T0136462

# COMMUNITY FORESTRY
# IN CANADA

# COMMUNITY FORESTRY
# IN CANADA
## Lessons from Policy and Practice

Edited by Sara Teitelbaum

**UBC**Press · Vancouver · Toronto

25 24 23 22 21 20 19 18 17 16     5 4 3 2 1

Printed in Canada on FSC-certified ancient-forest-free paper (100% post-consumer recycled) that is processed chlorine- and acid-free.

___

**Library and Archives Canada Cataloguing in Publication**

Community forestry in Canada : lessons from policy and practice / edited by Sara Teitelbaum

Includes bibliographical references and index
Issued in print and electronic formats
ISBN 978-0-7748-3188-8 (hardback). – ISBN 978-0-7748-3190-1 (pdf).
ISBN 978-0-7748-3191-8 (epub). – ISBN 978-0-7748-3192-5 (mobi)

1. Community forestry – Canada. 2. Forest policy – Canada. I. Teitelbaum, Sara, editor

| SD567.C644 2016 | 333.750971 | C2016-901668-4 |
| | | C2016-901669-2 |

___

Canadä

UBC Press gratefully acknowledges the financial support for our publishing program of the Government of Canada (through the Canada Book Fund), the Canada Council for the Arts, and the British Columbia Arts Council.

This book has been published with the help of a grant from the Canadian Federation for the Humanities and Social Sciences, through the Awards to Scholarly Publications Program, using funds provided by the Social Sciences and Humanities Research Council of Canada.

UBC Press
The University of British Columbia
2029 West Mall
Vancouver, BC V6T 1Z2
**www.ubcpress.ca**

# Contents

# Tables and Figures

**TABLES**

## FIGURES

# COMMUNITY FORESTRY
# IN CANADA

# Introduction
## A Shared Framework for the Analysis of Community Forestry in Canada

*Sara Teitelbaum*

This book is about community forestry in Canada. It focuses principally on those instances in which local communities have acquired
some rights and responsibilities for specific public forest lands in their
vicinity in order to achieve some collective benefit. In Canada, this type
of institutionalized arrangement remains on the margins of policy and
tenure development. The vast majority of public land is allocated to the
corporate sector through large industrial licences. However, since the
1990s, through a combination of public pressure and intermittent legal
reform, there has emerged a collection of initiatives on public land that
combine the qualities of formalized local governance with a place-
based approach. Both Aboriginal and non-Aboriginal actors have identified these initiatives as an important alternative on the forestry
landscape due to their potential to embrace a form of development that
is more adapted to local aims and conditions.

The qualities of these initiatives – including their relationship to
government policy, forms of collective action, and local patterns of development – is a central focus of this book. In bringing together a wide
group of researchers with experience related to community forestry,
this book sets out to provide a geographically inclusive and empirically
rich portrait of community forestry policy and practice in Canada. This
type of in-depth examination of community forestry, at both a macro
and a micro scale, allows for the emergence of a more nuanced and
representative portrait of community forestry. This book, the first to
capture a representation of community forestry from coast to coast, is
part of a growing research tradition that seeks to bridge the gap between empirical research and theoretical propositions associated with

community forestry. Most of this literature concerns cases in the Global South and includes numerous comparative case studies and meta-analyses, providing some of the best evidence for how community forestry implementation is faring (Glasmeier and Farrigan 2005; Pagdee et al. 2006; Charnley and Poe 2007; Bowler et al. 2012; Porter-Bolland et al. 2012; Hajjar, Kozak, and Innes 2012). While there are success stories, the general consensus is that adoption of community forestry in the Global South is hampered by constraints. These are apparent both at the community level, where the impediments include such things as limited capacity, internal conflict, and corruption, and at the state level, where there is clear evidence of an unwillingness on the part of governments to devolve substantive authority to the local level (Shackleton et al. 2002; Ribot 2010; Larson et al. 2010).

This level of detail and generalization has not yet emerged in the literature about community forestry in the Global North, including Canada, perhaps because of the absence of a critical mass of long-standing studies. The concept of "community forestry" is not well defined in the Global North, nor has it been implemented to the same extent. While examples are cited in countries such as Canada, the United States, Scotland, France, and Italy, these examples are better described as policy outliers than as the result of clear and sustained commitment on the part of national governments (Inglis 1999; Jeanrenaud 2001; Baker and Kusel 2003; Bullock and Hanna 2012; Lawrence et al. 2009). And yet in Canada, as in several other northern countries, there is a growing interest in community forestry and similar "hybrid" modes of governance, such as co-management and collaboration.[1] In Canada, this can be tied to a mounting social critique aimed at the perception of an entrenched pattern of industrial use and corporate control of public lands. The growing social mobilization around forests – which includes environmentalists, Aboriginal peoples, labour unions, and some rural organizations – has targeted a number of intersecting issues, including the perceived mismanagement of public forests, job losses, and the perception of a systematic pattern of exclusion of Aboriginal and local-level actors from forest-management decisions (Bernstein and Cashore 2000; Tindall, Trosper, and Perreault 2013). The unresolved nature of Aboriginal rights and title over public forest lands has become an issue of critical importance, with recent court rulings (e.g., *Tsilhqot'in Nation v. British Columbia*, 2014) pointing towards the need for a radical redefinition of relationships between governments and Aboriginal peoples in the matter of environmental governance. Overall, this growing politicization of forestry issues has opened up new space for discussions of alternative governance arrangements that include both Aboriginal and non-Aboriginal populations. Since the 1990s, there has been evidence of increased experimentation with participatory forms of forestry governance in Canada, including the Model Forest Program, co-management arrangements with Aboriginal peoples, a diversity of stakeholder advisory groups, third-party certification, and the development of community forestry tenures in several provinces (Parkins 2006; Teitelbaum, Beckley, and Nadeau 2006; Tollefson, Gale, and Haley 2008; Wyatt 2008; Howlett, Rayner, and Tollefson 2009).

In parallel with advancements at the policy and practical level, a distinct body of research literature has developed around community forestry in Canada. The early literature was instrumental in defining this approach and provided the first select surveys and descriptions of specific initiatives (Matakala and Duinker 1993; Allan and Frank 1994; Dunster 1994; Masse 1995; Beckley 1998; M'Gonigle, Egan, and Ambus 2001). This literature played an important role in exploring the potential of community forestry within a Canadian context, including enabling factors (Bouthillier and Dionne 1995; Duinker, Matakala, and Zhang 1991; Duinker et al. 1994; Harvey and Hillier 1994; Burda and M'.Gonigle 1996). It also helped to set out the underlying principles of community forestry in Canada, which have since become the barometer by which these approaches are understood and evaluated. These principles, which align closely with those expressed in the international literature, emphasize community forestry's potential to meaningfully involve local people in forest decisions as the key to ensuring appropriate and beneficial ecological and socio-economic outcomes for local communities. Thus, the early literature quickly established an association between community forestry and the principles of participatory democracy, community development, and ecological stewardship (Baker and Kusel 2003; Charnley and Poe 2007; Davis 2008).

The academic community has not only played an important role in generating knowledge about community forestry but has also been active in advocating for this approach among a broad set of publics (Duinker et al. 1994; Bouthillier and Dionne 1995; Beckley 1998; M'Gonigle, Egan, and Ambus 2001; Haley 2002; Ambus et al. 2007; Teitelbaum 2010; Smith, Palmer, and Shahi 2012). This work has included the organization of numerous conferences and workshops on the topic, which have brought together academics, policy makers, Aboriginal peoples, and practitioners for the purposes of learning and networking. Some academics have worked with non-governmental organizations on the production of educational materials, while others have collaborated directly with government on the design of community forestry policy. While these activities have most certainly helped move implementation forward, the combined research-advocacy role may also have contributed to what some have described as an idealistic tendency apparent in the community forestry literature (Beckley 1998; Bullock and Hanna 2012). In 1998, Beckley described the literature this way: "Much of the theoretical literature is written by proponents of these models who wish to see more widespread adoption of community forestry and co-management. Their discussions are more about what community forests or co-management could be or should be, rather than what the few nominal, empirical examples of these models actually are" (737).

While this depiction is no longer completely accurate, it is still possible to discern a certain tension between enthusiastic support and a more detached stance. The tendency for community forestry research to draw on qualitative methods – mainly interviews with key informants – may reinforce this tension. Few studies present measurable evidence of socio-economic and ecological outcomes; most prefer instead to rely on descriptive accounts. However, since the early 2000s, the theoretical frameworks associated with community forestry research have broadened considerably. Many new initiatives have been implemented, elevating the status of community forestry and creating new research interest, especially in British Columbia. The issue of devolution and the extent of power sharing with communities has been examined, revealing patterns of top-down governmental authority, particularly with regard to the more strategic dimensions of forest management (Chiasson, Andrew, and Leclerc 2008). Researchers have analyzed the economic trajectories of community forests through various lenses, including that of multi-functionality (Luckert 1999; Ambus et al. 2007; Chiasson, Leclerc, and Andres 2010). This work highlights the difficulties faced by community forests in breaking out of commodity-based approach to forest development, an approach that reflects the narrow economic base within many rural regions. Still others have analyzed community forestry through the theoretical framework of neoliberalism in order to clarify governmental motivations for policy reform and analyze the manoeuvrability of local forest-related objectives within the broader political climate (McCarthy 2006; Pinkerton et al. 2008). Some attention has also been given to the relationship between community forestry and the rights of Aboriginal peoples, revealing strong potential for commensurability, albeit under conditions of mutual respect and shared jurisdiction (e.g., Booth 1998; Bullock 2012; Smith,

Palmer, and Shahi 2012; Smith 2013). Overall, this research has helped connect community forestry scholarship to the broader fields of environmental governance. It has also helped identify a number of structural constraints to the full implementation of community forestry, including insufficient institutional arrangements, a lack of organizational resources, and a challenging economic context (McIlveen and Bradshaw 2005–6; Bullock, Hanna, and Slocombe 2009; McIlveen and Bradshaw 2009).

A smaller portion of the research examines conditions at the local level through in-depth case studies, pointing to the presence of an enduring empirical gap, as was highlighted by Beckley over a decade ago. One original contribution is Reed and McIlveen's (2006) case study of governance and civic science at the Burns Lake Community Forest in British Columbia. In examining governance practices, the authors observed a tendency for the organization to concentrate on capturing forestry expertise at the expense of a more pluralistic approach, illustrating "a tension between the traditional economic objectives of forestry and the expectation that community forests will also address a broader social agenda of taking care of local people" (602). This article spurred a lively debate on "success factors" in community forestry and the tension between inclusivity and effectiveness (Bradshaw 2007; Pagdee, Kim, and Daugherty 2007; Reed and McIlveen 2007). In another study, Davis (2008) examines the relationship between local control and ecosystem-based management, comparing case studies from British Columbia and Mexico. In a similar vein, Bullock and Hanna (2012) describe the challenges involved in adopting ecologically oriented practices at the Creston Valley Forest Corporation (CVFC) in British Columbia, pointing to insufficient local support, degraded site conditions, and weak provincial support.

This book is designed to fill key gaps within the Canadian research literature. It brings together twenty-eight researchers with expertise in the field of community forestry research in order to respond to three distinct objectives. The first is to provide a more complete portrait of community forestry policy and practice across *all* jurisdictions in Canada, through a series of regional portraits covering all ten provinces (Part 1). Thus far, the community forestry literature has focused almost exclusively on those jurisdictions with clear tenure arrangements, with a disproportionate amount of attention given to developments in British Columbia. Some provinces, such as Quebec, have a long tradition of collective action around forests but have generated few publications, especially in English (some exceptions include Chiasson and Leclerc 2013; Chiasson, Boucher, and Martin 2005). Still other regions, such as the Prairie provinces and the Atlantic provinces, are even further beneath the radar, perhaps because of an assumption that community forestry does not exist there. An important objective of this book, therefore, is to expand the horizons of community forestry research by spotlighting those regions that have not yet been sufficiently explored. What is the history of social mobilization around community-based forestry? What do patterns of governance look like? How are Aboriginal peoples engaged in forest governance? What has prevented or stimulated community forestry in particular regions? What have been the key drivers for community forestry development? These questions are equally relevant in those provinces that do have a documented tradition of community forestry practice as in those that don't, since this type of concise historical overview has rarely been produced. Indeed, this book represents the first attempt to synthesize and compare the evolution of community forestry across all Canadian provinces.

The second aim of this book is to address the empirical gap discussed above through the presentation of new case study work. The five chapters of Part 2 present original research findings from eleven different community forests across Canada. A critical mass of community forests now exists in Canada – more than one hundred overall, many with more than a decade of experience (Teitelbaum, Beckley, and Nadeau 2006). These provide an excellent opportunity for in-depth and comparative analysis, yet only a few of them have been the focus of academic research. The research reported in this volume enhances our knowledge of the range of experiences that exist in Canada and seeks to answer some key questions: What are the driving orientations of community forests? What kinds of benefits are being generated by community forests, and how are these benefits being distributed within communities? Is community forestry governance facilitating a shift towards more collaborative and inclusive forms of decision-making? Are community forests innovating with regard to ecological sustainability? The contributors to this book engage critically with these questions through new case study work in order to "bridge theory and practice" (Charnley and Poe 2007). These case studies draw on a common conceptual framework featuring four principles: participatory governance, rights, local benefits, and ecological stewardship. These principles are widely recognized as being underlying aspirations of community forestry, and as such, they provide a common foundation for the analysis of socio-economic and ecological outcomes.

The third objective of the book is to create a space for new reflections about community forestry, including its symbolic and practical importance within the mix of governance arrangements in Canada. As we have seen, community forestry has taken root in some parts of the country, yet its peripheral status has remained unchanged despite several decades of advocacy. Meanwhile, forestry reforms across the country indicate a continued preference for large-scale and globalized forms of production. This points to a potential incommensurability between the underlying ideological tenets of community forestry and the dominant economic model. In order to understand the full significance of community forestry, as a set of ideas *and* practices, we need to investigate the origin of this incommensurability. Can community forestry go beyond its current role? How do we navigate the tension between utopian ideals and pragmatic realities? Is community forestry really a viable option for Canada? If so, what is required for the strengthening of community forestry? The three chapters in Part 3 are conceptually rather than empirically driven, offering different perspectives on the future of community forestry and its transformative potential.

## Theoretical Framework

Four principles make up the conceptual framework for this book: participatory governance, rights, local benefits, and ecological stewardship. Expressed in different ways, they appear frequently as goals of community forestry, whether in the academic literature, in government policy, or in the mission statements of community forests themselves. Clearly, these are idealized terms; in adopting them as principles for this book, the idea is not to set unrealistic expectations for community forestry but rather to create a theoretical basis for exploring community forestry practices and outcomes. Indeed, these principles are unlikely to be present in equal force within every community forest initiative but will manifest with different degrees of importance. Krogman and Beckley (2002) propose a control-benefit continuum, which envisions a minimum threshold of local control and community benefit as being integral to the identity of a community forest. We might suppose something similar: that some minimum level of effort and achievement towards these four principles and goals must be in evidence for an endeavour to be identified as a community forest.

*Participatory governance* speaks to the proposition, described in the literature, of ensuring that local people have the opportunity to meaningfully participate in decision-making concerning forests in their region. This has been a key driver for community forestry in Canada and beyond, since community forestry has been envisioned as a means to correct the legacy of top-down and exclusionary forestry policies. However, the literature reveals that participatory governance requires much more than the simple transfer of power from central governments to communities; increasingly, studies document an absence of accountability at the local level due to corrupt or elitist practices within local institutions (Shackleton et al. 2002; Ribot 2010). Ensuring good governance therefore requires clear rules, accountability mechanisms, and redress mechanisms, as well as forms of community engagement (Tyler, Ambus, and Davis-Case 2007; World Resources Institute 2009; Secco, Pettenella, and Gatto 2011). This is reflected in Kearney et al.'s (2007, 2) definition of participatory governance: "the effort to achieve change through actions that are more effective and equitable than normally possible through representative government and bureaucratic administration by inviting citizens to a deep and sustained participation in decision-making." The analysis of participatory governance therefore speaks to a number of questions, including the following: Who decides? How is the community represented? How does the community participate? What are the accountability mechanisms, both upward and downward?

The second principle, *rights*, speaks to the level of authority that a community has over forest decisions (Fennell 2011). This is encapsulated by the concept of devolution, which can be described as the transfer of rights and responsibilities from central governments to local communities through legal arrangements such as tenures (Shackleton et al. 2002; Larson et al. 2010). While gaining legal rights is not a strict requirement for community forestry – there are examples in Canada where communities have few formal rights – it is widely seen as a facilitating step in gaining meaningful decision-making power and capturing benefits over the long term. Indeed, recent research indicates that positive results are more likely to be achieved if tenure reforms are fully implemented (Larson et al. 2010). Schlager and Ostrom (1992) provide a useful classification for the different degrees of rights transferred from the state to the local level. Rights at the operational level include the ability to access and withdraw the resource, while rights at the collective-choice level include management, exclusion, and alienation. These collective-choice rights represent more strategic dimensions, including decisions about how the resource should be used and managed. In the Canadian context, tenure arrangements often limit community forests to operational-level rights, an observation echoed in the international literature (Burda and M'Gonigle 1996; Clogg 1997; Ambus and Hoberg 2011; Cronkleton, Saigal, and Pulhin 2012). The issue of rights raises a number of questions: How much influence does the community have over decisions? Are these rights formalized through legal arrangements? Does the community have the right to establish rules concerning how the forest will be used?

The third principle, *local benefits*, relates to the aspiration that community forests generate various benefits for local communities (McDermott and Schreckenberg 2009; Colfer 2005). Some researchers have hypothesized that community forests are more likely than private forest companies to lead to enhanced benefits for communities, because private companies aim to generate profits for distant shareholders rather than to reinvest in the communities where they operate (Freudenburg 1992; Krogman and Beckley 2002). By contrast, descriptions of community forestry make reference to a type of locally centred economic development vision according to which keeping the wealth generated by the forest in the local area creates new opportunities, which in turn contribute to the revitalization of rural communities (Duinker et al. 1994; Burda and M'Gonigle 1996). The benefits associated with community forestry include economic ones such as the generation of employment, processing opportunities, and direct investments in services and infrastructures. However, benefits can also be socio-cultural (educational opportunities, trails, and recreational services, etc.) and ecological (water, wildlife, and forest protection; Colfer 2005). The idea that community forestry can help facilitate a diversification of activities, and thus movement away from a timber-production model, has often been asserted (Pinkerton and Benner 2013). Research indicates that many community forests adopt a mix of commodity-based and environmental-based goals (Luizza 2011). The concept of local benefits can be articulated through a number of questions: What are the activities of the community forest? What benefits do they bring to local people? How are benefits distributed among community members? Are inequalities being reinforced or lessened?

The fourth principle generally associated with community forestry is that of achieving higher levels of *ecological stewardship*, including protection of water quality, viewscapes, wildlife, and biodiversity (M'Gonigle, Egan, and Ambus 2001; Davis 2008; Furness and Nelson 2012). Indeed, the desire to counteract what have been characterized as the destructive forestry practices of industrial forestry has been an important point of mobilization for the community forestry move-ment. In Canada, some community forests have been given manage-ment responsibility over ecologically sensitive areas, precisely because of their particular orientation towards alternative harvesting tech-niques. However, there is also evidence that not all community forests are inherently "conservationist"; like other forestry actors, commun-ity forests face economic pressures as well as capacity limitations (Bradshaw 2003; Western, Wright, and Strum 1994). Given the small size of community forests, researchers have also raised the question of whether communities have the capacity to address landscape-level environmental issues such as biodiversity loss or climate change (Agrawal and Gibson 1999). Increasingly, community forestry scholars are drawing on interdisciplinary and systems-based approaches – such as complexity theory, resilience theory, and adaptive management – in order to present a more integrated perspective on how ecological dimensions intersect with broader issues of socio-economic and cul-tural sustainability (Folke et al. 2005; Gunderson 2003; Armitage et al. 2009). In this book, the issue of ecological stewardship is encapsulated in a number of questions: Is community forestry applying ecologically adapted management practices? Are there strong policies and norms in place to ensure oversight? Does the capacity exist to address landscape-level issues?

## Part 1: Creating a National Portrait

The first section of this book responds to the first objective: namely, to create a more complete portrait of community forestry policy and practices across all jurisdictions. This section features six regional portraits (Newfoundland, the Maritimes, Quebec, Ontario, the Prairie provinces, and British Columbia). Weaving together policy history, analysis of social movements, and description of initiatives, these chapters help us to understand what inroads community forestry has made in each province.

The national picture that emerges is marked by clear differences in the extent to which community forestry policies have been imple-mented from one province to the next. There are also, however, strong similarities across all jurisdictions with regard to their overarching political landscapes. Without exception, authors describe policy re-gimes characterized by a strong preference for building relationships with corporate-industrial actors rather than with non-traditional actors such as rural communities and Aboriginal peoples. This bilateralism is accompanied by a focus on prioritizing commercial harvesting over the range of other forest values. Erin Kelly and Sara Carson (Chapter 1) describe the dynamic within Newfoundland's forest managers: "The tendency at DNR Forestry was to optimize commercial harvest, which the department considered 'eroded' by competing uses, including domestic harvest, municipal watersheds, wildlife habitat, and cabin building."

That being said, a type of continuum can be distinguished: at one end of the spectrum are those provinces that have chosen to remain wedded to conventional public participation mechanisms (public review of plans, stakeholder advisory groups), and at the other end are those that have recognized community forestry legally through the creation of tenures. In between are provinces that have embraced what John Parkins and his coauthors (in Chapter 5) call "enhanced public input" or "community-based forestry." While these arrangements do not necessarily confer direct management rights or commercial benefits to communities, they nevertheless facilitate collaborative governance arrangements through models such as comanagement boards, partnership agreements, and regional corporations (see Chapters 4 and 5 for further details).

Table I describes the status of community forestry arrangements on public (Crown) land across Canada. Based on material presented in Chapters 1 through 6, with the addition of a few statistics (Canada, NRCAN 2014), Table I reveals mixed progress towards implementing community forestry. Four out of ten provinces have established community forestry on area-based tenures: British Columbia, Ontario, Quebec, and Nova Scotia. Of these four, British Columbia and Quebec have the highest concentrations of community forestry initiatives.

The chapters in Part 1 provide important insights as to the reasons for these marked regional differences in the level of community forestry implementation. Social mobilization and public pressure emerge as enabling factors in several provinces where community forests are well established. In Chapter 6, Lisa Ambus presents a historical overview of British Columbia's first community forestry tenure, the Community Forest Agreement (CFA), tying its creation to the conflictual political climate surrounding forestry at the time. "The mid-1990s saw the emergence of a community forestry movement consisting of a loose coalition of communities and other groups, which aimed to capitalize on potential opportunities for alternative and community-based approaches to forestry as government sought out policy solutions to

**TABLE I Status of community forestry on public (Crown) land in Canada**

| Province | Crown forests (provincial and federal, as % of forest lands) | Progress towards community forestry | Approximate number of initiatives on Crown land | Name of tenure or enabling legislation |
|---|---|---|---|---|
| Newfoundland | 99% | Absent | 0 | NA |
| Nova Scotia | 32% | Emergent | 1 | Forest utilization licence agreement (standard Crown licence) |
| New Brunswick | 50% | Absent | 0 | NA |
| Prince Edward Island | 9% | Absent | 0 | NA |
| Quebec | 89% | Established | 67 | Forest management contract Territorial management agreement |
| Ontario | 92% | Established | 3 | Algonquin Forestry Authority Act Sustainable forest licence (standard Crown licence) |
| Manitoba | 97% | Absent | 0 | |
| Saskatchewan | 94% | Absent | 0 | |
| Alberta | 97% | Absent | 0 | |
| British Columbia | 97% | Established | 49 | community forest agreement |

Source: Crown forest statistics drawn from Canada, CFS, NRCAN (2005).

the War in the Woods." Ambus describes an evolving dynamic of civil society support, which has since coalesced in the creation of a provincial association that is uniquely positioned to provide a common voice for community forestry actors in British Columbia. Similarly, in Chapter 3, Solange Nadeau and Sara Teitelbaum describe a long tradition of collective action originating within rural communities in Quebec. These collective efforts, described by the authors as a response to economic decline and centrist policies, spawned a diversity of community-based forestry models, including joint management groups, cooperatives, and tenant farms. While few of these models were successful in gaining direct access to public lands, the authors describe a parallel development in the form of small-scale tenures allocated to municipal and, to a lesser extent, Aboriginal communities. Nadeau and Teitelbaum's chapter provides a rich account of endogenous forest development strategies in Quebec and their unsettled relationship to forest policy – the first such account produced for an English-speaking audience.

The chapters in Part 1 also tell a story of compromise. The translation of ideals into practice, in all those jurisdictions that have established community forests, has been arduous, in part because of what Ambus calls a situation of "constrained devolution," meaning the limited transfer of management rights:

> With the institutionalization of community forestry in the CFA, the vision of the community forest movement became bound within a set of rules determined, ultimately, by the provincial government. As a result, CFA holders struggle to work within the limitations of their management mandates and strive to balance the high, and sometimes contradictory, expectations of diverse local stakeholders.

Ambus's observation is echoed in Nadeau and Teitelbaum's exposé of Quebec, where the implementation of community forestry tenures has been restricted to a narrow set of actors and involves only small parcels of unallocated forest lands rather than a wholesale redistribution of industrial forestry allocations.

The stories of other provinces, including Ontario and the Prairies, follow a different path, one that arguably aligns more closely with the concept of "enhanced participation." A distinguishing feature of policy in these provinces is a preference for wedding governance innovations to large industrial tenures. In Chapter 4, Lynn Palmer, Peggy Smith, and Chander Shahi draw on complexity theory to describe the different stages of policy-making in Ontario and the place of community forestry within each historical period. The chapter reveals repeated moments of tenure reform and governance experimentation that resulted in some large-scale examples of community forestry, such as Westwind Forest Stewardship and Algonquin Forest Authority, and recent partnerships between First Nations and municipalities. Overall, however, the authors describe insufficient change: "The rigidity of the command-and-control approach meant the system was able to accommodate only minor variations ... As a result, all community forestry attempts, whether enduring or not, have amounted to localized experiments only, failing to transform the system." Along similar lines, in Chapter 5, John Parkins, Ryan Bullock, Bram Noble, and Maureen Reed explore the landscape of governance arrangements in Manitoba, Saskatchewan, and Alberta, provinces that are virtually unmentioned in the community forestry literature. The authors identify three initiatives that encompass some of the qualities of community forestry, several of which are agreements established with Aboriginal communities. However, the authors stop short of characterizing these as community forests because of an absence of shared jurisdiction. They speculate that in the Prairies, where forestry is a more recent development and has been less controversial than in neighbouring British Columbia, provincial governments have opted for a strategy of large-scale industrial development combined with enhanced community participation because such an arrangement "appeared to offer government the best of both worlds: that is, the potential benefits of larger, international companies (such as access to capital and markets) while ensuring that the interests of local, northern, and Aboriginal communities were also included in decisions. Given this policy thrust, governments and communities were much less interested in smaller-scale, community-oriented development."

The distribution of public-private property rights has created a different context for the implementation of community forestry in the Atlantic provinces. The Maritime provinces (New Brunswick, Nova Scotia, and Prince Edward Island) have a much higher proportion of private land, held under small-scale property ownership. In Chapter 2, Thomas Beckley posits a novel thesis to explain the lack of progress towards community forestry. He attributes this deficit to the high proportion of small private forest holdings, which meet many of the same socio-economic objectives as community forests, but at an individual level. "If we scale this up one level of social organization (Beckley 1998) – that is, if we consider that communities comprise individuals – we can argue that much of the forest land in the Maritimes is managed for community benefits, which is a cornerstone of community forestry. This ownership pattern is virtually non-existent in Newfoundland, however, where there are no institutionalized examples of community forestry." As Erin Kelly and Sara Carson point out in Chapter 1, community-based management can take different forms, as evidenced by the strong tradition of subsistence activities in Newfoundland. However, the authors point to an entrenched pattern of disregard for citizen concerns in planning characterized by "one-size-fits-all rules and planning, limited access and decision-making power for local users, and top-down fixes for local resource needs." Drawing on the example of the Great Northern Peninsula, Kelly and Carson describe the opportunities for renewal in the form of community forestry partnerships between government and local development organizations.

Finally, the chapters in Part 1 emphasize the growing connection between community forestry and Aboriginal rights. Across all the provinces in which community forests have been implemented, we see examples of small- and large-scale tenures accorded to Aboriginal communities. In Chapter 4, Palmer, Smith, and Shahi describe recent tenure reforms in Ontario that include clear orientations towards enhanced Aboriginal and local involvement – in some cases, through new municipal-Aboriginal cooperation. Comanagement arrangements with Aboriginal communities in Saskatchewan are also profiled in Chapter 5. It is clear that community forestry, in the sense defined in this book, has strong appeal for Aboriginal communities because of the focus on rights, benefits, and participation. However, chapter authors agree that in practice, these arrangements are falling short of these three principles because of an absence of shared jurisdictional authority, a key step towards the implementation of Aboriginal and treaty rights.

## Part 2: Bridging Theory and Practice through Case Studies

This section responds to the second objective of this book: to address the gap in empirical work through the presentation of new case studies. The chapters in Part 2 describe eleven community forestry initiatives in the provinces of British Columbia, Ontario, and Quebec. These initiatives are, for the most part, small-scale and encompass a variety of organizational structures within unique cultural and environmental contexts. Some, such as Burns Lake Community Forest in

British Columbia (Chapter 7), are well-established organizations that have achieved a certain level of financial stability and success through integration into the traditional forest economy. Others, such as the cooperatives in Quebec described by Édith Leclerc and Guy Chiasson (Chapter 9), represent short-lived attempts to provide employment opportunities in geographically peripheral regions. Still others, such as the Common Ground initiative in Kenora, Ontario, presented by James Robson and his coauthors (Chapter 8), are emerging attempts to build cross-cultural collaboration in an urban forest setting. Collectively, these cases point to the diversity of social values driving community forestry implementation.

Governance is a pervasive theme within this book, which provides new evidence concerning the long-standing debate about the level of inclusivity and participation manifested by community forests (Reed and McIlveen 2006; Bradshaw 2007; Pagdee, Kim, and Daugherty 2007). Previous research has highlighted the tension between expert-driven governance approaches and more participatory models. McIlveen and Bradshaw (2005–6) argue that it may be unrealistic to expect community forests to out-perform private forest companies in public participation, given the former's small size and lack of institutional support. The case studies in this section reinforce the notion that community forestry cannot be equated with the drive towards broad-based community engagement. In Chapter 11, Sara Teitelbaum compares participatory governance practices in four community forests in three provinces (Ontario, Quebec, and British Columbia), revealing very different scenarios. Two of these organizations demonstrate very strong practices, including community surveys, open houses, public advisory groups, and so on, while the other two rely simply on community representation through a formal board of directors. This chapter brings an additional dimension to the issue of participatory governance by highlighting a potential connection between citizen engagement and the presence of a more diversified vision of forest use. Teitelbaum posits that non-timber activities create a natural bridge towards community engagement.

This resonates with the findings of Robson and colleagues in Chapter 8, who describe a local forest governance initiative in an urban, post-extractive setting where the priority is placed on ensuring cross-cultural communication and collaboration.

Another set of governance issues converges around the topic of rights and levels of decision-making influence. While it is well established that community forests, both in Canada and internationally, are characterized by what Lisa Ambus (Chapter 6) calls a situation of "constrained devolution," the case studies provide important lessons concerning the different ways in which this is manifested and the types of strategies that community forests adopt in order to mitigate its effects. For example, in Chapter 10, Lauren Rethoret, Murray Rutherford, and Evelyn Pinkerton describe two community forests in British Columbia, Harrop-Procter and Creston, that are pursuing source water protection as a priority management objective. Drawing on a combination of interviews, government documents, forestry plans, and board meeting minutes, the authors describe positive results in achieving objectives around water protection. However, they also describe a tug-of-war between top-down regulation and local approaches:

> Harrop-Procter was refused the right to manage its watersheds without cutting timber. Creston's right to manage wildfire risk in one of its watersheds was limited by legislation that prescribes standards for restocking designed primarily to sustain timber harvests, leaving little room to implement alternative ecological objectives. Both of these limitations on local discretion arise from the provincial government's focus on maintaining the overall level of timber harvest and economic return from BC's forestry land base, a long-time driver of forest policy in the province.

This chapter is unique in that it represents one of very few Canadian case studies that specifically target environmental outcomes.

The chapters in this section also point to a fragility related to the broad economic context within which community forests are embedded. The case studies reveal that community forests are deeply inscribed into regional patterns of economic development – an environment characterized by globalization, neoliberal policies, and market instability. Within this context, scale becomes an important factor. In Chapter 7, Kirsten McIlveen and Michelle Rhodes present a fascinating study of the Burns Lake Community Forest and its response to the mountain pine beetle outbreak, a landscape-level threat in northern British Columbia. Drawing on research from two different periods, the authors provide an illustration of how, despite strong experience and organizational capacity (community support, leadership, and expertise), the Burns Lake Community Forest is limited both by the scale of its operations and land base and by the market conditions that prevail in the region, which restrict opportunities for diversification. A similar dynamic exists among community forests in rural Quebec, described by

Édith Leclerc and Guy Chiasson in Chapter 9. Theirs is an original contribution in that it presents some of the first descriptions of local governance initiatives that did not persist. Of the four case studies analyzed, two (Coopérative forestière de Beaucanton and Coopérative de solidarité de Duhamel) ran into difficulties early on because of a number of intersecting factors, including a lack of qualified labour, technical expertise, and financial capital. This chapter highlights the precarious nature of community forest organizations and the pivotal issue of community capacity, especially for those organizations that operate in geographically remote locations without the benefit of tenure.

One final topic, rarely explored in the community forest literature, deserves attention: the role that community forests play in building connections, identity, and a sense of place. In Chapter 8, James Robson and his coauthors describe Common Ground, an emerging collaborative governance initiative in Kenora, Ontario. The case study describes a post-industrial forest in an urban setting that has become a de facto recreation and spiritual commons for both First Nation and non-First Nation residents. This research reveals strong connections between local people and these forests and explores the history of First Nations' use of, and then exclusion from, these lands. The implementation of local governance therefore requires careful attention to local values, a shared understanding of place, and jointly established rules that respect different cultural traditions and resource use patterns. Chapter 8 is an important complement to the other case studies in this book, reminding us that governance means much more than ensuring democratic representation and effective administration; rather, it can be a platform for collaboration and can help to "construct shared experiences and meanings that transcend social and cultural differences."

## Part 3: Casting Forward

This section of the book addresses the third objective: to create a space for new reflections about community forestry, including its symbolic and practical significance and its future role in Canada. Can community forestry expand beyond its current role? How do we navigate the tension between utopian ideals and pragmatic realities? Is community forestry really a viable option for Canada? If so, what is required for strengthened forms of community forestry?

The first two chapters in Part 3 focus on the need for deep institutional reform, which requires a fundamental shift in how governments understand and engage with community forestry. In Chapter 13, Peter Duinker and Kris MacLellan take us on a journey into the future, positing different scenarios for Canada's forests in the year 2050. Within the context of the Forest Futures Project, a two-year participatory research endeavour, the authors find "a strong leaning across Canada for a future characterized by more community-controlled forests." On this basis, they construct an argument in favour of a formalized strategy of adaptive policy development on the part of governments. Drawing on the example of Nova Scotia, the authors describe their vision for a robust community forestry policy based on the principle of institutional experimentation and evaluation. This approach is premised on a strong and stable enabling policy framework, a reliance on distributed leadership tapping into the creative energy of local people, the application of diverse governance models (a range of sizes, organizational structures, and objectives), and "a boldness to abandon prevailing conventions and step smartly, with eyes wide open, into novel territory."

In Chapter 12, Erik Leslie, a practitioner with the Harrop-Procter Community Forest in British Columbia, builds his argument around the need for community forests to acquire more comprehensive sets of rights. Tracing the development of community forestry policy in British Columbia, Leslie observes only "small de facto steps" in the direction of enhanced operational management independence rather than a transfer of higher-level management rights. In his view, providing communities with more strategic decision-making regarding resources would not only facilitate innovative practices but also help ensure citizen engagement. "When community members feel like they actually have meaningful rights and ultimate control, a subtle but profound societal shift occurs and new approaches can emerge." Leslie illustrates this argument through a skilful interweaving of common property theory and a narrative account, drawing on the experiences of the Harrop-Procter Community Forest and negotiations around the determination of the annual allowable cut.

In the final chapter of this book, Ryan Bullock and Maureen Reed propose a vast reorganization of the relationship between communities and forests. Confronting the "myth of self-reliance," a cultural fiction based on "self-reliant northerners and isolated communities eking out an existence in rugged Canadian landscapes, surrounded by pristine environs and natural resources containing immeasurable wealth," the authors advance their own conceptualization of resource communities drawing on a concept that they call the "company-community system." This system is characterized by a core-periphery dependence that yields insufficient diversification, social support, and community infrastructure. The authors propose a shift from the company-community system to a "community-forest system," which is based on endogenous and decentralized development. The authors' vision foresees greater integration among the many localized community forestry movements emerging in Canada and the creation of a functional community forestry network to coordinate actions across different scales.

## Conclusion

This book brings together many of the leading researchers of community forestry in the Canadian context. Drawing on a common theoretical framework based on four principles – participatory governance, rights, local benefits, and ecological stewardship – the authors seek to bridge theory and practice in order to build a more empirically based portrait of community forestry across all Canadian provinces. The picture that emerges is one of a governance approach that has found broad consensus among civil society groups because of its ability to provide a tangible counterpoint to the corporate-industrial model. However, the attempt to fully realize this approach has been a struggle. The translation of ideas into practice has proven challenging, in large part because of the structure of the existing forest sector and a lack of clear political support from governments. These barriers place an unfair burden on community forests, since they are expected to be both economically competitive, albeit with a much smaller land base, and innovative, as prescribed by their mandate. The case studies in this book provide fresh evidence for how this balancing act is going. Some community forest initiatives, particularly those that target social or ecological innovation, have fared well, creating new opportunities for collaboration, intercultural communication, and strengthened environmental stewardship. For others, the convergence of structural barriers and low levels of community capacity has led to failure. The key to the future of community forestry on public land lies in deep institutional reform, which would provide communities not only with greater influence over forest management but also with the necessary support to ensure ongoing progress.

## Notes

1 Lemos and Agrawal (2006) describe two types of hybrid governance: soft governance strategies, which rely on self-regulatory processes, and cogovernance, which relies on partnerships across state-market-society divisions.

## References

Agrawal, A., and C.C. Gibson. 1999. "Enchantment and disenchantment: The role of community in natural resource conservation." *World Development* 27 (4): 629–49. http://dx.doi.org/10.1016/j.worlddev.2005.07.013.

Allan, K., and D. Frank. 1994. "Community forests in British Columbia: Models that work." *Forestry Chronicle* 70 (6): 721–24. http://dx.doi.org/10.5558/tfc70721-6.

Ambus, L., D. Davis-Case, D. Mitchell, and S. Tyler. 2007. "Strength in diversity: Market opportunities and benefits from small forest tenures." *BC Journal of Ecosystems and Management* 8 (2): 88–100.

Ambus, L., and G. Hoberg. 2011. "The evolution of devolution: A critical analysis of the community forest agreement in British Columbia." *Society and Natural Resources* 24 (9): 933–50. http://dx.doi.org/10.1080/08941920.2010.520078.

Armitage, D.R., R. Plummer, F. Berkes, R.I. Arthur, A.T. Charles, I.J. Davidson-Hunt, A.P. Diduck, et al. 2009. "Adaptive co-management for social-ecological complexity." *Frontiers in Ecology and the Environment* 7 (2): 95-102.

Baker, M., and J. Kusel. 2003. *Community forestry in the United States: Learning from the past, crafting the future.* Washington, DC: Island Press.

Beckley, T.M. 1998. "Moving toward consensus-based forest management: A comparison of industrial, co-managed, community, and small private forests in Canada." *Forestry Chronicle* 74 (5): 736–44. http://dx.doi.org/10.5558/tfc74736-5.

Bernstein, S., and B. Cashore. 2000. "Globalization, four paths of internationalization, and domestic policy change: The case of ecoforestry in British Columbia, Canada." *Canadian Journal of Political Science* 33 (1): 67–99. http://dx.doi.org/10.1017/S0008423900000044.

Booth, A.L. 1998. "Putting 'forestry' and 'community' into First Nations' resource management." *Forestry Chronicle* 74 (3): 347–52. http://dx.doi.org/10.5558/tfc74347-3.

Bouthillier, L., and H. Dionne. 1995. *La forêt à habiter: La notion de forêt habitée et ses critères de mise en œuvre – Rapport final au Service canadien des forêts (Région du Québec).* Rimouski: Université du Québec à Rimouski et Université Laval.

Bowler, D.E., L.M. Buyung-Ali, J.R. Healey, J.P.G. Jones, T.M. Knight, and A.S. Pullin. 2012. "Does community forest management provide global environmental benefits and improve local welfare?" *Frontiers in Ecology and the Environment* 10 (1): 29–36. http://dx.doi.org/10.1890/110040.

Bradshaw, B. 2003. "Questioning the credibility and capacity of community-based resource management." *The Canadian Geographer* 47 (2): 137–50. http://dx.doi.org/10.1111/1541-0064.t01-1-00001.

–. 2007. "On definitions of 'success' and contingencies affecting success in community forestry: A response to Reed and McIlveen (2006) and Pagdee et al. (2006)." *Society and Natural Resources* 20 (8): 751–53. http://dx.doi.org/10.1080/08941920701429155.

Bullock, R. 2012. "Reframing forest-based development as First Nation–municipal collaboration: Lessons from Lake Superior's north shore." *Journal of Aboriginal Economic Development* 7 (2): 78–89.

Bullock, R., and K. Hanna. 2012. *Community forestry: Local values, conflict, and forest governance.* New York: Cambridge University Press. http://dx.doi.org/10.1017/CBO9780511978678.

Bullock, R., K. Hanna, and S. Slocombe. 2009. "Learning from community forestry experience: Challenges and lessons from British Columbia." *Forestry Chronicle* 85 (2): 293–304. http://dx.doi.org/10.5558/tfc85293-2.

Burda, C., and M. M'Gonigle. 1996. "Tree farm or community forest?" *Making Waves* 7 (4): 16–21.

Canada. CFS (Canadian Forest Services). NRCAN (Natural Resources Canada). 2005. *The state of Canada's forests 2004-2005: The boreal forest.* Ottawa: CFS, NRCAN. http://cfs.nrcan.gc.ca/pubwarehouse/pdfs/25648.pdf.

–. NRCAN (Natural Resources Canada). 2014. *State of Canada's forest report.* Ottawa: NRCAN.

Charnley, S., and M.R. Poe. 2007. "Community forestry in theory and practice: Where are we now?" *Annual Review of Anthropology* 36 (1): 301–36. http://dx.doi.org/10.1146/annurev.anthro.35.081705.123143.

Chiasson, G., C. Andrew, and E. Leclerc. 2008. "Territorialiser la gouvernance du développement : Réflexions à partir de deux territoires forestiers." *Canadian Journal of Regional Science* 31 (3): 489-506.

Chiasson, G., J.L. Boucher, and T. Martin. 2005. "La forêt plurielle: Nouveau mode de gestion et d'utilisation de la forêt, le cas de la Forêt de l'Aigle." *Vertigo* 6 (2): 115-25.

Chiasson, G., and E. Leclerc. 2013. *La gouvernance locale des forêts publiques québécoises: Une avenue de développement des régions périphériques?* Montréal: Presses de l'Université du Québec.

Clogg, J. 1997. *Tenure reform for ecologically and socially responsible forest use in British Columbia.* North York, ON: Faculty of Environmental Studies, York University.

Colfer, C.J.P. 2005. *The equitable forest: Diversity, community, and resource management.* Washington, DC: Resources for the Future and CIFOR.

Cronkleton, P., S. Saigal, and J.M. Pulhin. 2012. "Co-management in community forestry: How the partial devolution of management rights creates challenges for forest communities." *Conservation and Society* 10 (2): 91–102. http://dx.doi.org/10.4103/0972-4923.97481.

Davis, E.J. 2008. "New promises, new possibilities? Comparing community forestry in Canada and Mexico." *BC Journal of Ecosystems and Management* 9 (2): 11–25.

Duinker, P.N., P. Matakala, F. Chege, and L. Bouthillier. 1994. "Community forestry in Canada: An overview." *Forestry Chronicle* 70 (6): 711–20. http://dx.doi.org/10.5558/tfc70711-6.

Duinker, P., P. Matakala, and D. Zhang. 1991. "Community forestry and its implications for northern Ontario." *Forestry Chronicle* 67 (2): 131–35. http://dx.doi.org/10.5558/tfc67131-2.

Dunster, J. 1994. "Managing forests for forest communities: A new way to do forestry." *International Journal of Ecoforestry* 10 (1): 43–47.

Fennell, L.A. 2011. "Ostrom's law: Property rights in the commons." *International Journal of the Commons* 5 (1): 9–27.

Folke, C., T. Hahn, P. Olsson, and J. Norberg. 2005. "Adaptive governance of socio-ecological systems." *Annual Review of Environment and Resources* 30: 441-73.

Freudenburg, W.R. 1992. "Addictive economies: Extractive industries and vulnerable localities in a changing world economy." *Rural Sociology* 57 (3): 305–32. http://dx.doi.org/10.1111/j.1549-0831.1992.tb00467.x.

Furness, E., and H. Nelson. 2012. "Community forest organizations and adaptation to climate change in British Columbia." *Forestry Chronicle* 88 (5): 519–24. http://dx.doi.org/10.5558/tfc2012-099.

Glasmeier, A.K., and T. Farrigan. 2005. "Understanding community forestry: A qualitative meta-study of the concept, the process, and its potential for poverty alleviation in the United States case." *Geographical Journal* 171 (1): 56–69. http://dx.doi.org/10.1111/j.1475-4959.2005:00149.x.

Gunderson, L.H. 2003. "Adaptive dancing: Interactions between social resilience and ecological crises." In *Navigating social-ecological systems: Building resilience for complexity and change,* edited by F. Berkes, J. Colding, and C. Folke, 33-52. Cambridge, UK: Cambridge University Press.

Hajjar, R.F., R.A. Kozak, and J.I. Innes. 2012. "Decentralization leading to real decision-making power for forest-dependent communities: Case studies from Mexico and Brazil." *Ecology and Society* 17 (1): 12. http://dx.doi.org/10.5751/ES-04570-170112.

Haley, D. 2002. "Community forests in British Columbia: The past is prologue." *Forests, Trees, People Newsletter* 46: 54–61.

Harvey, S., and B. Hillier. 1994. "Community forestry in Ontario." *Forestry Chronicle* 70 (6): 725–30. http://dx.doi.org/10.5558/tfc70725-6.

Howlett, M., J. Rayner, and C. Tollefson. 2009. "From government to governance in forest planning? Lessons from the case of the British Columbia Great Bear Rainforest initiative." *Forest Policy and Economics* 11 (5–6): 383–91. http://dx.doi.org/10.1016/j.forpol.2009.01.003.

Inglis, A.S. 1999. "Implications of devolution for participatory forestry in Scotland." *Unasylva* 199 (50): 45–50.

Jeanrenaud, S. 2001. *Communities and forest management in Western Europe: A regional profile of the Working Group on Community Involvement in Forest Management.* Gland, Switzerland: International Union for the Conservation of Nature.

Kearney, J., F. Berkes, A. Charles, E. Pinkerton, and M. Wiber. 2007. "The role of participatory governance and community-based management in integrated coastal and ocean management in Canada." *Coastal Management* 35 (1): 79–104. http://dx.doi.org/10.1080/10.1080/08920750600970511.

Krogman, N., and T. Beckley. 2002. "Corporate 'bail-outs and local 'buyouts': Pathways to community forestry?" *Society and Natural Resources* 15 (2): 109–27. http://dx.doi.org/10.1080/089419202753403300.

Larson, A.M., D. Barry, G. Ram Dahal, and C.J.P. Colfer. 2010. *Forests for people: Community rights and forest tenure reform.* London, UK: Earthscan.

Lawrence, A., B. Anglezarke, B. Frost, P. Nolan, and R. Owen. 2009. "What does community forestry mean in a devolved Great Britain?" *International Forestry Review* 11 (2): 281–97. http://dx.doi.org/10.1505/ifor.11.2.281.

Lemos, M.C., and A. Agrawal. 2006. "Environmental governance." *Annual Review of Environment and Resources* 31 (1): 297–325. http://dx.doi.org/10.1146/annurev.energy.31.042605.135621.

Luckert, M.K. 1999. "Are community forests the key to sustainable forest management? Some economic considerations." *Forestry Chronicle* 75 (5): 789–92. http://dx.doi.org/10.5558/tfc75789-5.

Luizza, M.W. 2011. "Community forestry in Canada and the United States: Patterns of governance along the public/private continuum." Presentation to the Colorado Conference on Earth Systems Governance: Crossing Boundaries and Building Bridges, 17–20 May 2011, Fort Collins, Colorado State University.

M'Gonigle, M., B. Egan, and L. Ambus. 2001. *The community ecosystem trust: A new model for developing sustainability.* Victoria, BC: POLIS Project on Ecological Governance, University of Victoria.

Masse, S. 1995. *Community forestry: Concept, application, and issues.* Ste-Foye, QC: Natural Resources Canada, Canadian Forest Service.

Matakala, P.W., and P. Duinker. 1993. "Community forestry as a forest-land management option in Ontario." In *Forest dependent communities: Challenges and opportunities,* edited by D. Bruce and M. Whitla, 26–58. Sackville, NB: Rural and Small Town Research and Studies Program, Mount Allison University.

McCarthy, J. 2006. "Neoliberalism and the politics of alternatives: Community forestry in British Columbia and the United States." *Annals of the Association of American Geographers* 96 (1): 84–104. http://dx.doi.org/10.1111/j.1467-8306.2006.00500.x.

McDermott, C., and K. Schreckenberg. 2009. "Equity in community forestry: Insights from North and South." *International Forestry Review* 11 (2): 157-70.

McIlveen, K., and B. Bradshaw. 2005-6. "A preliminary review of British Columbia's Community Forest Pilot Project." *Western Geography* 15–16: 68–84.

–. 2009. "Community forestry in British Columbia, Canada: The role of local community support and participation." *Local Environment* 14 (2): 193–205. http://dx.doi.org/10.1080/13549830802522087.

Pagdee, A., Y.S. Kim, and P.J. Daugherty. 2006. "What makes community forest management successful: A meta-study from community forests throughout the world." *Society and Natural Resources* 19 (1): 33–52. http://dx.doi.org/10.1080/08941920500323260.

–. 2007. "A response to Bradshaw's commentary article: On definitions of 'success' and contingencies affecting success in community forestry." *Society and Natural Resources* 20 (8): 759–60. http://dx.doi.org/10.1080/0894192070 1429189.

Parkins, J.R. 2006. "De-centering environmental governance: A short history and analysis of democratic processes in the forest sector of Alberta, Canada." *Policy Sciences* 39 (2): 183–202. http://dx.doi.org/10.1007/s11077-006-9015-6.

Pinkerton, E., and J. Benner. 2013. "Small sawmills persevere while the majors close: Evaluating resilience and desirable timber allocation in British Columbia, Canada." *Ecology and Society* 18 (2): 34. http://www.ecologyandsociety.org/vol18/iss2/art34/.

Pinkerton, E., R. Heaslip, J. Silver, and K. Furman. 2008. "Finding 'space' for co-management of forests within the neoliberal paradigm: Rights, strategies, and tools for asserting a local agenda." *Human Ecology* 36 (3): 343–55. http://dx.doi.org/10.1007/s10745-008-9167-4.

Porter-Bolland, L., E.A. Ellis, M.R. Guariguata, I. Ruiz-Mallén, S. Negrete-Yankelevich, and V. Reyes-Garcia. 2012. "Community managed forests and forest protected areas: An assessment of their conservation effectiveness across the tropics." *Forest Ecology and Management* 268: 6–17. http://dx.doi.org/10.1016/j.foreco.2011.05.034.

Reed, M., and K. McIlveen. 2006. "Toward a pluralistic civic science? Assessing community forestry." *Society and Natural Resources* 19 (7): 591–607. http://dx.doi.org/10.1080/08941920600742344.

–. 2007. "Other voices from the neighbourhood: Reconsidering success in community forestry – A response to Bradshaw's commentary paper: 'On definitions of "success" and contingencies affecting success in community forestry.'" *Society and Natural Resources* 20 (8): 755–58. http://dx.doi.org/10.1080/08941920701429304.

Ribot, J. 2010. "Forestry and democratic decentralization in Sub-Saharan Africa: A rough review." In *Governing Africa's forests in a globalized world*, edited by L. German, A. Karsenty, and A. Tiani, 29–55. London, UK: Earthscan.

Schlager, E., and E. Ostrom. 1992. "Property-rights regimes and natural resources: A conceptual analysis." *Land Economics* 68 (3): 249–62. http://dx.doi.org/10.2307/3146375.

Secco, L., D. Pettenella, and P. Gatto. 2011. "Forestry governance and collective learning process in Italy: Likelihood or utopia?" *Forest Policy and Economics* 13 (2): 104–12. http://dx.doi.org/10.1016/j.forpol.2010.04.002.

Shackleton, S., B. Campbell, E. Wollenberg, and D. Edmunda. 2002. "Devolution and community-based natural resource management: Creating space for local people to participate and benefit?" *ODI Natural Resources Perspectives* 76: 1–4.

Smith, M.A. 2013. "Natural resource co-management with aboriginal peoples in Canada: Coexistence or assimilation?" In *Aboriginal peoples and forest lands in Canada*, edited by D.B. Tindall, R.L. Trosper, and P. Perreault, 89-113. Vancouver: UBC Press.

Smith, M.A., L. Palmer, and C. Shahi. 2012. "We are all treaty people: The foundation for community forestry in Northern Ontario." In *Pulp friction: Communities and the forest industry in a global perspective*, edited by R.N. Harpelle and M.S. Beaulieu, 100–20. Northern and Regional Studies Series 21. Thunder Bay, ON: Centre for Northern Studies, Lakehead University.

Teitelbaum, S. 2010. *Local people managing local forests: An information guide to community forestry in Quebec*. Nicolet, QC: Solidarité rurale du Québec.

Teitelbaum, S., T. Beckley, and S. Nadeau. 2006. "A national portrait of community forestry on public land in Canada." *Forestry Chronicle* 82 (3): 416–28. http://dx.doi.org/10.5558/tfc82416-3.

Tindall, D.B., R.L. Trosper, and P. Perreault. 2013. *Aboriginal peoples and forest lands in Canada*. Vancouver: UBC Press.

Tollefson, C., F. Gale, and D. Haley. 2008. *Setting the standard: Certification, governance, and the Forest Stewardship Council*. Vancouver: UBC Press.

Tyler, S., L. Ambus, and D. Davis-Case. 2007. "Governance and management of small forest tenures in British Columbia." *Journal of Ecosystems and Management* 8 (2): 67–79.

Western, D., R.M. Wright, and S.C. Strum. 1994. *Natural connections: Perspectives in community-based conservation*. Washington, DC: Island Press.

World Resources Institute. 2009. *The governance of forests toolkit: A draft framework of indicators for assessing governance of the forest sector*. Washington, DC: Governance of Forests Initiative, World Resources Institute.

Wyatt, S. 2008. "First Nations, forest lands, and 'Aboriginal forestry' in Canada: From exclusion to comanagement and beyond." *Canadian Journal of Forest Research* 38 (2): 171–80. http://dx.doi.org/10.1139/X07-214.

# PART 1
## Regional Portraits

Chapter 1    **The Roots of Community Forestry**
Subsistence and Regional Development
in Newfoundland

*Erin C. Kelly and Sara Carson*

*Isolation and hardship bred an overpowering sense of
place. Newfoundlanders belong to a series of widening
circles: to their family; to their parish; to their faith;
to their hamlet, bay, stretch of coastline, and, above all,
to their Island.*

Community forestry generally includes devolved control over natural
resources, accrual of benefits to local communities, and management
based on ecological sustainability (Charnley and Poe 2007), but the
forms taken by community forestry are locally distinct and depend on
the circumstances of a place. Newfoundland is a place with a long history of common-pool management of resources but without formalized
community tenure rights.[1] In this chapter, we identify customary access and rights as the "roots" of community forestry in Newfoundland
and describe a recent effort to create a community forest tenure in the
remote Great Northern Peninsula (GNP), a region struggling to capture economic and social benefits from its natural resources following
the collapse of the cod fishery in the early 1990s and years of instability and contraction in the forest sector. The GNP is typical of much of
Newfoundland: it has an economically and culturally vital subsistence
economy, high levels of rural out-migration, declining traditional industries (fisheries and forestry), and geographic isolation.

In this chapter, we review community forest literature regarding
devolution and forest access and then provide an overview of forestry

in Newfoundland from a historical and political perspective. We explore reasons why community forestry has not yet been successfully implemented on the island and examine the proposed GNP Community Forest, its proponents, and current forest users in the region as a case study of possible community forest tenure. Through the GNP Community Forest case study, we consider the potential impacts of community forestry in Newfoundland – particularly how a community forest could combine regional and local economic development with natural resource management, provide a platform for resolving land-use conflicts, and integrate existing forest users and government agencies with forest planning processes.

## Methods

In the summers of 2011 and 2012, we developed the GNP case study because of the region's struggling forest sector and early discussions among community leaders and economic development groups concerning community forestry. We conducted semi-structured interviews with residents of the GNP (n = 29). These interviewees comprised forest managers and industry employees (4), logging contractors (7), economic development officers (4), community leaders (7), outfitters (4), and other community members (3). Interview topics centred on questions about the changing role of forestry in the region – its history and possible futures, forest planning and management, and economic constraints and opportunities in forestry and other sectors. We also interviewed forest policy experts, academics, and forest managers across the province (n = 65) to provide context for the case. These interviews centred on forest policy in the province and the role of communities and the public in forest management. One of us (Erin Kelly) also attended two meetings in 2011 regarding the possible creation of a community forest in the GNP and subsequently helped community forest proponents craft a statement in support of a pilot community forest project.

## Literature Review: Impacts of Community Forestry

Community forests devolve some decision-making authority to a group of citizens to shape management and determine distribution of benefits (Brendler and Carey 1998; Ambus and Hoberg 2011). Devolution, or vesting power or rights directly in local or non-state groups, differs from decentralization, which involves granting management functions to lower branches of government (Charnley and Poe 2007). Ideally, community forestry not only relocates the nexus of power but also involves a more diverse array of voices and broad community representation (Reed and McIlveen 2006).

Forest community access includes subsistence activities, which are important in rural communities across Canada (see Duinker et al. 1994; Beckley and Hirsch 1997; Beckley 1998). Subsistence activities, which utilize natural resources to meet material and cultural needs outside the formal market, forge links between local communities and forests, and subsistence users may regard themselves as legitimate stewards of resources regardless of formal management authority (Emery and Pierce 2005).

Many rural residents therefore already have substantial access to forests and considerable knowledge about their resources; we argue that this de facto commons arrangement may be strengthened and formalized through community forestry. A traditional view of the commons asserts that overexploitation of shared natural resources is inevitable and that privatization or state management are the only viable solutions (Hardin 1968). But a large body of evidence suggests that resource users have a wide array of common-pool governance structures, often resulting in sustainable, efficient, equitable resource use (Ostrom 1999). Formalizing commons arrangements through clearly delineated management and tenure rights may empower local communities and give local residents a sense of ownership over the fate of nearby landscapes (Kellert et al. 2000).

Community forestry is a form of commons governance that grants local residents decision-making power; as such, it has a host of purported benefits, including economic benefits, as communities retain more revenues and reinvest in local projects (Dunster 1994; Krogman and Beckley 2002; Teitelbaum 2009) and "set the pace of [their] own development" (Varghese et al. 2006, 506); political benefits, through collaborative decision-making that responds to local needs and contributes to the reinvigoration of local democracy (Beckley 1998; Robinson, Robson, and Rollins 2001; Teitelbaum 2009); social benefits, as communities build capacity and more equitable distribution mechanisms (Baker and Kusel 2003); and ecological benefits, as communities oversee appropriately scaled, place-based forest management (Bray et al. 2003). We are focused on the possibilities of these benefits for a place like the GNP, though we recognize that benefits may not materialize or may not be realized concurrently. Rather, community forest tenure creation may follow the model of building a house, in which social change is the foundation of the house and other benefits (especially ecological) are added at a later date (McDermott 2009). We focus on potential social, political, and economic implications of formalizing community forest tenure in Newfoundland.

### Forestry in Newfoundland: Parallel Tenures

From the sixteenth century through the ninteenth century, the forests of Newfoundland supplemented the fisheries, which formed the primary industry of the island. Beginning in the early twentieth century, the pulp and paper industry rose to prominence, and at its peak in the 1930s, it produced 53 percent of total goods exported from Newfoundland (Munro 1978). Tenure thus evolved along two paths: a "three-mile commons," or "fisherman's reserve," along the coast, supporting the fisheries and domestic forest uses, and government-supported industrial tenures targeting inland forest resource development (Munro 1978). Even after the dissolution of the formal three-mile commons in the mid-twentieth century, the coastal forests surrounding rural communities continued to provide wood, primarily for domestic harvesting and locally owned sawmills.

The coastal lands, designated "unalienated" Crown lands because they had no pulp and paper leases, provided the setting for a subsistence economy that has remained central to community well-being in Newfoundland (Den Otter and Beckley 2002). As fisheries, mines, and logging camps opened and closed in rural communities, workers maintained diversified livelihoods by combining seasonal fishing and forestry income with subsistence activities such as domestic fuelwood and sawlog harvesting, moose hunting, snaring, fishing, berry and mushroom picking, and gardening (Omohundro 1995). Subsistence activities declined through the twentieth century as the formal economy grew but rebounded in the late twentieth century, in part because of a revitalization of traditional Newfoundland culture (Omohundro 1995) and "as a recreation, a regional mark of distinction, a bank of useful skills, an expression of self-esteem, a way to stretch limited cash, and an insurance against sudden drops in a household's income" (xviii). A cultural sense of entitlement to the forest, including the right to harvest trees and to build hunting cabins, shaped aspects of forest access and governance across the island.

Access to forests well beyond the bounds of the three-mile commons grew with technology and an expanding road network. Until technological changes occurred around the 1970s – including the expansion of the road network and broader access to technology such as vehicles, ATVs, and snowmobiles – domestic users were largely confined to unalienated Crown lands. This was due to the proximity of unalienated lands to towns, though according to one interviewee, some industry-leased lands were gated and even had guards. With expanded physical access, Newfoundlanders harvested domestic wood and built cabins across the island, including on leased industry lands.

For most of the twentieth century – and arguably through the current day – the needs of the pulp and paper industry determined forest policies and management priorities throughout the province, as implemented by the Department of Natural Resources, Forestry Branch (DNR Forestry). The provincial government supported the industry through guaranteed loans and grants, generous tax incentives, road-building and silvicultural assistance, and control over hydropower resources (Ommer 2007). On pulp and paper–tenured lands, which covered two-thirds of the island, industry enjoyed relative autonomy and decision-making authority (APEC 2008). Though pulp and paper companies technically owned only the trees, they created management plans and granted cabin permits, effectively "regulat[ing] internal use patterns" of the landscape (Schlager and Ostrom 1992, 251) and determining the end uses and beneficiaries of forest utilization. The province continued to support the pulp and paper industry with subsidies and other assistance, despite the closure of two of the island's three mills after 2005, as well as capacity reduction and tenure relinquishments at the one remaining mill, Corner Brook Pulp and Paper (CBPP; Auditor General of NL 2011; Wernerheim and Long 2011). As a result of mill closures and land relinquishments, pulp and paper leases in 2011 amounted to less than one-third of the island, while the remainder was unalienated Crown lands, which were managed by DNR Forestry for sawmill and domestic use. The provincial government tendered an Expression of Interest for many of the relinquished pulp and paper lands immediately following the closure of a major mill in 2009, demonstrating the government's commitment to maintain forest land in commodity production. Until the recent decline of the pulp and paper industry, chronic wood deficits on the island because of pulp and paper–fibre needs allowed for little flexibility in forest planning.

Most planning, and most forestry expertise within DNR Forestry, was narrowly defined by the goal of optimizing fibre allocation for the pulp and paper industry. In addition, the government created forest tenure systems that favoured the pulp and paper industry: long-term tenure agreements were only granted to pulp and paper companies, while all other forest users, including local sawmills, relied on unalienated Crown lands permits or exchange agreements with the pulp and paper industry to procure wood.

While DNR Forestry has favoured the pulp and paper industry, it has had an uneasy relationship with the three-mile commons and with domestic wood cutters in general. For example, in the early twentieth century, DNR Forestry allowed industrial pulp and paper encroachment in the three-mile commons, against the wishes of many subsistence users (Cadigan 2006). From 1984 to 1986, DNR Forestry created the only community forest model on record in Newfoundland to address the perceived problem of "uncontrolled indiscriminate domestic cutting" (Roy 1989, 345). The community forest was created on 550 hectares just north of Gros Morne National Park in the GNP. Residents of the nearby community were consulted but were not granted control over management of the forest. Rather, DNR Forestry paid domestic harvesters to cut according to the government department's specifications. DNR Forestry has thus displayed uneasiness about the concept of allowing communities control over nearby forest lands.

### Forest Policies Shift

In the 1990s, DNR Forestry shifted its policy objectives from timber-based sustained-yield management to more diverse objectives such as ecosystem-based management and inclusive public participation (Nazir and Moores 2001). But mechanisms for implementing these new objectives remained unclear, and a gap between policy objectives and forest planning became evident (Auditor General of NL 2011).

The tendency at DNR Forestry was to optimize commercial harvest, which the department considered "eroded" by competing uses, including domestic harvest, municipal watersheds, wildlife habitat, and cabin building. In the words of the 2008 five-year management plan for the GNP, "The land base available for forest activity is constantly being eroded by other [non-commercial] users" (Newfoundland and Labrador, DNR Forestry 2008, 48). In other words, rather than integrating competing uses, forest planners in Newfoundland appeared to view different forest uses as mutually exclusive.

As part of its five-year planning process, DNR Forestry held public meetings in various communities in order to gain public input. According to most interviewees, including employees of DNR Forestry, residents had little substantive input into forestry plans. At meetings, residents were presented with maps of harvest areas that had already been determined and justified through internal DNR Forestry planning exercises. Participants were allowed to comment on proposed harvest areas only, but they were permitted no input into management priorities in general. Most meetings were poorly attended and most interviewees – both in the GNP and in other regions of Newfoundland – felt that forest plans were completed before their input was sought. One outfitter on the GNP who had travelled hours to attend a meeting said, "You could travel [to meetings] and voice your opinion on the plan, but it didn't matter."[2]

Regional DNR Forestry staff were similarly frustrated: one manager indicated that "local values" were complicating forest planning and management and that people came to the meetings with "an agenda" that he could not resolve. Government agencies dealing with other natural resource and land-use issues – such as matters related to wildlife, parks and natural areas, agriculture, and mining – had little input into five-year plans, and collaboration was limited. This resulted in the establishment of separate silos, with each agency attending to its own mandate. This policy structure was ineffective for reconciling diverse land-use objectives and for incorporating local voices into planning.

### Setting the Stage: The Great Northern Peninsula

GNP is a sparsely populated peninsula that is approximately eighty kilometres wide and extends three hundred kilometres from its southern end at Gros Morne National Park to its northern tip. The Long Range Mountains form the spine of the peninsula, separating the west coast, which contains most of the GNP's communities and the only major road (Hwy 430), from the isolated east coast. Forests of the GNP are boreal and dominated by balsam fir trees. The GNP has a short growing season, and productive forest stands are naturally highly fragmented and interspersed with scrub, bogs, and rocky highlands (Newfoundland and Labrador, DNR Forestry 2008).

Two mills – located in Grand Falls (built in 1909) and Corner Brook (built in 1925) – had substantial land claims on the interior forests of the GNP.[3] The Grand Falls mill closed in 2009, and Corner Brook Pulp and Paper relinquished its lands on the GNP in 2010, leaving unalienated Crown land as the sole tenure in the region. Forestry revenues in the GNP shrank from $19 million in 2008 to $9 million in 2010, while

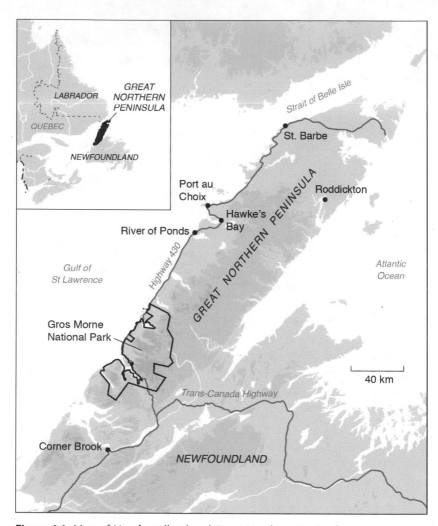

**Figure 1.1** Map of Newfoundland and Great Northern Peninsula.
*Source:* Original map, data from Government of Newfoundland. Recreated by Eric Leinberger.

the labour force declined from 140 forest workers in 2003 to 75 in 2010 (RED Ochre Board 2011).

The proposed GNP Community Forest would extend along the west coast of the GNP, comprising approximately 100,000 hectares of land adjacent to twenty-one communities, five of which are incorporated (see Figure 1.1). This area, including the human communities, is referred to as the St. Barbe Development Association (SBDA) Region.

TABLE 1.1 Demographic information for the communities of the
SBDA Region and Canada as a whole

|  |  | Community forest region of GNP | Canada |
|---|---|---|---|
| Population | 2006 | 5,502 | 31,612,897 |
|  | 2001 | 6,183 | 30,007,094 |
|  | Change from 2001–06 | –11% | + 5.4% |
| Median household income (after taxes), 2005 |  | $36,968 | $55,111 |
| Adults aged 25–64, 2006 | % with no high school diploma | 45.2% | 15.4% |
|  | % with a university degree | 7.6% | 22.9% |

Source: Statistics Canada (2006).

Newfoundland was long one of the most rural and economically depressed provinces of Canada. Though some parts of Newfoundland have recently benefited from offshore oil revenues, demographic trends for the community forest region as measured by Statistics Canada reveal population decline and low income and education levels (Table 1.1).

However, some researchers have found that demographic data do not capture well-being in rural Newfoundland, because of the importance of the informal subsistence economy (Den Otter and Beckley 2002).

## Proponents of the Community Forest

As the once-dominant pulp and paper industry diminished across the province and eventually abandoned some regions altogether, including the GNP, a forest tenure and management vacuum arose. A group of regional leaders in the GNP, representing a number of diverse institutions, stepped up and submitted a proposal for a community forest pilot project to the minister of Natural Resources in November 2011.[4] These leaders included city council members and mayors of the incorporated municipalities, employees of three rural economic and business development organizations, and locally based employees of many provincial agencies, including DNR Forestry and several provincial economic development agencies.

Many of the economic development agencies interested in community forestry were the legacies of rural development efforts begun in the 1960s that were meant to spur sustainable growth in small, isolated communities (Blake 2003). One prominent group created in 1974, the SBDA, proposed to spearhead the effort towards community forestry. The SBDA had some experience with forestry projects, including organizing a silviculture training program and several forest restoration projects, but it lacked the authority to manage land use.

Meanwhile, DNR Forestry, which had forest-management authority and expertise, did *not* have an explicit rural development mandate. It therefore did not prioritize different forest uses from the perspective of maximizing employment or returning benefits to local communities. In 2011, commercial logging was carried out by seventeen independent contractors remaining in the region, many of whom were not operating full time. For this project, we contacted seven of those contractors, all of whom expressed pessimism and resignation about the decline of commercial forestry. They had significant debt and were wary of investing more into an industry that was in a downward spiral. As one contractor said, "Kruger [CBPP] asked we [*sic*] to get another [harvester] and then they left. We had to do something to pay the banks so we started cutting for different people. Then the price went down and costs went up."[5] This contractor spoke of moving to Alberta for work, following the trend of many residents of the GNP who moved away to find jobs elsewhere. Although two small commercial sawmills remained in the GNP, the pulp and paper–dominated tenure and management system that had long been supported by DNR Forestry disappeared.

In meetings and interviews, project participants described the traditional forest sector as disconnected from community well-being and spoke of community forestry as an avenue for resizing forestry within the region and maintaining benefits for local communities. In the words of one mayor, former forestry projects were "too big" and community forestry "would be smaller pieces, a self-sustained, small industry." An economic development officer identified the need to transform forestry by developing tourism and non-timber forest product markets:

> The traditional rural business plan is simplistic: you have x resource and x capacity, and you need a certain amount of money for machinery, and you take resources to market. But now somebody is

talking about ecosystem interpretation, somebody about mountain biking, somebody about logging, somebody about mushrooms. There are multiple potential revenue streams and a very cluttered business plan.[6]

Interviewees frequently described the traditional forest industry as "outside" the region. In contrast, the various economic and natural resource groups enjoyed local support and representation. The SBDA, for example, had twenty of the twenty-one communities in the region represented on its board and was created for communities to "tak[e] more control themselves over their economic destinies" (Blake 2003, 207). This objective meshed with the idea of relocating forest access – especially the flows of benefits – to a local scale. As one economic development officer noted, "A community forest can't walk away."

## The Forest Users

After the pulp and paper industry left the GNP in 2010, forest uses in the region were dominated by subsistence and recreational activities. In the district that includes the proposed community forest, planned domestic harvests constituted more than one-third of the total planned harvest (see Table 1.2), and actual domestic harvests were an even larger percentage.[7] As many as 80 percent of households used firewood as their primary heat source (Omohundro 1994); In all, more than 2,800 domestic harvesting permits were issued in the GNP in 2011.

TABLE 1.2   Harvest information for Forest Management District 17, which includes most of the proposed community forest

|  | District 17* |
| --- | --- |
| Total land | 587,076 ha |
| Productive forest | 203,792 ha |
| Total harvest scheduled (2008–12) | 291,197 m³ |
| Total domestic harvest scheduled (2008–12) | 111,413 m³ |
| Domestic as proportion of total (planned) harvest | 38% |

\* The proposed community forest is almost entirely within District 17, which covers the western part of the GNP. District 18 contains both remaining sawmills and a pellet mill, and lies north and east of District 17. The districts share a forest manager, and planning processes are combined for the two districts.
*Source:* Anderson (2011).

Other subsistence activities included hunting and cabin-building, with about nine thousand moose licences issued in the GNP (Newfoundland and Labrador, Environment and Conservation 2011) and exceptional access for cabin building. Though cabin building required a permit, if a cabin was built on any forest access road without permission, the cabin owner paid an "illegal occupation fee" of only $500, and all cabin owners, legal or illegal, paid $100 in annual land rent. Long-standing tensions existed between domestic harvesters and DNR Forestry, which criticized domestic fuelwood and sawlog harvesters as wasteful or inefficient and at odds with commercial forestry (Sinclair and Kean 2006).

Though regulations existed – domestic harvests required a permit, and wood removal was generally limited to non-commercial species (hardwoods, larch) in cutover stands or on designated domestic harvest units – interviewees were divided about the sufficiency of regulations and enforcement. One economic development officer said that domestic forest harvesting resulted in high-grading, with the best trees cut out and the worst trees left behind, and that "domestic firewood monitoring needs to be stepped up ... Now domestic wood cutters just get a pamphlet; it's very unregulated." An alternative perspective came from a mayor, who said, "Some conditions they [DNR Forestry] impose on people impose hardship ... We can give the domestic cutter what he wants; he can't hurt the forest – there's only so much he can use."

Despite these differences of opinion, all interviewees supported the important economic and cultural role of domestic forest use. As one mayor stated, "The most value in forests for rural Newfoundland is when the local man can go to the woods and build his home." The interviews clearly showed that domestic use is an important component of Newfoundlanders' connection to place; as one economic development officer stated, "It's a connection to place, to their homes, that makes people want to make this community forest idea work."[8]

## Discussion: Envisioning a Community Forest Approach in the GNP

As outlined above, Newfoundland has a long history of commons rights and subsistence uses of the forest. It also has a regional leadership structure that includes a number of economic development agencies and natural resource management agencies. Formalizing local rights to the commons through a community forest may provide some solutions to the challenges outlined above, including economic development, land-use conflicts, and lack of local involvement in making decisions related to forest uses.

We consider the creation of a community forest as an experiment in governance, in accordance with the adaptive management framework outlined in the province's 2003 *Sustainable Forest Management Strategy* (Newfoundland and Labrador, DNR Forestry 2003). The GNP Community Forest serves as an experiment in implementing a complex adaptive system, wherein rules may be changed according to the needs of users, as described by Ostrom (1999).

This governance experiment, however, is not guaranteed to work. In many jurisdictions, community forestry has encountered problems that are often related to a lack of central government support or a reluctance to grant control to communities (Charnley and Poe 2007; Ambus and Hoberg 2011). The Government of Newfoundland has demonstrated a resistance to ceding control over its lands, especially if potential industrial development opportunities could be threatened. The provincial government has continued to support the pulp and paper industry, despite warnings of its continuing decline (Milley 2008). For now, long-term tenure rights on Crown lands link forest tenure to wood-processing capacity and are granted only to pulp and paper companies.

Capacity obstacles within the province could also impair the establishment of a successful community forest tenure. These obstacles include human resource concerns – in particular, declining human capital as a result of out-migration and a perceived lack of entrepreneurial motivation – and forest industry concerns. A lack of forest product diversification is of special concern, since the market for pulp has deteriorated and there is a dearth of value-added processing facilities in the province (Carson 2013). Without viable economic prospects, a community forest tenure would not be likely to succeed.

In the following sections, we outline some of the potential benefits of community forestry, while also pointing to the very real constraints facing the implementation of this tenure.

## Integrating Economic and Natural Resource Mandates

Newfoundland has a wealth of both local and provincial economic development groups, many of which emerged from a rural development movement in the late 1960s and early 1970s that focused on small-scale enterprise and took a cue from Third World "small is beautiful" development literature (House 2003, 228). These groups, including the SBDA, have persisted in many rural regions, focusing on job creation and capacity building. By contrast, natural resource agencies of Newfoundland, particularly DNR Forestry, have focused on land management and resource sustainability and have contributed to an overriding economic policy within the province of supporting large-scale industry development. These distinct governmental objectives have not been well integrated in the past, as evidenced by forestry policies that prioritized the needs of pulp and paper companies over local employment (Omohundro 1994) and, more recently, have been slow to respond to the complete abandonment of the region by industry. Forestry development has frequently been misaligned with the development needs of rural Newfoundland (Byron 2003).

Community forestry may address this disjuncture between rural economic revitalization and forestry. Both the various economic development boards and the district DNR Forestry office are regionally based, decentralized provincial government agencies that could facilitate the devolution of power and decision-making to a community forest authority (CFA), shifting power from distant provincial offices to a regionally based, broadly representative authority.

The CFA could be responsible for allocating timber for commercial uses, in line with the traditional mandates of DNR Forestry, but could also oversee a locally grounded approach to forest resources. The CFA, working across the established silos of different agencies and community groups, could realign forest-based economic opportunities to fit with regional economic development plans. Different objectives and economic visions for the forest could then be prioritized among the various communities and stakeholder groups, including outfitters, subsistence users, recreationists, and commercial wood users, and revenues could be generated for use in the local economy, providing opportunities for further local development (Krogman and Beckley 2002; Varghese et al. 2006). The community forest tenure could therefore expand forest utilization to include a host of "alternative" economic projects, grounded in community expectations and needs (McCarthy 2006).

The pulp and paper industry left the GNP because the region was no longer economically viable for its operations. The provincial government has not demonstrated a high capacity for industrial diversification, instead subsidizing an aging and deteriorating pulp industry (Milley 2008). Community forest supporters would need to think beyond traditional economic models to nurture entrepreneurial businesses that have a high risk of failure, but who would take on this challenge is unknown. The region has experienced high levels of outmigration – especially of young, motivated people (McGinn 2010). The remaining residents of the GNP tend to be older and may have little interest in investing time and money in an economically and politically risky forestry business.

## Resolving Land-Use Conflicts at the Local Level

Land-use conflicts in Newfoundland reflect the diversity of forest uses in the province, especially the importance of subsistence and recreational uses. The traditional decision-making system linked the economic destiny and forest planning within the GNP to the plans of the pulp and paper industry, a system that proved unable to accommodate many alternative forest uses and dissenting voices. Restructuring public participation and interagency collaboration under a community forest could give various government agencies, not limited to DNR Forestry, a clearly delineated region for experimenting with land-use planning and conflict resolution. It would provide a platform for open discussion of management priorities, as they exist at the local level (Robinson, Robson, and Rollins 2001). A community forest can provide a forum for the exchange of ideas and information, and "mutual understanding and trust may follow" (Bullock and Hanna 2008, 80).

The uncertainty surrounding current forest-related decision-making, in large part due to the departure of the pulp and paper industry, offers a window of opportunity. The province itself has identified the need for more inclusive public participation processes, wherein residents and agency members identify forest objectives, participate in workshops, and continuously evaluate and comonitor forest projects and compare desired forest conditions with forest-management objectives (Newfoundland and Labrador, DNR Forestry 1995, 17–18). These activities could be pursued at a local level and be coordinated by the CFA, rather than being entirely the responsibility of DNR Forestry.

## Devolving Decision-Making: Bringing in the Forest Users

Including the many subsistence and recreational forest users in planning could give residents both a voice and a stake in how the forests are managed (Charnley and Poe 2007). Many subsistence users may already have a sense of ownership over the forest (Emery and Pierce 2005), but access to the forest has long been restricted to use rights – domestic wood harvests, hunting, and cabin building – both in the GNP and across the province. The traditional commons, still an integral part of the cultural fabric of Newfoundland, could be formalized through a more systematic planning system, which is one component of community-based management (Kellert et al. 2000).

The land use conflicts discussed in the previous section have proven frustrating for both forest managers and other stakeholders; devolving decision-making authority regarding the uses of the forest and distribution of benefits may increase social acceptability (Robinson, Robson, and Rollins 2001). For example, the remnant forest industry may gain the support of conservation groups as different stakeholders gain knowledge of industry needs through planning. Harvesting could be controlled by residents with a keen interest in supporting an array of commercial operations. It is important to be aware, however, of the experience of British Columbia, where the establishment of community forest tenures has not led to meaningful devolution of decision-making authority (Ambus and Hoberg 2011).

Basing forest planning and management within the region itself may create more place-based accountability and bring local knowledge generated by thousands of forest users to the planning table (Ostrom 1999). Currently, Newfoundland's public participation processes resemble mere tokenism, whereby the public has some opportunity to voice concerns and the rights and options of the public are identified, but the public voice has very little influence over management and planning (Arnstein 1969). The role of the CFA would not only be to identify and help negotiate conflicts by coordinating with leaders and representatives of government agencies; it would also be expected to engage with citizens and to incorporate a wide range of ideas and interests. This could be a very difficult prospect for agency scientists, who would have to recognize and work with diverse opinions and different types of knowledge and expertise (Reed and McIlveen 2006).

A relevant constraint is that the communities of rural Newfoundland evolved from isolated fishing villages; a lingering effect of this is a tendency for residents to identify with their towns rather than their regions (House 2003). A regional forester stated in an interview, "Regional cooperation is a huge issue. Currently, there is [sic] lots of little communities working against each other and fighting for the same projects." Local inequalities may be exacerbated by the devolution of control, a commonly identified challenge for community forests

(Agrawal and Gibson 1999). Inclusiveness would need to be built into the structure of the community forest.

Ostrom (1999) demonstrates that the commons can be governed effectively by local users. Forest management in Newfoundland displays many of the problems of one-size-fits-all rules and planning, limited access and decision-making power for local users, and top-down fixes for local resource needs. Community forestry has the potential to fix these shortcomings by bringing people together to solve common problems, creating a structure for sharing information, and devising common rules.

## Conclusion

The possible benefits of community forestry discussed above are inextricably linked: through the involvement of existing forest users, land-use conflicts could be openly discussed and economic priorities set at a regional level. This could also translate into lessons for other sectors. Researchers have demonstrated the capacity of community groups to build on successes in natural resource management in order to address other challenges and to bring together residents of a region around a common cause (Baker and Kusel 2003).

The community forest tenure may provide precedence for community control over development across the island and in different sectors. The roots of community forestry – especially the importance of the subsistence economy – exist in almost every region of Newfoundland, and the traditional forest products industry has collapsed in many areas outside the GNP, creating opportunities for new tenures. Other natural resource sectors, particularly the fisheries, could look to community forestry as an example and as a source of capacity.

Relocating decision-making authority to local users carries risks and will certainly require immense political will. But as this book demonstrates, there are successful models across Canada, and it may only be a matter of time before Newfoundlanders construct their own version of a community forest.

### Notes
*Epigraph:* R. Gwyn, *Smallwood: The Unlikely Revolutionary* (Toronto: McClelland and Stewart, 1972).

1  The province is called Newfoundland and Labrador. Because this paper focuses on a region of the island of Newfoundland, we refer to Newfoundland, not Newfoundland and Labrador.

2  Interview with outfitter, June 8, 2011.
3  Originally constructed by a consortium of investors, the Corner Brook mill was purchased by International Paper, then Bowater's Pulp and Paper, then Corner Brook Pulp and Paper Limited (owned by Kruger). CBPP is the only remaining pulp and paper mill on the island.
4  At the time of writing, this proposal has not yet gained traction.
5  Interview with logging contractor, July 4, 2011.
6  Interview with economic development officer, June 24, 2011.
7  Domestic harvests were a larger percentage because planned commercial harvests were curtailed with industry decline, and an unknown number of domestic users harvested wood.
8  Ibid., March 9, 2011.

### References

Agrawal, A., and C.C. Gibson. 1999. "Enchantment and disenchantment: The role of community in natural resource conservation." *World Development* 27 (4): 629-49. http://dx.doi.org/10.1016/j.worlddev.2005.07.013.

Ambus, L., and G. Hoberg. 2011. "The evolution of devolution: A critical analysis of the community forest agreement in British Columbia." *Society and Natural Resources* 24 (9): 933–50. http://dx.doi.org/10.1080/08941920.2010.520078.

Anderson, A. 2011. *Domestic fuel wood feasibility study.* Corner Brook, NL: Anderson and Yates Forest Consultants.

APEC (Atlantic Provinces Economic Council). 2008. *Building competitiveness in Atlantic Canada's forest industries: A strategy for future prosperity.* Halifax, NS: APEC.

Arnstein, S.R. 1969. "A ladder of citizen participation." *Journal of the American Institute of Planners* 35 (4): 216–24. http://dx.doi.org/10.1080/019443669 08977225.

Auditor General of NL. 2011. *Annual Report, part 2.14: Forest management.* St. John's: Auditor General of Newfoundland and Labrador.

Baker, M., and J. Kusel. 2003. *Community forestry in the United States: Learning from the past, crafting the future.* Washington, DC: Island Press.

Beckley, T. 1998. "Moving toward consensus-based forest management: Comparison of industrial, co-managed, community, and small private forests in Canada." *Forestry Chronicle* 74 (5): 736–44. http://dx.doi.org/10.5558/tfc74736-5.

Beckley, T., and B. Hirsch. 1997. *Subsistence and non-industrial forest use in the Lower Liard Valley.* Information Report NOR-X-352. Edmonton, AB: Natural Resources Canada, Canadian Forest Service, Northern Forestry Centre.

Blake, R.B. 2003. *Regional and rural development strategies in Canada: The search for solutions.* St. John's, NL: Royal Commission on Renewing and Strengthening Our Place in Canada. http://www.gov.nl.ca/publicat/royalcomm/research/blake.pdf.

Bray, D.B., L. Merino-Perez, P. Negreros-Castillo, G. Segura-Warnholtz, J.M. Torres-Rojo, and H.F.M. Vester. 2003. "Mexico's community-managed forests as a global model for sustainable landscapes." *Conservation Biology* 17 (3): 672–77. http://dx.doi.org/10.1046/j.1523-1739.2003.01639.x.

Brendler, T., and H. Carey. 1998. "Community forestry defined." *Journal of Forestry* 96 (3): 21–23.

Bullock, R., and K. Hanna. 2008. "Community forestry: Mitigating or creating conflict in British Columbia." *Society and Natural Resources* 21 (7): 77–85.

Byron, R. 2003. *Retrenchment and regeneration in rural Newfoundland*. Toronto: University of Toronto Press.

Cadigan, S. 2006. "Recognizing the commons in coastal forests: The three-mile limit in Newfoundland, 1875–1939." *Newfoundland and Labrador Studies* 21 (2): 209–33.

Carson, S. 2013. "An investigation into the potential for community forestry in Newfoundland." Master's thesis, Department of Forestry and Environmental Management, University of New Brunswick, Fredericton.

Charnley, S., and M.R. Poe. 2007. "Community forestry in theory and practice: Where are we now?" *Annual Review of Anthropology* 36 (1): 301–36. http://dx. doi.org/10.1146/annurev.anthro.35.081705.123143.

Den Otter, M., and T. Beckley. 2002. *This is paradise: Community sustainability indicators for the Western Newfoundland Model Forest*. Information Report M-X-216E. Fredericton, NB: Natural Resources Canada, Atlantic Forestry Centre.

Duinker, P., P. Matakala, F. Chege, and L. Bouthillier. 1994. "Community forests in Canada: An overview." *Forestry Chronicle* 70 (6): 711–20. http://dx.doi.org/10.5558/tfc70711-6.

Dunster, J. 1994. "Managing forests for forest communities: A new way to do forestry." *International Journal of Ecoforestry* 10 (1): 43–46.

Emery, M., and A.R. Pierce. 2005. "Interrupting the telos: Locating subsistence in US forests." *Environment and Planning* 37 (6): 981–93. http://dx.doi.org/10.1068/a36263.

Hardin, G. 1968. "The tragedy of the commons." *Science* 162 (3859): 1243–48. http://dx.doi.org/10.1126/science.162.3859.1243.

House, J. 2003. "Does community really matter in Newfoundland and Labrador?" In *Retrenchment and regeneration in rural Newfoundland*, edited by R. Byron, 226–67. Toronto: University of Toronto Press.

Kellert, S.R., J.N. Mehta, S.A. Ebbin, and L.L. Lichtenfeld. 2000. "Community natural resource management: Promise, rhetoric, and reality." *Society and Natural Resources* 13 (8): 705–15. http://dx.doi.org/10.1080/089419200750035575.

Krogman, N., and T. Beckley. 2002. "Corporate 'bailouts' and local 'buyouts': Pathways to community forestry?" *Society and Natural Resources* 15 (2): 109–27. http://dx.doi.org/10.1080/089419202753403300.

McCarthy, J. 2006. "Rural geography: Alternative rural economies – The search for alterity in forests, fisheries, food, and fair trade." *Progress in Human Geography* 30 (6): 803–11. http://dx.doi.org/10.1177/0309132506071530.

McDermott, M. 2009. "Locating benefits: Decision-spaces, resource access, and equity in US community-based forestry." *Geoforum* 40 (2): 249–59. http://dx.doi.org/10.1016/j.geoforum.2008.10.004.

McGinn, J. 2010. "Rural depopulation in Newfoundland and Labrador: Attitudes of young people and the impact of new industry development." Master's thesis, School of Environmental Sciences, University of Ulster, Coleraine, UK.

Milley, P. 2008. *Newfoundland forest sector strategy: Final report*. Submitted to Forestry Services Branch, Department of Natural Resources, Government of Newfoundland. Halifax, NS: Halifax Global Inc.

Munro, J. 1978. "Public timber allocation policy in Newfoundland." PhD diss., Department of Forestry, University of British Columbia, Vancouver.

Nazir, M., and L. Moores. 2001. "Forest policy in Newfoundland and Labrador." *Forestry Chronicle* 77 (1): 61–63. http://dx.doi.org/10.5558/tfc77061-1.

Newfoundland and Labrador. DNR Forestry (Department of Natural Resources, Forestry). 1995. *Environmental preview report: Proposed adaptive management process*. St. John's, NL: Newfoundland Forest Service.

–. DNR Forestry (Department of Natural Resources, Forestry). 2003. *Provincial sustainable forest management strategy*. Corner Brook, NL: Department of Natural Resources.

–. DNR Forestry (Department of Natural Resources, Forestry). 2008. *Districts 17 and 18 five year management plan, 2008–2012*. Corner Brook, NL: Department of Natural Resources.

–. Environment and Conservation. 2011. *Hunting and trapping guide 2011–2012*. St. John's, NL: Department of Environment and Conservation.

Ommer, R. 2007. *Coasts under stress: Restructuring and social-ecological health*. Montreal and Kingston: McGill-Queen's University Press.

Omohundro, J. 1994. *Rough food: The seasons of subsistence in northern Newfoundland*. St. John's, NL: Institute of Social and Economic Research, Memorial University.

Omohundro, J. 1995. "Living off the land." In *Living on the edge: The Great Northern Peninsula of Newfoundland*, edited by L. Felt and P. Sinclair, 103–27. St. John's, NL: Institute of Social and Economic Research, Memorial University.

Ostrom, E. 1999. "Coping with tragedies of the commons." *Annual Review of Political Science* 2 (1): 493–535. http://dx.doi.org/10.1146/annurev.polisci. 2.1.493.

RED Ochre Board. 2011. *Strategic economic plan (2011–2014)*. Parsons Pond, NL: RED Ochre Regional Board Inc., Board of Directors.

Reed, M., and K. McIlveen. 2006. "Toward a pluralistic civic science? Assessing community forestry." *Society and Natural Resources* 19 (7): 591–607. http://dx.doi.org/10.1080/08941920600742344.

Robinson, D., M. Robson, and R. Rollins. 2001. "Towards increased citizen influence in Canadian forest management." *Environments* 29 (2): 21–41.

Roy, M. 1989. "Guided change through community forestry: A case study in forest management unit 17 – Newfoundland." *Forestry Chronicle* 65 (5): 344–47. http://dx.doi.org/10.5558/tfc65344-5.

Schlager, E., and E. Ostrom. 1992. "Property-rights regimes and natural resources: A conceptual analysis." *Land Economics* 68 (3): 249–62. http://dx.doi.org/10.2307/3146375.

Sinclair, P.R., and R.W. Kean. 2006. "Forest politics: Contested issues and governance in forest management for Newfoundland's Great Northern Peninsula."
*Newfoundland and Labrador Studies* 21 (2): 1719–26.

Statistics Canada. 2006. 2006 Census of population. http://www12.statcan.gc.
ca/census-recensement/2006/index-eng.cfm.

Teitelbaum, S. 2009. "An evaluation of the socio-economic outcomes of community forestry in the Canadian context." PhD diss., Department of Forestry and
Environmental Management, University of New Brunswick, Fredericton.

Varghese, J., N. Krogman, T. Beckley, and S. Nadeau. 2006. "Critical analysis of the
relationships between local ownership and community resiliency." *Rural
Sociology* 71 (3): 505–27. http://dx.doi.org/10.1526/003601106778070653.

Wernerheim, M., and B. Long. 2011. *Commercial forestry at a cross-roads: Emerging
trends in the forest sector of Newfoundland and Labrador.* St. John's, NL: Harris
Centre, Memorial University.

## Chapter 2

**Community Forestry in the Maritimes**
Long-Standing Debates and Recent
Developments

*Thomas Beckley*

The main purpose of this chapter is to review the history of the community forest movement in the Maritime provinces and explain the
lack of on-the-ground experience in that region relative to other jurisdictions in Canada. I give special emphasis to New Brunswick and
Nova Scotia, since Prince Edward Island's experience is scant. The
chapter has four interconnected pieces. I begin by explaining a subtle
but important difference between the broad category of forest communities and the much more specific case of community forests. I then
describe the policy history around the idea of community forestry
in New Brunswick and Nova Scotia. Following this context, I offer a
theory about why demand for and practice in community forestry has
been somewhat less in the Maritimes compared to the rest of Canada. I
end the chapter by suggesting an approach for conducting community
forest policy experiments in the region. That approach is adaptive
management.

   The perspective I present here comes from more than a decade of
participation in research and policy debates concerning community
forestry in New Brunswick. I believe that we have not sufficiently
explored the topic or experimented enough with pilots or models to
wholly embrace or reject community forests as a potential management model. I have promoted the idea of testing various models of
community forestry in the Maritimes. As an academic, I have participated in social science research to gauge the social acceptability of
Crown land tenure reform, supervised graduate and postgraduate students who worked on community forestry, contributed to proposals

for pilot projects, and made presentations (by invitation only) to government officials.

## Forest Communities and Community Forests: What Is the Difference?

The geographical region that we now refer to as the Maritime provinces has featured human forest communities for millennia. Historically, Mi'kmaq and Maliseet communities relied on forest resources for their livelihoods. They spent considerable time transforming forest resources into their material culture: food, shelter, clothing, tools, and so on. The forest naturally featured prominently in the rich stories and spiritual dimensions of these cultures as well. Given that most of the region was forested before European colonization, these Indigenous communities may fairly be characterized as forest communities. While less of their material culture is derived from local forests today than in the past, forests still provide medicines, some food, and tremendous cultural value for these long-time residents of the Maritimes.

When Europeans first arrived on North American shores, they were primarily interested in fish, not forests. Once settlement became an agenda of colonial powers, forest resources began to feature more prominently in their affairs. Settlement occurred on the coasts and rivers first, but virtually all Maritime communities were cut from forests, were surrounded by forests, and, to a significant degree, relied on forest resources (Wynn 1981). As the economy developed and matured over the centuries, many of these communities became single industry towns, and most continued to rely on forests to a degree. In the last half-century, we have tended to measure forest dependence as the percentage of the industrial employment base that relies to a significant degree on forest employment. According to Dave Watson, an analyst from the Canadian Forest Service who has been examining custom runs of Statistics Canada data on forest dependence for nearly twenty years (White and Watson 2004, Stedman et al. 2007), in 2006, New Brunswick had 284 census subdivisions. Of these, 39 were heavily forest dependent, with over one-third of their economic base in the forest sector. An additional 84 census subdivisions were moderately dependent, with between 10 and 33 percent of their economies rooted in forest work. In the same census year, Nova Scotia had 22 census subdivisions in the moderately dependent category and 6 in the heavily dependent range. PEI had 10 communities in the moderately forest dependent category and zero that were heavily forest dependent (pers. comm., David Watson, 2013). By these measures, New Brunswick has the highest level of forest dependence in the Maritimes and the largest number of communities that depend primarily on forests for their livelihoods. In reality, measures of jobs and industrial forest dependence are only a part of the story and provide, at best, a partial definition of forest communities. Many places, despite the number of mill or silviculture jobs, are forest communities because they are surrounded by forests and their residents use the forest for their livelihoods, play, and subsistence. As in precontact times, the culture, stories, and spirituality of these places is wrapped up with human interaction with the surrounding forest landscape.

The term *forest communities*, then, refers to human communities that are geographically in and around forests, that rely on forests, and for whom forests are an important part of the culture. *Community forests*, on the other hand, refers to unique institutional arrangements that have some key features. These are explained in more detail in other contributions to this volume (see especially the introductory chapter), but for the sake of quick review, community forests typically include the following four elements:

- Local benefits: the economic and non-economic benefits of forest management should accrue primarily to local residents.
- Participatory governance or local decision-making: participation in the objectives for forest management should be skewed in favour of local residents.
- Respect for diverse rights holders in a multiple-use forestry context: management objectives should include the full suite of market, non-market, and ecosystem service values that flow from forests.
- Ecological stewardship: management should involve a heightened attention and commitment to the long-term maintenance and stewardship of this full suite of forest values.

These tenets are fairly simple and have been derived explicitly in contrast to what many feel are characteristics of status quo Crown land management across Canada: leakage of local economic benefits, dominance of extra-local interests in decision-making, a disproportionate emphasis on fibre values over all other forest values, and the prioritizing of short-term profit taking and political manoeuvring over long-term stewardship.

Community forests remain a contentious policy issue, in part because of misunderstandings about what community forests are or what people think they ought to be. On the one hand, despite increasing experience with the phenomenon, many foresters, politicians, policy analysts, economists, and particularly business people view community forests with extreme skepticism and believe that community forests would bring ruin to the forest industry and usher in an era of haphazard, ill-conceived, and inefficient forest management. On the other hand, some community developers, environmentalists, activists, and academics are persistent and uncritical boosters of community forests and argue that the simple act of placing responsibility for forest management into local hands will create a new dawn for forestry, with more jobs, better provision of non-timber values, and more democracy. It is likely that neither of these extreme positions is an accurate depiction of what would ensue with a more widespread application of community forest-management regimes. However, these perceptions have hampered the development of constructive, critical debate about how a more community-oriented management model might be structured. What is clear is that when it comes to discussions of community forests or tenure reform, representatives from both sides of the debate feel that there is a lot at stake.

Forest communities, then, comprise the hundreds of Maritime communities that are surrounded by forests. For thousands of years, First Nation peoples and subsequent waves of settler societies have relied on those forests for livelihoods, sustenance, and important cultural and spiritual values. Community forests, however, are characterized by a distinct institutional arrangement that devolves significant goal setting and decision-making to the community level. The explicit intent of community forestry is to enhance community benefits, whether social, economic, or ecological. In a media scrum following his testimony before the Legislative Select Committee on Wood Supply in 2003, Jim Irving, CEO of JDI Ltd., and Kevin Matthews, a filmmaker, had the following exchange:

*Kevin Matthews*: In terms of where to go, I am wondering what you think of community-based forestry.
*Jim Irving*: Well, to me, community-based forestry – we have community-based forests today.
*Kevin Matthews*: Well, I am talking about communities controlling that resource directly.

*Jim Irving*: I will tell you, as I mentioned inside, it is a very complex, sophisticated, capital-intensive business. This can't be run on a year-by-year basis. We are talking about a twenty-five-year, a fifty-year horizon, and we need to have the technology in place to make sure we manage it. And we can't be up and down with the turns in the economy. We have to be consistent about it. I don't think putting the local communities in charge of managing forestry, as complex and as complicated as it is, is the right thing to do. These communities are totally dependent, around New Brunswick, on forest products. And you have a number of them: Chipman, Juniper, Edmundston, Dalhousie, Saint John. Forest products are a major part of the community. So I don't think trying to go, trying to put the timberlands in the hands of the community is the right thing at all, because of the complexity of it. (Matthews 2004)

Mr. Irving's statement speaks directly to the misunderstanding that many people, and especially those involved in the status quo industrial fibre management system, have about community forestry. He appears to believe that community forests are simply communities surrounded by forests or communities with a significant number of people employed in the forest sector. Instead, community forests are about access to and governance of forests. They are about empowerment. The community forest, as an institutional form, is about who gets to set the management objectives for Crown forests. Currently, it is definitely not the case that the residents of those communities named by Mr. Irving decide, or even have much voice in, the fate of the surrounding public forests.

## Community Forest Experience in the Maritime Provinces

This section reviews community forest development in the Maritimes, highlighting the development and discussion of the *idea* of community forestry rather than on-the-ground experience. With the exception of some urban forests (e.g., the forest lands owned and managed by cities such as Moncton, Fredericton, and Saint John), there has been no significant experience with actual community forest models in the Maritimes until very recently. Furthermore, virtually no significant discussion of community forestry as a management option has occurred in Prince Edward Island, largely because 90 percent of the island's forests are privately owned and are already managed by local residents for local benefits. Therefore, I review two decades of community forest debate and discussion in New Brunswick and the much shorter history of community forestry developments in Nova Scotia.

## Modern Community Forest Experience in New Brunswick

For the last fifteen to twenty years, community forestry has been emerging in New Brunswick as both a policy idea and a grassroots movement, with an interesting interplay between these dimensions. The first modern references to community forestry in the province appeared in the mid-1990s in the form of two reports by Matthew Betts and David Coon (Betts and Coon 1996), under the auspices of the Conservation Council of New Brunswick. Since that time, community forestry has been a part of most policy conversations that deal with Crown land, although it appears that government and industry rarely wish to engage in these conversations.

In 1999, in an unprecedented move, the government of New Brunswick purchased 157,000 hectares of land from the US-based firm Georgia Pacific (GP). GP owned a mill across the US border in Baileyville, Maine, and the company's forest holdings on the New Brunswick side were primarily managed to provide fibre to that processing facility. The purchase of this land was unique in that most prior large acquisitions of land by the government (federal or provincial) had been done through "eminent domain" – the government's right to expropriate land, with compensation, for public use – as was the case with Base Gagetown, both national parks in New Brunswick, and the properties adjacent to the Saint John River required for the construction of Mactaquac Lake, the headpond behind the Mactaquac Dam, built in the 1960s. The history of land tenure in New Brunswick was primarily the opposite, with various governments (first Britain, then Nova Scotia, then New Brunswick) divesting their public (Crown) land assets to settlers, military veterans, railroad companies, and others, who would take it over and pay taxes on it (Wynn 1981). The reversal of that trend – the government acquiring land and thus removing it from the tax base – was unique. This action by government created "new" Crown land for the first time since the modern Crown tenure legislation had been put in place, and it sent the imaginations of community forest proponents running with regard to possibilities for alternative management and tenure structures.

Two different proposals emerged for community forests on the former GP land. One was from residents of the town of McAdam. It was spearheaded by the mayor and several people involved with the forest sector in that area. The other initiative came from a coalition of individuals from the YSC (York, Sunbury, Charlotte) Forest Products Marketing Board (a woodlot owners' collective); the Falls Brook Centre (an environmental non-government organization, or ENGO); and social ociontists in the Canadian Forest Service (including me). This second group completed a feasibility study and secured federal funding for a pilot project, but when the group asked the provincial government for sixty thousand acres on which to run the pilot, they were denied access to any land, and because of this denial, the funds for the pilot project had to be returned to Ottawa. The provincial government also denied the McAdam proposal. After setting aside a portion of the newly acquired land for conservation and another portion for negotiations with First Nations, the government redistributed the fibre on the remaining land to existing Crown licensees. This significant acquisition of new Crown land in the late 1990s was New Brunswick's best opportunity for some experimentation with community forests, but the government of the day made it clear that they were not interested in such experimentation.

The New Brunswick government certainly had ideas for how it might use that land, but community forest experiments did not appear to be one of them. At that time (the late 1990s), industry was feeling insecure about its fibre supply because of the two government processes mentioned above: the Protected Natural Areas Strategy (PNAS) and negotiations that would see harvest rights to a certain percentage of Crown land flow through First Nations. The PNAS process was identifying several thousand hectares of Crown land to be set aside for biodiversity and to meet post-Rio conservation commitments. In the end, close to thirty thousand hectares of the newly acquired GP land was designated as protected through the PNAS process.

As the protection of these lands was being formalized and finalized from 2000 to 2002 with the culmination in the Protected Natural Areas Act in 2003, industry launched a lobbying campaign to secure more robust timber rights from Crown land. Part of that lobbying process involved commissioning a report from the forest consulting firm Jaakko Pöyry in 2002. The government paid for a portion of that report so that the results might be made public. Essentially, the report benchmarked forest management in New Brunswick against management regimes in Maine, Ontario, and Finland. Following the release of the Jaakko Pöyry report, industry, conservationists, individual rural communities, and the private woodlot sector intensified their lobbying efforts. This particular iteration of the forest policy debate in New Brunswick became formalized when the legislature struck a multi-party Legislative Select Committee on Wood Supply (LSCWS) to hear testimony regarding policy alternatives and proposals for the management of Crown land. Many of the presenters – from ENGO representatives, to private wood-lot operators, to ordinary citizens – cited the idea of community forests and suggested that the large forest companies had too much power and were not managing Crown land in the best interest of the public. After the formal public hearing process, the LSCWS continued with a series of invited presentations, including one by me, on community forestry because committee members wanted to learn more about the concept and how it might work in New Brunswick.

Ultimately, the Legislative Select Committee's report rejected the recommendation of the Jaakko Pöyry report that a timber objective be determined for Crown land. However, the committee also rejected the idea that community forests should be established on Crown land:

> The Committee does not recommend establishment of community forests on New Brunswick Crown land. Such a form of forest tenure was advocated by some at the hearings on the grounds that the current system allows too little public influence over management objectives, provides too few local employment opportunities, and stifles opportunity for innovative value-added and non-timber based economic enterprises. These three claims have merit and the Committee has attempted to address each in its recommendations. However, it has done so in a way which does not undermine the strengths of the Crown Lands and Forests Act and which does not introduce inherent difficulties it envisions with implementing community forests in New Brunswick. (LSCWS 2004, iii)

It appears the committee wished to end the debate over community forestry by promising to address the issues that lead to demand for community forests through other means. That didn't really occur, however, despite the fact that interest in the concept of community forests was considerably heightened in 2004, with the release of a National Film Board documentary titled *The Forbidden Forest*. The film features two New Brunswickers – Jean Guy Comeau, a woodlot owner, mill worker, and long-time community forest advocate from the Miramichi, and Francis Wishart, a wealthy gentleman farmer and artist from France with deep New Brunswick roots and a passion for his patch of the Acadian forest in northern New Brunswick. The film is a critique of the existing tenure and management system for Crown land in the province, as well as an advocacy piece for the idea of community forestry. It portrays the two protagonists visiting community forests in Germany and discussing the potential application of the concept in New Brunswick. A shortened version of the film was aired on David Suzuki's *The Nature of Things*. The film and television program were widely viewed and discussed.

One of the government's commitments in the wake of the LSCWS report was to do a better job of public engagement in the management of Crown land. The New Brunswick Department of Natural Resources, in partnership with university and Canadian Forest Service social scientists, undertook a province-wide public opinion survey about Crown land management. One of the questions asked respondents who they would like to see given an opportunity to manage Crown land if one of the major licensees walked away from its licence and no other industrial owner expressed interest in managing the land base. Respondents were listed multiple groups, but interestingly, 58 percent of respondents said they would like to see environmental groups have an opportunity to manage Crown land, 46 percent expressed a desire to see local communities be given a chance, and 38 percent suggested that woodlot owner organizations be granted access. Support for status quo management was low under this hypothetical scenario. The response category "forest companies that currently have rights to Crown wood" came in a distant fifth, with only 21 percent of respondents supporting this choice (Nadeau et al. 2007). The stark contrast between the survey results and status-quo management of Crown forests produced a chilling effect on future public engagement regarding Crown land management in New Brunswick (Beckley 2014). While the environmental community frequently references this survey in policy debates, the government has virtually abandoned any meaningful consultations regarding forest policy since that time – even though a wholesale change in the Crown land management system was established in 2014 (Beckley 2014).

Currently, a few initiatives in New Brunswick could formally be considered community forests or have aspirations to become community forests. One is the 15,000 acres managed as a municipal forest in Moncton, New Brunswick's second-largest city. It was identified by Teitelbaum, Beckley, and Nadeau's 2006 survey as the only area that met the definition of "community forest" at the time of that study. It is a forest that is managed by municipal employees for diverse, multiple objectives including, but not primarily, timber. Watershed protection is the most important management objective for that forest, and recreation and even non-timber forest products factor into the planning for, and management of, that forest.

Another emerging (or potential) initiative is located in the community of Upper Miramichi, a rural community with a new municipal designation made possible by recent legislation. The jurisdiction was formed through the merger of sixteen previously unorganized or loosely organized local service districts. Upper Miramichi now has elected officials, a new town office, and responsibility for its own budget. As this new community develops its capacity and takes responsibility for its own economic self-determination, it is interested in exploring the idea of greater control over the one thousand square kilometres of Crown land within its boundaries. To date, the community has no formal tenure or access to this land as it is licensed to J.D. Irving Limited, though many use it for recreation, non-timber products, and non-forest business opportunities such as guided hunting and fishing. The aspiration of community forest proponents in Upper Miramichi is to eventually gain more direct access to the fibre resources on Crown land within their territory.

**Modern Community Forest Experience in Nova Scotia**
The experience with community forestry in Nova Scotia is much more limited than in New Brunswick, but events are occurring at an extremely rapid pace relative to its neighbouring province. In August 2011, the Nova Scotia Department of Natural Resources published *The Path We Share: A Natural Resources Strategy for Nova Scotia, 2011–2020.* The document arrived in the aftermath of some heated political rhetoric regarding the future direction for management of Nova Scotia's public and private forests. The debate was similar to the debate in New Brunswick in that it focused on the degree to which forest policy would support the goals of industry (favourable stumpage rates for industry from Crown land, subsidized silviculture that favours softwood species, etc.) versus prioritizing environmental values such as ecological integrity and biodiversity. The report was intended to chart the future course for forest management in the province. In the section on shared stewardship, the government made a commitment to "explore ways to establish and operate working community forests on Crown land" (Nova Scotia Department of Natural Resources 2011, 38).

In the years prior to *The Path We Share*, the major industrial players in Nova Scotia were struggling. Bowater Mersey Paper Company's operation in Liverpool had seen down time and layoffs, and was well known to be on thin ice financially when it was purchased by Resolute

Forest Products in 2011. Despite considerable government financial support and promises on the part of Resolute to "run the mill for at least 5 years," (Jackson 2011), in July 2012, the relatively new owners of the mill announced its indefinite closure. At the other end of the province, the New Page mill in Port Hawkesbury went into receivership and was shuttered in August 2011. The government stepped in, and for about a year, the mill was managed by existing mill staff with government oversight. The government wanted to keep the Port Hawkesbury mill on "hot idle" while they shopped the asset (and the Crown licence that came with it) to potential new owners. Both of these cases entailed massive government intervention and financial support and only served to ratchet up the debate about the viability of the pulp and paper industry in Nova Scotia (Canadian Press 2012).

As had occurred in New Brunswick, when Nova Scotia's forest industry appeared to be in a financial crisis, people started to imagine what "post-industrial" forestry might look like. After years of seeing job declines but no significant reduction in harvest levels, some citizens warmed up to the idea of different management models such as community forestry. Though community forestry entered the conversation later in Nova Scotia than in New Brunswick, it quickly picked up momentum.

In June 2012, Dalhousie University and the Nova Forest Alliance released a community forest discussion paper (MacLellan and Duinker 2012). When Resolute Forest Products announced the indefinite closure of its facility in Liverpool and its intent to sell off its Nova Scotia assets, many of the region's residents expressed immediate interest in having the province purchase the freehold land. Some hoped that the province would then reallocate and relicense those lands, or a portion of them, to new, more community-oriented licensees. Ultimately, that is what happened. In December 2012, the province closed the deal with Resolute and purchased the entire operation (a newsprint mill, a sawmill, a hydroelectric facility, and 220,000 hectares of woodland). The purchase price was $1.00, but for that, the government accepted all of Resolute's liabilities, including over $100 million in pension liabilities (Canadian Press 2012).

At about the same time, the Nova Scotia government put out a call for community forest pilot projects. Seven different organizations or partnerships of organizations met the 31 January 2013 deadline and submitted proposals: two ENGOs, a woodlot owners' organization, a First Nation band, a municipality, and two partnerships with existing industrial forestry players (one a sawmill consortium and one led by a contractor). In April 2013, all seven applicants were told that they did not fully meet the criteria set forth by government. A new call for proposals was made, open to the seven original applicants but also any additional applicants. This call, however, required applicants to limit their proposals to fifteen thousand hectares (pers. comm., Peter Duinker, 19 May 2013). The deadline for new applications was set for July 2013. In October 2013, the government announced that it was entering negotiations for a single pilot project involving fifteen thousand hectares with the Medway Community Forest Co-operative (Nova Scotia, Natural Resources 2013; Scrine 2014).

The initial excitement over the possibility of community forestry in the region has been severely tempered. The scope of community forest activity in the region appears to be considerably less than what was initially implied or suggested. Not only has the number of proposals in the running declined, but the limit of fifteen thousand hectares means that a relatively small area will be under this management model. The Medway Co-op initially asked for sixty thousand hectares. In the meantime, most of the Crown wood from the former Bowater/Resolute land appears to be headed to existing industrial producers (Lindsey 2014).

In both Nova Scotia and New Brunswick, community forestry discussions have emerged from a sense that the large-scale industrial tenure model has some serious problems. The long-term viability of these facilities appears to be questionable. New Brunswick has seen the closure of five major facilities in the last decade (in Dalhousie, Miramichi [two], Bathurst, and Nackawic). Others have been in perennial financial difficulty. Two of Nova Scotia's three major pulp mills have closed in the last five years and have been bought outright by government or repurchased by new investors with what many view as "sweetheart" deals in terms of stumpage rates, energy costs, and so on. Despite having some structural similarities with forest industries in places like Quebec, British Columbia, and Ontario, the Maritime provinces have very little experience with Crown land community forests as actual policy or management experiments. The next section presents an argument as to why that may be the case.

## Supply and Demand for Tenured Reform towards Community Forests

We typically think of supply and demand as economic concepts that refer to commodities or services, more or less tangible items that are bought, sold, or traded in markets. But supply and demand may also apply to ideas and concepts, and community forestry is one such idea. In comparison to British Columbia, Ontario, and Quebec, there has been very little experimentation in the Maritime provinces with community forests on public land. According to Teitelbaum, Beckley, and Nadeau (2006), 105 of 106 community forests exist in the three jurisdictions listed above. I contend that demand for community forests in the Maritimes is strongly related to the degree to which the socioeconomic elements that characterize community forestry are already provided, not through Crown land management but through the higher proportion of small forest holdings that are managed by non-industrial private interests. In Nova Scotia, the first provincially sanctioned community forest experiment in the Maritimes were announced in October 2013. New Brunswickers have been talking about and lobbying for community forests on Crown land, at least as pilots or experiments, for nearly twenty years, yet none exist on the ground. With 90 percent of its land being private and with little processing capacity, Prince Edward Island has had little experience or demand for community forests. While there is a rich tradition and long history of forest communities in the Maritimes, the region has had little practical experience and, until recently, little demand for community forests as a unique institutional management arrangement or tenure type. I believe that this is due to some historically contingent reasons.

One of the interesting and unique features of the policy debate over community forests in the Maritimes is that significant elements of community forests already exist on private forest lands. The unique tenure structure in the Maritimes is, in my view, a causal factor in the lower incidence of community forest initiatives in the region. Table 2.1 provides a review of the distribution of privately controlled productive forest lands in Canada (excluding the northern territories).

The Maritimes have a unique land ownership pattern relative to the rest of Canada, with private forests representing the predominant tenure type (Nova Scotia, Prince Edward Island) or being on par with Crown forests (New Brunswick). Nova Scotia and Prince Edward Island are particularly heavily weighted to private owners compared to Crown land, and more of that private land is in the hands of non-industrial owners. Nova Scotia has approximately thirty thousand woodlot owners, and 68 percent of the province's land base is privately owned. Prince Edward Island has fourteen thousand owners, and 87 percent of productive forest is privately owned with virtually none of it owned by forest corporations (referred to as industrial freehold;

TABLE 2.1   Private woodlot distribution across Canada's provinces

| | Productive forest area in private woodlots (%) | Number of woodlot owners | Average size of woodlots (ha) | Rural/ urban population (%) | Population owning woodlots (%) |
|---|---|---|---|---|---|
| Prince Edward Island | 87 | 14,000 | 17 | 53/47 | 10 |
| Nova Scotia | 57 | 30,000 | 57 | 43/57 | 3.2 |
| New Brunswick | 30 | 41,900 | 45 | 48/52 | 5.5 |
| Quebec | 13 | 125,000 | 54 | 19/81 | 1.5 |
| Ontario | 12 | 169,000 | 28 | 14/86 | 1.3 |
| Alberta | 7 | 17,500 | 88 | 17/83 | .04 |
| Manitoba | 7 | 13,500 | 73 | 28/72 | 1.1 |
| British Columbia | 3 | 22,500 | 53 | 14/86 | 0.5 |
| Saskatchewan | 3 | 15,000 | 27 | 33/67 | 1.4 |
| Newfoundland and Labrador | 1 | 4,000 | 9 | 41/59 | 0.7 |
| Canada | 9.5 | 451,000 | 43 | 19/81 | 1.3 |

Source: Dansereau and deMarsh (2003).

Dansereau and deMarsh 2003). Private land in New Brunswick falls into two distinct categories; private industrial land that comprises about 40 percent of New Brunswick's private land (industry-owned) and private woodlots that make up about 60 percent of New Brunswick's private land. As a percent of the total land in New Brunswick, industrial freehold comprises about 20 percent of the total forest land in New Brunswick while private woodlot owners hold 30 percent. New Brunswick has almost forty-two thousand individual owners who manage 1.7 million hectares of land (Nadeau et al. 2012, 6). Canada as a whole, by contrast, has 94 percent of its total forest land under government tenure.

These statistics matter when you consider the four themes that generally constitute the definition of community forestry: local benefits, local decision-making, rights of multiple users, and ecological stewardship. Most forest land in Nova Scotia and Prince Edward Island, and a third of it in New Brunswick, is managed expressly for individual benefits: that is, for the benefit of the individual private owners and their families. If we scale this up one level of social organization (Beckley

1998) – that is, if we consider that communities comprise individuals – we can argue that much of the forest land in the Maritimes is managed for community benefits, which is a cornerstone of community forestry. In other words, community benefits accrue through the collective actions of individual forest owners' decisions. Landowners decide how to allocate their land across a range of potential uses – from recreation to wildlife, from timber to non-timber products, or as investments or legacy projects.

A similar logic applies to the second characteristic of community forests: locals have at least some degree of control over planning and management decision-making regarding local forests. Once again, this is certainly the case in the Maritimes, since 30 percent of the forest land of New Brunswick, roughly 57 percent of Nova Scotia, and 87 percent of Prince Edward Island is managed at the most local level possible by individuals or families. While there is no explicit "communal" or community dimension to this management, the region has a tradition of fairly strong private property rights. The state does regulate against uses that are counter to the public good, but for the most part, the laws align with the social norms of private landowners (Quartuch and Beckley 2014) and respect the rights of individuals to manage their own land. Again, scaled up to the community level, local land is managed in individual units but in accord with community values.

With respect to rights of multiple users, even though much land is private, most woodlot owners have generous trespass policies with regard to hunting and recreational use. Neighbours and friends of woodlot owners, if not the entire community, benefit from the diversity of uses and products the woodlot sector in the Maritimes provides, such as maple sugar products, fiddleheads, Christmas trees, pulpwood, firewood, sawlogs, fenceposts, recreation, and wildlife. One could argue that far more human benefits are derived from private forest land than from Crown land, if for no other reason than that this land is in direct proximity to rural residents who, through ownership, have the authority to derive (and share) such benefits from the land.

Finally, with regard to ecological stewardship, privately owned forests in the Maritimes demonstrate a wide range of management practices, from the very best to some of the worst. A select cadre of owners are extremely knowledgeable, intimately connected to their land, and highly motivated to manage sustainably for a wide range of forest values. Some private woodlot owners, however, view their land as having little more than revenue-generating potential, and there is less stringent enforcement of environmental laws on private land than on public land. Roughly 9 percent of private woodlot owners in New Brunswick reported having no interest in harvesting timber products in the future, so that land could be considered to be managed ecologically (Nadeau et al. 2012).

My thesis, then, is that demand for community forestry is proportional to the supply of two "commodities": community benefits and community involvement in decision-making for the Maritime provinces' overall forest estate. It makes sense that of the three Maritime provinces, demand for community forests would be highest in New Brunswick, which has the smallest proportion of land held by private woodlot owners. The twenty years of active debate over community forests and consistent lobbying for community forest pilots or experiments on Crown land in New Brunswick appear to support the hypothesis that demand for community forestry will be highest where there is the lowest proportion of private land. There has been considerably less demand for community forestry in Nova Scotia, the jurisdiction with the second highest proportion of Crown land, and virtually none in Prince Edward Island, which is almost entirely private. Over the last thirty years, there has been a concentration and rationalization of mills in New Brunswick.

With respect to Crown land, for the better part of two centuries, there was little demand for community forests because two of the four elements of community forestry were also in place on Crown land (local benefits and local decision-making) and a third (ecological stewardship) had not yet been conceived. The local forests were managed by local timber barons, who doled out patronage in an arbitrary and unequal manner. Throughout the nineteenth and early twentieth centuries, forests were locally managed, albeit with a very top-down structure. Local benefits did accrue, though they were unevenly distributed. There was not much expectation of democratic participation in resource management. Such expectations only emerged with the rise of the welfare state and mass media outlets that report routinely and regularly on the activities of government. The old patronage system was tolerable to local communities as long as benefits trickled down to them. Again, until recent decades, in most corners of the Maritimes, sawmills employed local people, bought local wood, and provided spinoff benefits to local communities. Even in the modern pulp and paper era, there was a feeling that pulp and timber giants such as Repap and Irving were essentially "home grown" capitalists and that they would look after New Brunswickers' interests, reinvest in the province, and continue to provide local employment. Nova Scotia had a different tradition, with most of its large forestry players (e.g., Oxford Paper, Stora, Kimberly Clark) coming from "away." Furthermore, the Nova Scotia mills came later than most of the large mills in New Brunswick, as the era of paternalistic capitalism was winding down.

In both New Brunswick and Nova Scotia, the dominant paradigm has been "social contract" forestry, characterized by institutional relationships whereby access to land and resources is granted to (mostly) private firms at quite favourable rates with the tacit understanding that the quid pro quo for those firms is to fund substantial payrolls and be responsible local corporate citizens (Wang 2005). Given increasingly tighter financial margins for international forest-products players and the continuous substitution of capital for labour (e.g., greater productivity through technology and thus fewer jobs), the social contract has become a bit frayed. From the 1970s through to the early 2000s, as mills were upgraded and refurbished, capacity increased but the productivity increases resulted in fewer jobs. The same process, of course, was going on with work in the forest itself, as bucksaws gave way to chainsaws, and chainsaws to harvesters. It is in this context that reformers are calling for a new social contract called "community forestry." The hope is that returns to communities and taxpayers will be greater under a system that maximizes employment while maintaining profitability rather than one that chases profit above all else.

After two decades of debate and discussion, community forestry has not made significant inroads in the Maritimes, although Nova Scotia has embraced it, albeit in a small way. Industrial players have managed to continue to make their model work, primarily through lobbying to keep stumpage rates low and by successfully depressing wood prices for private owners. They have survived on various handouts and bailouts from governments, with the same "too big to fail" logic that led to bank bailouts in the United States after the 2008 financial crash. The argument from government is that these heavily forest-dependent regions are too reliant on their major employers and the jobs they provide to allow these corporate operations to fail. Yet fail they have. New Brunswick has seen the closure of four major mills in the last half decade, and in Nova Scotia, two of the province's three major pulp plants have closed. And with the exception of J.D. Irving Limited, there has been a constant turnover in ownership of these facilities.

## Policy Experiments, Adaptive Management, and the Future of Community Forestry in the Maritimes

The following section offers a suggested approach regarding parameters for establishing and monitoring community-based forest management experiments in the Maritimes. Despite years of debate, we are no further along in terms of knowing what benefits and services a community forest might provide, because no systematic experiments have been done to reveal the potential of such a model. Where community forests have been tried – most notably, in British Columbia, Ontario, and Quebec – the monitoring and assessment systems have been too weak to offer policy-makers much useful information on what societal benefits are gained or lost through the implementation of community-based management systems. Only recently have academics and practitioners begun to discuss how one might evaluate community forests or pilot projects in community forestry (Reed and McIlveen 2006; Teitelbaum 2014).

Advocates of community forests have ideas about the advantages of a community-based management structure over status quo industrial tenures. Establishing community forests under adaptive management principles would allow for rigorous evaluation of the efficacy of these institutional experiments. One key feature of adaptive management is the posing of hypotheses for testing (Duinker and Trevisan 2003). Advocates of community forests have implicit assumptions about the superiority of this model compared to status quo industrial fibre management. Establishing such institutions under a systematic adaptive management framework would turn these implicit assumptions into explicit, formal hypotheses. These hypotheses could then be tested by running pilot community forest projects as "policy experiments," with the intent of answering research questions that have arisen in the debate over the efficacy of community forestry. Because governments are often risk averse, entering into wholesale Crown land policy reform that includes alternative tenure systems is frightening to many officials, both elected representatives and members of the civil service. Policy experiments would allow governments to answer some key questions in a limited, controlled framework, with considerable oversight, clear performance targets, and predetermined evaluation criteria to determine the success of the experiment. Such a framework for policy experimentation would potentially answer questions that have remained unanswered for two decades. One side of the community forestry debate says, "It would never work, and it could never produce the same degree of economic benefit as does the current system," while the other side says, "Community forests would result in happier, more engaged citizens, more local investment of profits from forest management, and a healthier forest." Until we try, and unless we establish the necessary monitoring and assessment protocols, we will never discover which of these perspectives is correct, even if we do implement community forest experiments in the region.

Under an adaptive management, policy experiment framework, the government would determine how many community forest pilots would be established, how large they would be, and what governance structures might be acceptable (e.g., local corporations, cooperatives, or community organizations). These organizations would need to generate a business plan and, ultimately, a forest-management plan. One of the fears that is frequently expressed by detractors of community forests is that communities will lack the capacity and expertise to manage the forest. However, just like a for-profit company, the directors of a community forest pilot project would simply have to hire such expertise from the ranks of professional foresters and environmental managers.

A key feature of adaptive management is defining indicators to measure performance and establishing the direction of trend that is desired for these indicators (Teitelbaum 2014). An important point to consider is that in terms of jobs and income, community forests may not perform as well as industrial firms that manage primarily for fibre and for profit have historically performed. However, there is mounting evidence that such firms will not perform as well in the future on key indicators of social concern, such as job numbers, as they have in the past. This is, in part, due to a subsidization of the industrial fibre system by provincial governments. The key would be to set up the adaptive management assessment criteria in such a way as to be able to take those subsidies into account in the comparison. However, at the core of the community forest proponents' critique of the status quo model of forest management is the notion that efficiency (in a narrow economic sense) and profit have trumped equity, local environmental quality, and the rights of citizens to participate in setting goals for Crown land. Advocates of community forestry have claimed that community-based management may deliver on issues of ecological integrity and social justice (Hammond 1991). An unrealistic expectation of a community forest pilot project in the Miramichi might be that it would provide two thousand high-paying, unionized jobs (which was the maximum number employed in the heyday of Repap's operation). Similarly, it is unlikely that whatever community forest phoenix may rise from the ashes of the Mersey mill in Nova Scotia will rival that mill's former employment or profit levels. It should be noted, however, that those high numbers were not sustainable and that they involved a degree of "mining" of both the forest and the public purse. As Nova Scotia moves forward with plans to establish community forestry pilots, an adaptive management evaluative framework could go a long way in helping policy makers understand whether the pilots are meeting their targets and objectives regarding social, ecological, and economic indicators.

## Conclusion

In the late 1990s, Dr. Luc Bouthillier of Laval University said, "Community forestry should not be built on the sweepings after the party." The "party," of course, referred to exploitive harvest regimes and low stumpage rates. The "sweepings" referred to the remaining marginal, poorly stocked land with poor soils or in places too remote to make economic sense for large, corporate players to manage once an initial harvest was made. Dr. Bouthillier feared that these were the only places where the government would allow community forest experimentation to happen; in allocating land in this way, community forests would be set up to fail. He was basically asking for a level playing field so that community forest tenures could compete fairly with industrial tenures. Community forest advocates do not want community forests to compete head to head with experienced and efficient industrial players in the production of internationally traded commodities; rather, they want to see whether community forest tenures could remain economically viable while producing a more socially acceptable suite of goods and services.

To summarize, the story of community forestry in the Maritimes is one of possibility, not practice. There are historically contingent reasons for this. Over the years, demand for community forests has been greater in places within the region where the proportion of small private ownership is low relative to Crown and industrial freehold land. The interest in community forestry stems, in part, from frustration in watching resources that are extracted locally go far down the road for processing. As the logs leave the region, so do profits and the possibility of making more out of the wood through value-added processing. Greater economic benefits, employment, and working up the value chain are important parts of the community forest ideal, but they are not the whole story. Community forestry is more about governance and the opportunity to have a say than it is about the opportunity to profit or capture economic benefits. Those two elements – benefits and decision-making authority – are not mutually exclusive, however, and many community forest proponents believe that they are in fact mutually supportive.

Time will tell if community forestry will take root in the Maritimes. Recent developments, however, have not been encouraging to community forest proponents. In 2014, the New Brunswick government initiated a plan to place the Crown wood supply under even greater control by large-scale private industry (Beckley, 2014). The government has entered into contracts that will make it extremely difficult and costly to reallocate Crown forest to other users such as community forest pilots. The contracts will be in place for twenty-five years, and some experts believe that they will be difficult or extremely expensive to reverse. In Nova Scotia, the promise of community forestry in the Mersey region is still alive, but expectations have been scaled back considerably. For now, the prospect of multiple community forests with diverse partners and potentially different models seems to be on hold. One project, Medway Community Forest, is going forward. Undoubtedly, many in government, industry, and civil society are waiting to see how this experiment fares.

**References**

Beckley, T.M. 2014. "Public engagement, planning, and politics in the forest sector in New Brunswick, 1997-2014." *Journal of New Brunswick Studies* 5: 41-65.

–. 1998. "The nestedness of forest-dependence: A conceptual framework and empirical exploration." *Society and Natural Resources* 11 (2): 101–20. http://dx.doi.org/10.1080/08941929809381066.

Betts, Matthew, and David Coon. 1996. *Working with the Woods: Restoring Forests and Community in New Brunswick.* Fredericton, NB: Conservation Council of New Brunswick.

Canadian Press. 2012. "Province buys Bowater lands." *CBC News*, 10 December. http://www.cbc.ca/news/canada/nova-scotia/province-buys-bowater-lands-1.1186704.

Dansereau, J.P., and P. deMarsh. 2003. "A portrait of Canadian woodlot owners in 2003." *Forestry Chronicle* 79 (4): 774–78. http://dx.doi.org/10.5558/tfc79774-4.

Duinker, P.N., and L. Trevisan. 2003. "Adaptive management: Prospects and progress in Canadian forests." In *Towards Sustainable Management of the Boreal Forest*, edited by P.J. Burton, C. Messier, D.W. Smith, and W.L. Adamowicz, 857–92. Ottawa: National Research Council of Canada.

Duinker, Peter. 2013. Professor, School of Resources and Environmental Studies (personal communications, 19 May 2013).

Hammond, Herb. 1991. *Seeing the forest among the trees: The case for wholistic forest use.* Winlaw, BC: Canada Polestar Calendars.

Jackson, D. 2011. "Resolute boss confident plan will keep Bowater mill running." *Chronicle Herald,* 6 December. http://thechronicleherald.ca/novascotia/40013 -resolute-boss-confident-plan-will-keep-bowater-mill-running.

Lindsey, D. 2014. "Northern pulp, western Crown: The politics of wood supply and public consultation." *Atlantic Forestry Review* 20 (4): 20–21.

LSCWS (Legislative Select Committee on Wood Supply). 2004. *Final report on wood supply in New Brunswick.* Fredericton: Legislative Assembly of New Brunswick.

MacLellan, L.K., and P.N. Duinker. 2012. *Community Forests: A Discussion Paper for Nova Scotians.* Stewiacke, NS: Nova Forest Alliance, and Halifax: School for Resource and Environmental Studies, Dalhousie University.

Matthews, K. (director). 2004. *Forbidden Forest.* Documentary film. Ottawa: National Film Board of Canada.

Nadeau, S., T. Beckley, E. Huddart-Kennedy, B. McFarlane, and S. Wyatt. 2007. "Public views on forest management in New Brunswick: Report from a provincial survey." Information Report M-X-222E. Fredericton, NB: Canadian Forest Service.

Nadeau, S., T. Beckley, M. McKendy, and H. Keess. 2012. *A snapshot of New Brunswick non-industrial forest owners in 2011: Attitudes, behaviour, stewardship, and future prospects.* Fredericton: Department of Natural Resources, Government of New Brunswick.

Nova Scotia. Department of Natural Resources. 2011. *The path we share: A natural resources strategy for Nova Scotia, 2011-2020.* Nova Scotia Department of Natural Resources.

–. 2013. "Province moving forward on community forests." Press release, 18 October. http://novascotia.ca/news/release/?id=20131018004.

Quartuch, M., and T.M. Beckley. 2014. "Carrots and sticks: New Brunswick and Maine forest landowner perceptions toward incentives and regulations." *Environmental Management* 53 (1): 202–18. http://dx.doi.org/10.1007/s00267 -013-0200-z.

Reed, M., and K. McIlveen. 2006. "Toward a pluralistic civic science? Assessing community forestry." *Society and Natural Resources* 19 (7): 591–607. http:// dx.doi.org/10.1080/08941920600742344.

Scrine, J. 2014. "N.S. community forest gets green light." *Atlantic Forestry Review* 20 (3): 24–26.

Stedman, R. C., W. White, M. Patriquin, and D. Watson. 2007. Measuring community forest-sector dependence: Does method matter? *Society and Natural Resources* 20: 629-46.

Teitelbaum, S. 2014. "Criteria and indicators for the assessment of community forestry outcomes: A comparative analysis from Canada." *Journal of Environmental Management* 132: 257–67. http://dx.doi.org/10.1016/j.jenvman.2013. 11.013.

Teitelbaum, S., T. Beckley, and S. Nadeau. 2006. "A national portrait of community forestry on public land in Canada." *Forestry Chronicle* 82 (3): 416–28. http://dx.doi.org/10.5558/tfc82416-3.

Wang, S. 2005. "Managing Canada's forests under a new social contract." *Forestry Chronicle* 81 (4): 486–90. http://dx.doi.org/10.5558/tfc81486-4.

Watson, David. 2013. Field Economist, Natural Resources Canada (personal communication, 22 May 2013).

White, W. and D. Watson. 2004. "Natural resource based communities in Canada: An analysis based on the 1996 Canada Census." Internal report produced for the WINS initiative of Natural Resources Canada, Canadian Forest Service. Northern Forestry Centre: Edmonton, AB.

Wynn, G. 1981. *Timber colony: A historical geography of early nineteenth-century New Brunswick*. Toronto: University of Toronto Press.

Chapter 3 **Community Forestry in Quebec**
A Search for Alternative Forest
Governance Models

*Solange Nadeau and Sara Teitelbaum*

In this chapter, we examine the development and evolution of community forestry in Quebec. The main focus is on public forests, which are a provincial responsibility, managed under a distinct legal and regulatory framework. We do not intend to provide a comprehensive analysis of all community forestry models in the province but rather to highlight how community forest initiatives have become interwoven with a wider social movement, rooted in rural communities, that has actively challenged provincial forest policy for more than a century. We discuss grassroots initiatives that have emerged in Quebec, as well as the evolution of policy, both of which have helped enlarge the spectrum of community forestry approaches. As is suggested by Baker and Kusel (2003), we conceive of community forestry in this chapter as an attempt to reorganize the web of relationships that exists among people, their communities, the forest, and the wider political and economic system. In Quebec, this impetus towards reorganization has involved both Aboriginal and non-Aboriginal peoples, but often on parallel (albeit intersecting) paths. This chapter focuses more on rural "settler" communities than on Aboriginal communities; however, we wish to recognize and acknowledge the long history of Aboriginal mobilization to develop forest governance models that align with Aboriginal and treaty rights, values, and aspirations. Aboriginal initiatives – whether in the form of modern-day treaties (the James Bay and Northern Quebec Agreement); specific agreements such as the Paix des Braves; or community endeavours such as those spearheaded by the Lac Barrière, Pikogan, and Kitcisakik Algonquins, the

Weymontaci Atikamekw, the Waswanipi Crees, the Mashteuiatsh and Essipit Innus, and the Wendake Hurons – have characteristics in common with other community forestry initiatives but are also distinct in origin and political aspirations (e.g., Wyatt 2004; Waswanipi Cree Model Forest 2007; Jacqmain 2008; Saint-Arnaud 2009; Beaudoin, St-George, and Wyatt 2012; Beaudoin 2014).

## The Emergence of Community Forestry as an Alternative Governance Model

The proposal that rural communities should play a central role in forest-management decisions has been a recurrent theme in forest policy discussions in Quebec. It reflects a tension between two visions of development. One vision has been the predominant approach applied in Quebec: it favours a pattern of development where industrial forest companies are entrusted with the rights and responsibilities for managing public forest lands. The other, less conspicuous vision favours the empowerment of local communities for the same purpose. There are clear differences in the fundamental orientation of these two models: the former is primarily concerned with building a strong forest industry and generating jobs whereas the latter seeks to develop a suite of forest resources through innovations in local governance and institution building at local and regional levels. Although the former model has received much stronger institutional support in Quebec, and also in the rest of Canada, there is nonetheless an important story to be told in Quebec concerning the push for local and participatory management.

According to Howlett and Rayner (2001), the tenure system in Canada has had difficulty adapting to needs other than traditional timber uses. The importance of adaptation was raised early on by Esdras Minville, a well-known professor in the business school at the Université de Montréal. Minville's analysis of the crisis that followed the economic crash of 1929 led him to question the fundamental tenets of the economic system in Quebec and to recommend the establishment of a more collaborative system (Paradis 1980). Later, Minville published a seminal text on Quebec's forests, where he describes what he refers to as "the social problem of the forest" (Minville 1944). In short, he criticizes the forest policy regime for placing too much emphasis on the generation of economic benefits to the detriment of a suite of social values that are of direct importance to local communities. In Minville's time, public forests were primarily allocated to private companies, the

TABLE 3.1 Portrait of Quebec's historical cantonal reserves

|  | Number of cantonal forest reserves | Area cover (ha) | Sawlog production (m³) | Firewood production (m³) |
|---|---|---|---|---|
| 1924–25 | 26 | 80,549 | 10,896 | 9,947 |
| 1944–45 | 183 | 625,178 | 64,332 | 28,806 |

Source: AIFQ (1949).

only exception being a system of cantonal reserves within which locals were allowed to harvest wood for fuel and construction but were allotted no role in management (Bouthillier 2001; see Table 3.1). Minville highlighted the importance of reinforcing this type of community-oriented reserve with a stronger focus on local governance, local entrepreneurship, and resource diversification. In his seminal report, Minville made a number of recommendations, including the following:

- using different forest reserves (district forests, forest reserves) jointly to protect rights of existing businesses while making forests available to meet the needs of local populations
- organizing timber harvesting not as a single activity but in coordination with management of other local resources to ensure more continuity for local populations
- establishing timber harvesting on a sustained basis to avoid resource depletion
- promoting the creation of cooperatives or small businesses rather than relying on transient lumberjacks paid on the basis of piecework.

Minville is also known for his role in spearheading the creation of the first forestry cooperative in the 1930s in the Gaspésie region, the Société agro-forestière de Grande-Vallée. Under the administration of the cooperative, a collective of fifty family run forest lots was established on four hundred square kilometres of public land. The cooperative took care of managing the forest and sawmilling, storing, and marketing the wood, and families split their year between agriculture and timber harvesting (Teitelbaum 2009). Several of Minville's ideas were echoed in one of the first vision statements for community forestry, published by the Quebec Association of Registered Professional Foresters, or QARPF (AIFQ 1949). The QARPF advocated for the creation of a network of community forests, located within twenty miles of village or town centres on land owned and managed by local administrations (towns, villages, parishes, school districts, etc.), with a mandate to contribute to the well-being of the local population (Guertin 1997). Indeed, the QARFP also made reference to the cantonal reserves, suggesting that a joint governance scheme involving both government and communities would result in better management outcomes and offset the mismanagement that had beset some of these reserves.

The period after the Second World War saw important changes in the structure and productive capacity of the forest sector in Quebec, including a shift from seasonal to full-time labour. This period saw the expansion of the cooperative model, which emerged during the 1930s and 1940s in the Gaspésie and Abitibi regions in response to the difficult working conditions experienced by settlers working seasonally for forestry companies. The postwar period saw new opportunities for cooperatives, as large companies facing labour shortages proved willing to subcontract work to these smaller companies. However, as the cooperative model spread, those in industry became increasingly concerned that workers would gain too much influence, thereby threatening their financial bottom line. This led industry to change their subcontracting practices, resulting in less work for cooperatives (Vincent 1995).

However, this reversal did not diminish the cooperative idea put forward by Minville and others. Indeed, the tradition of the forestry cooperative persisted and multiplied, despite a difficult economic context. As indicated in Box 3.1, at its height in the 1970s, there were 167 forest cooperatives in Quebec. This number has since decreased, primarily because of consolidation and challenges related to policy and economic conditions in the forest sector. Today, cooperatives act primarily as contractors to the forest industry on public land in activities such as harvesting and silviculture, but many have also diversified their activities towards plant production, timber processing, and the harvesting and processing of non-timber forest products. Thus, although forestry cooperatives have not succeeded in gaining direct tenure rights to public lands, they represent an important model of collective action in the forest sector and remain a key actor within regional economic development (Lessard 2012; Chiasson and Andrew 2013).

### Challenges, Conflicts, and the Emergence of New Ideas

The 1960s saw a policy shift in Quebec, as government sought to take control of the levers of development. Rural development proved to be a

|||||||||||||||||||||||||||||||||||||||||||||||||||||||||||||||||||||||||||||||||||||||||||||||||||||||||||||

**BOX 3.1  A Short History of Forestry Cooperatives in Quebec**

- 1946: Quebec has 41 forestry cooperatives under two categories: cooperatives harvesting timber to sell to mills or to process internally (21) and cooperatives harvesting and trucking as subcontractors to forestry companies (20).
- 1970: The number of forestry cooperatives has increased to 167.
- 1977: Provincial policy for forestry cooperatives is introduced to help consolidate forestry cooperatives.
- 1985: A provincial federation to represent and promote forestry cooperatives is created.
- 1980–90: The province negotiates contracts with forestry cooperatives for up to 50 percent of forest-management activities in regions where they operate.
- 1990s: Forestry cooperatives become more involved in wood processing through direct mill ownership or through partnerships with forest companies.
- 2005: Quebec has 43 forestry cooperatives with 3,135 members and an annual turnover of $309 million.
- 2000 onward: There is increased interest, locally and internationally, in integrated resource management, forest certification, and participation in pilot projects. A new approach emerges called the "solidarity cooperative," which extends the concept of membership to include not only workers but also forest users and organizational supporters.

|||||||||||||||||||||||||||||||||||||||||||||||||||||||||||||||||||||||||||||||||||||||||||||||||||||||||||||

*Source:* FQCF (2015).

major challenge, since many areas were facing serious demographic shifts and economic decline. Public hearings held for a forest policy review in 1965 revealed considerable public dissatisfaction with the government's practice of allocating public forests exclusively to large corporations (Québec, MTF 1965). Once more, the QARPF advocated for greater community involvement, this time proposing that government create a network of municipal forests (Guertin 1997).

For its part, government opted to implement a strategy of centralized rural planning that aimed to raise living standards in outlying areas to the provincial average (Linteau et al. 1989). In the eastern part of the province, a policy of regionalization was proposed, which involved closing many small parishes and relocating residents to larger centres. In 1969, a first group of ten villages labelled "economically unprofitable and socially unviable" was targeted (Deschênes and Roy 1994). However, as plans unfolded, resistance grew among local residents, who held public protests and organized a wide mobilization, resulting in the creation of a veritable social movement called Opérations dignité. This movement, active in the early 1970s, sought to strengthen local economies through, among other things, the reappropriation of natural resources such as forests. Communities and farmers' organizations petitioned the government for control over public forests in proximity to local communities, which they called "inhabited forests" (UCC 1969; UCC and FPBQ 1971). Although the government did not grant them any substantial new authority, rural communities nonetheless undertook an important reflection on local governance and became innovators with regard to the application of alternative management approaches. For example, tenant-farming projects were first implemented in the municipality of Sainte-Paule (Bas-Saint-Laurent), once in the 1970s and again in the 1990s (Otis 1989; Masse 1995). The tenant-farms model is based on the idea of allocating forest parcels to individual entrepreneurs, who manage and commercialize forest products, with some collective responsibilities.

Another model that developed in the early 1970s, this time in neighbouring Gaspésie, is the joint management organization (see Box 3.2).

||||||||||||||||||||||||||||||||||||||||||||||||||||||||||||||||||||||||||||||||||||||||||||||||

**BOX 3.2   Joint Forestry Ventures in Quebec**

Joint venture organizations have a membership composed of private woodlot owners who buy shares in the organization and thereby benefit from the right to vote and from the services provided by the organization.

- 1971: Quebec's first joint venture is created.
- 2007: Quebec has 44 joint ventures across the province, with 25,770 woodlot owner members managing a total of 1.3 million hectares. These organizations employ 112 professional foresters and biologists, 487 forest technicians, and 2,500 silviculture workers.
- 2015: Several joint venture organizations are active on public land. Several hold forest management contracts with the government to manage small public forest lands. Others have contracts with regional county municipalities to manage public intramunicipal lands.

||||||||||||||||||||||||||||||||||||||||||||||||||||||||||||||||||||||||||||||||||||||||||||||||

*Source:* RESAM (2012).

The first of its kind was the Groupement forestier de Ristigouche, formed by a group of citizens in l'Ascension-de-Patapédia (Deschênes and Roy 1994). Under this model, small-scale private landowners consolidate their forest holdings, while retaining full ownership rights, in order to improve efficiency through collective management and commercialization of forest products. In the wake of Opérations dignité, the joint management model became a popular approach across eastern Quebec and beyond. This approach became an important way for private forest owners to consolidate expertise and enhance financial return, thereby empowering local actors within a provincial forest regime where large industrial interests still dominated on public lands. As the joint management organizations gained experience, some also sought to expand their activities onto public forest lands located within municipal boundaries (Guertin 1997). However, as was the case with forestry cooperatives, it proved very difficult to break into the tenure system.

**Efforts to Reframe Forest Governance**

While the provincial government attempted a significant forest reform in the mid-1970s, which would have represented a departure from the industrial concession-based system, this reform stalled. Although an initial act adopted in 1974 had brought the management of public forests under the government's responsibility, the new Forest Act passed in 1986 did not go as far and retained the model of industry-led management, but with additional requirements for public participation and greater consideration of multiple uses of the forest (Désy et al. 1995; Dubois 1995; Duchesneau 2004). The 1986 Forest Act opened the door to public involvement by subjecting all public forests to land-use planning exercises conducted by regional governments known as regional county municipalities, or RCMs (Bouthillier 2001). The new law provided no specific opportunities for community-based governance; however, it did retain existing programs to support small private forest owners and forestry cooperatives.

Whereas the 1986 Forest Act addressed the needs of a key industry for many rural areas, the challenges faced by those areas went well beyond their industrial fabric. Studies documented serious deficiencies in the social and demographic conditions of rural areas and stressed the need to rethink rural development (Conseil des affaires sociales 1989). This analysis was echoed by academics, who highlighted the importance of adopting a holistic view of rural development and promoting strategies that would enhance local self-sufficiency through community development and local governance (Vachon 1993; Désy et al. 1995).

The reflection on rural areas highlighted the diversity of rural realities and the need for a policy framework that would enable communities to shape their own economic, social, cultural, and political destinies (Vachon 1991). On the forestry front, the shift from an industry-oriented model to a community-based model was, once again, recognized as holding strong potential.

By the early 1990s, the idea of devolving rights and responsibilities related to public forest lands to rural communities had become part of the discourse of a wide range of actors. The activities of Solidarité rurale du Québec, a coalition of rural organizations, as well as public participation processes within forest planning provided new platforms for citizens to articulate alternative visions for rural development and to advocate for greater community control over forest management. These proposals were put forward during the Forest Protection Commission hearings in 1991, which resulted in the commission's recommendation that government amend provincial legislation to allow local communities to gain access to forest tenures (BAPE 1991). The combination of strong rural institutions with widespread public debate helped to build a movement around community forestry that is unique to Quebec.

The government undertook a number of modest reforms in the 1990s, which opened the door to the first legally recognized community forests in the province. Although these reforms were not labelled as community forests in 1993, the provincial Forest Act was amended in order to introduce a new tenure, the forest management contract (FMC), which allows individuals or organizations to obtain forest management rights on a specific parcel of public forest, without the usual obligation to own a processing facility (Roy 2006). A similar volume-based tenure called the forest management agreement (FMA) followed in 2001. The FMC and FMA were designed for smaller-scale players such as municipalities, non-profits, First Nations, and other local organizations interested in participating in timber harvesting and forest management (Québec, MRNF 2009). However, the creation of these tenures was not accompanied by any redistribution of existing forest allocations; thus, their implementation remained limited. In 2009, a total of sixty-eight FMCs covered an area equivalent to 1,275,732 hectares, and eight FMAs held rights over 2 percent of the timber from public lands that was allocated on a volume basis (Québec, MRNF 2009).

During the mid-1990s, the provincial government also opted to de-centralize some of its forest-management authorities to RCMs through the establishment of territorial management agreements, or TMAs (see Box 3.3). These agreements allow RCMs to take over responsibility for intramunicipal lots, which consist of scattered parcels of public land within municipal boundaries that are not already allocated to indus-try. This tenure gives the RCMs rights that go beyond forest manage-ment, including leasing or selling some land as well as enforcing rules and regulations (Teitelbaum, Beckley, and Nadeau 2006). Between 1997 and 2004, the government concluded twenty-two TMAs with RCMs, covering approximately 500,000 hectares of forest. Some RCMs opted to manage these lands themselves and acquired the services of professional foresters. Others transferred this responsibility to other groups in the community, such as forestry cooperatives, non-profit cor-porations, or joint management organizations. These policy reforms,

|||||||||||||||||||||||||||||||||||||||||||||||||||||||||||||||||||||||||||||||||||||||||||||||||||||||||||

**BOX 3.3  Territorial Management Agreements (TMAs)**

- TMAs are agreements attributed to RCMs by the provincial government to manage scattered public lands (often referred to as intramunicipal lands) within their boundaries.
- Agreements give RCMs management authority over most resources, as well as control of land use and land protection.
- Decision-making is ultimately the responsibility of the mayors. The tenure conditions include an obligation to form an advisory committee made up of forestry stakeholders in the region.
- Many RCMs contract out management responsibilities to other community-based organizations such as joint ventures, forestry cooperatives, or not-for-profit corporations.
- Despite their common goal of contributing to regional economic development, RCMs have embraced a variety of goals and approaches. Some are primarily timber oriented, while others are developing other forest resources and activities.
- RCMs are allowed to sign FMCs with community-based organizations for the management of these lands.
- In 2004, a total of 22 TMAs covered approximately 500,000 hectares of land.

|||||||||||||||||||||||||||||||||||||||||||||||||||||||||||||||||||||||||||||||||||||||||||||||||||||||||||

*Source:* Teitelbaum, Beckley, and Nadeau (2006); Roy (2006).

although not part of a wider community forestry strategy per se, did facilitate the emergence of numerous small-scale initiatives in Quebec, most of which involved the participation of regional municipalities.

In addition to providing funds to help TMA start-ups, the government introduced the Forest Resource Development Program in 1995, a cost-sharing program to support work conducted by tenure holders and local initiatives on either public or private forest lands. The program supported work related to silviculture, wildlife habitats, recreation, education, and environmental issues. Many funded projects were partnerships: for example, a partnership involving First Nations, industry, and the federal government conducted work related to timber production, recreation, wildlife, and cultural values (Gagnon 2002).

In the mid-1990s, the provincial government also took steps towards the implementation of the "inhabited forest," mandating a working group to suggest a framework for the implementation of the concept of inhabited forest and to facilitate the emergence of pilot projects to

|||||||||||||||||||||||||||||||||||||||||||||||||||||||||||||||||||||||||||||||||||||||||||||||||||||

**BOX 3.4   Inhabited Forest Pilot Projects in Quebec**

- *Chibougamau*: 103,000 ha in Nord-du-Québec
- *Comité de gestion intégrée des ressources des Bois-Francs coopérative*: 2,939 ha in Centre-du-Québec
- *Coopérative forestière de Ferland-Boileau*: 40,100 ha in Saguenay
- *Expérience pilote de gestion décentralisée et de mise en valeur des ressources forestières*: 74,000 ha in Bas-Saint-Laurent
- *Fermes forestières de la MRC de Matane*: 8,130 ha in Bas-Saint-Laurent
- *Forêt de l'Aigle*: 14,000 ha in Outaouais
- *Forêt du Massif*: 4,500 ha in Québec
- *Forêt Habitée de la Chute-Saint-Philippe*: 11,000 ha in Laurentides
- *Forêt Habitée du Mont Gosford*: 6,000 ha in Estrie
- *Forêt Habitée Iberville*: 25,497 ha in Côte-Nord
- *Forêt Ouareau*: 14,900 ha in Lanaudière
- *Habitafor*: 25,500 ha in Gaspésie
- *Parc régional du Massif du Sud*: 10,300 ha in Chaudière-Appalaches
- *Production et aménagement des ressources collectives récréo-forestier de Saint-Mathieu*: 12,704 ha in Mauricie
- *Roulec 95 Inc.*: 37,429 ha in Abitibi-Témiscamingue

|||||||||||||||||||||||||||||||||||||||||||||||||||||||||||||||||||||||||||||||||||||||||||||||||||||

*Source:* Corporation de gestion de la Forêt de l'Aigle (2000).

validate the conditions of implementation of this concept (Groupe de travail interministériel sur la forêt habitée 1996). "Inhabited Forest" was the name given to a social movement that had manifested itself through the years via numerous local initiatives, such as Opérations dignité, and that promoted the occupancy and use of forest lands by rural communities to ensure their sustainability as a natural eco-system and as a place to live (Bouthillier and Dionne 1995). However, as was pointed out by Bouthillier and Dionne (1995), in order to make this vision a reality, better institutional arrangements were clearly needed (the most important being a tenure) to support the adoption of local governance and ecosystem-based management. The government opted to set up a series of pilot projects, fourteen in total, designed to investigate the potential of different models of local engagement (see Box 3.4). These fourteen initiatives covered a range of organizational arrangements, decision-making structures, goals, and practices. Many of the project proponents came from the municipal milieu, but there was also a strong emphasis on bringing together other forest stakehold-ers, including private landowners, industrial forest companies, recrea-tion organizations, joint management organizations, and cooperatives. The pilot projects yielded mixed results (Bérard 2000). Those with ex-clusive rights to timber harvesting fared better, such as the Forêt de l'Aigle (Outaouais) and the Forêt habitée du Mont Gosford (Estrie). After five years, support for the pilot projects was discontinued.

The Model Forest Program, a federally funded initiative, was also active in supporting local organizations, including two in Quebec, dur-ing this period. In the Bas-Saint-Laurent region, a proposal was de-veloped to implement a tenant-farming project, which became part of the Model Forest Network through the Bas-Saint-Laurent Model Forest. This project, owned by Abitibi-Consolidated, was implemented on pri-vate forests and proved quite successful, until the company opted to end the lease in the mid-2000s. The Model Forest Program also sup-ported the development of the Waswanipi Cree Model Forest, which worked at developing a collaborative approach between Cree land users and forest managers (Waswanipi Cree Model Forest 2007).

Finally, there were also communities that, in the absence of institu-tional support, opted to negotiate directly with industrial tenure hold-ers in order to get involved with or take charge of management of public lands adjacent to their villages (Guertin 1997; Mercier 2002; Hébert-Sherman 2011). Those initiatives, most of them spearheaded by municipal corporations, exist in the Lac-Saint-Jean region, where close to a dozen organizations are involved in forest management through collaborative arrangements with industry. Examples include the Corporation d'aménagement de la forêt de Normandin, which has spearheaded the combination of blueberry and timber production on some forests, and the Coopérative de solidarité forestière de la Rivière aux Saumons, which has undertaken a comprehensive multi-resource planning process. These initiatives have persevered and have sought more secure tenure under the most recent policy regime, but thus far without success.

Despite commitments to develop a policy on the inhabited forest, no further steps were taken to enshrine this concept in law. Revisions to provincial forest policy during this time, which included public consultations, revealed divergent views regarding the inhabited forest. The majority of forest companies voiced their opposition to the allocation of more timber to nonindustrial interests, while many organizations and local interest groups expressed strong support for devolution of rights to communities (Québec, MRN 1999). The result of these consultations, Bill 136, was silent on the question of the inhabited forest, leading actors like the Union of Quebec Municipalities (Union des municipalités du Québec, or UMQ) to infer a shift within the ministère des Ressources naturelles towards a more centralized vision of forest management, a move that the UMQ saw as contrary to the government's broader goal of promoting sustainable regional development (UMQ 2000).

In the midst of this policy review, managers of one of the inhabited forest pilot projects organized the first ever conference on this topic to assess the situation and reiterate the need for government commitment to greater devolution of management responsibility towards local communities. The conference brought together more than two hundred participants from government, industry, municipalities, academia, and nongovernmental organizations. It led to a common position, encapsulated in the Maniwaki Declaration, which pointed out that the current forest regime only partially responded to communities' aspirations and that what was needed was the creation of a legal framework for the inhabited forest that would confer more power to local communities (Corporation de gestion de la Forêt de l'Aigle 2000).

### Community Forestry: Are We There Yet?

The 2000s proved a tumultuous time for the forest sector in Quebec. The release of Richard Desjardins's documentary *L'erreur boréale* (1999) and a critical report from Quebec's Auditor General (*Vérificateur générale* 2003) fuelled public concerns over forest management – more specifically, about the availability and sustainability of the resource. Forestry issues became highly politicized, and both government and industry, now on the defensive, sought ways to abate public concern. In 2003, the Commission for the Study of Public Forest Management (known as the Coulombe Commission) was put in place, with the mandate to analyze the state of Quebec's public forest management in order to recommend improvements on how to enhance social acceptability and environmental sustainability. The final report, which covered a broad range of issues, did acknowledge long-standing pressures for community forests from rural communities and recommended that the government support the establishment of inhabited forest projects as well as the creation of new tenures better adapted to meeting community needs (Coulombe Commission 2004). This recommendation was echoed during the subsequent *Sommet sur l'avenir du secteur forestier québécois* (Summit on the Future of the Quebec Forest Sector), which brought together a vast number of forestry-oriented organizations and in which the closing statement called for a review of tenure allocation practices in order to promote a diversification of approaches (Summit Partners 2007). The closing statement of the summit used the term *forêt de proximité* (local forests), which was, in essence, a new name for an old concept, but with the explicit recognition of geographical proximity.

After the Coulombe Commission, the provincial government undertook a major overhaul of forest policy, which culminated in a new forest law, the Sustainable Forest Development Act (SFDA), adopted in 2013. The law incorporates many of the recommendations of the Coulombe Commission and represents an important shift in the way forests are governed and managed in the province. This shift includes the application of a number of key approaches, including ecosystem-based management, integrated resource management, and regionalization of forestry planning. In the new regime, forest-management responsibilities will be shared among government, regional organizations, First Nations, and other users of the forest, but the roles of the non-government groups still seem largely limited to consultation (Laplante 2010). However, in sharp contrast to the previous policy regime, under the SFDA, the ministère des Ressources naturelles, rather than the forest industry, is in charge of forest planning. The concept of "local forests," introduced at the summit in 2007, was part of the discussion from the outset and was included in Bill 37, which became the new Sustainable Forest Development Act that came into force in April 2013.

With regard to the local forests, the SFDA committed the government to developing a local forest policy and to holding public consultations on the topic. In summer 2011, the government released a consultation document on local forests that reiterated many of the ideas expressed in earlier policy discussions. Local forests were envisioned as a tool to support the socio-economic development of rural and Indigenous communities through local decision-making and local hiring. Territories would be defined according to geographic proximity to communities, as well as sense of connection to specific territories (Québec, MRNF 2011). The government envisioned the conversion of existing forest management contracts and forest management agreements to local forests, as well as the creation of ten to fifteen new initiatives across Quebec in the first round. The groups eligible to apply include Native band councils, RCMs, and municipalities. Public responses to this proposition were, for the most part, favourable, with some reservations – for example, regarding implications for Indigenous rights. In the period leading up to the release of the new forest law, several municipalities and RCMs began preparing proposals for new or expanded local forests.

However, since the release of the new law, there has been very little movement on the local forests file. The local forest policy has not yet been released, with some projecting a moratorium of two years before action will be taken (FQM 2013). This has raised the frustration level for local actors, many of whom were anticipating a rapid transfer of existing small-scale tenures to local forests and the creation of new projects. Particular concerns have been expressed for forest lands that have been identified as suitable candidates for local forests but that are currently under corporately held tenures. Most recently, the government made a commitment to develop four pilot projects, with no firm date attached (Québec 2013).

## Conclusion

In Quebec, attempts to reorganize the web of relationships in the forest sector, creating space for enhanced community involvement, has been ongoing for more than half a century. These efforts have been made within the context of what Minville (1944) describes as a forest governance system with a near-exclusive focus on economic benefits, without sufficient attention and valorization of other forest values. As demonstrated in this chapter, this model has regularly been contested by local communities in Quebec, some of whom have sought to develop initiatives that counteract the corporate model. Thus, despite a tenure regime that links forestry rights to the ownership of processing facilities, communities have succeeded in carving out a space, albeit a small one, within the patchwork of tenure arrangements in the province. Forestry cooperatives, joint ventures, and inhabited forest projects are all lasting examples of initiatives that aim to enhance local management and decision-making and to capture economic benefits locally. However, these initiatives have suffered from a lack of legal recognition and institutional support, which has contributed to a certain fragility and has left them on the sidelines of major forest sector developments, such as Quebec's new policy regime (Blais and Boucher 2013). The lack of government support for community forestry should perhaps not come as a surprise, considering the much stronger precedent in Quebec of prioritizing corporate approaches to forest development. For their part, community forests often espouse a different philosophy, focused more on adopting practices of participatory governance, local management, and local development. Carving out this space within an economic context that has been difficult for all players has not been an easy endeavour, and not all community forests have been successful. However, this does not appear to have diminished enthusiasm for community forestry. The recent local forests policy, although not yet fully developed, has generated a great deal of excitement within rural communities. As with past initiatives, the local forests approach has fed the imagination of Quebec's rural communities, who appear poised to play a greater role in forest-management decisions in order to make forests a key asset in meeting their economic, social, and environmental needs and aspirations.

## References

AIFQ (Association des ingénieurs forestiers du Québec). 1949. *Le problème du Québec forestier.* Montréal: Éditions Fides.

Baker, M., and J. Kusel. 2003. *Community forestry in the United States: Learning from the past, crafting the future.* Washington, DC: Island Press.

BAPE (Bureau des audiences sur l'environnement). 1991. *Des forêts en santé: Rapport de la Commission sur la protection des forêts.* Bureau des audiences publiques du Québec, Commission sur la protection des forêts. Québec: Les Publications du Québec.

Beaudoin, J.M. 2014. "Growing deep roots: Learning from the Essipit's culturally adapted model of Aboriginal forestry." PhD diss., Department of Forest Resources Management, University of British Columbia, Vancouver.

Beaudoin, J.M., G. St-Georges, and S. Wyatt. 2012. "Valeurs autochtones et modèles forestiers: Le cas de la Première Nation des Innus d'Essipit." *Recherches Amérindiennes au Québec* 42 (2–3): 97–109. http://dx.doi.org/10.7202/1024105ar.

Bérard, L. 2000. "Les projets témoins de Forêt habitée: Forces et faiblesses." In *Corporation de gestion de la Forêt de l'Aigle, Colloque sur la forêt habitée*, 17-25.

Blais, R., and J.L. Boucher. 2013. "Les temps des régimes forestiers au Québec." In *La gouvernance locale des forêts publiques québécoises: Une avenue de développement des régions périphériques?*, edited by G. Chiasson and É. Leclerc, 33–63. Montréal: Presses du l'Université du Québec.

Bouthillier, L. 2001. "Quebec: Consolidation and the movement towards sustainability." In *Canadian forest policy: Adapting to change*, edited by M. Howlett, 237–78. Toronto: University of Toronto Press.

Bouthillier, L., and H. Dionne. 1995. *La forêt à habiter: La notion de "forêt habitée" et ses critères de mise en œuvre*. Rimouski: Université du Québec à Rimouski et Université Laval.

Chiasson, G., and C. Andrew. 2013. "Les coopératives forestières ou les difficultés du développement 'à la périphérie de la périphérie.'" In *La gouvernance locale des forêts publiques québécoises: Une avenue de développement des régions périphériques?*, edited by G. Chiasson and É. Leclerc, 147–68. Montréal: Presses de l'Université du Québec.

Conseil des affaires sociales. 1989. *Deux Québec dans un: Rapport sur le développement social et démographique*. Boucherville, QC: Gaëtan Morin Éditeur.

Corporation de gestion de la Forêt de l'Aigle. 2000. *Colloque sur la forêt habitée: Nouveaux modes d'exploitation et d'aménagement des forêts au Québec?* Conference proceedings, 18–20 October, Maniwaki, QC.

Coulombe Commission (Commission d'étude sur la gestion de la forêt publique québécoise). 2004. *Rapport de la Commission d'étude sur la gestion de la forêt publique québécoise*. Québec, QC.

Deschênes, M.A., and G. Roy. 1994. *Le JAL: Trajectoire d'une expérience de développement local*. Groupe de recherche interdisciplinaire en développement de l'Est du Québec. Rimouski: Université du Québec à Rimouski.

Désy, J., G. Bélanger, C. Brisson, L. Fraser, G. Tremblay, and S. Tremblay. 1995. *Des forêts pour les hommes et les arbres*. Laval, QC: Méridien.

Dubois, P. 1995. *Les vrais maîtres de la forêt québécoise*. Montréal: Écosociété.

Duchesneau, M. 2004. *Gestion de la forêt publique et modes d'allocation de la matière ligneuse avant 1986*. Report prepared for Commission d'étude sur la gestion de la forêt publique québécoise, Québec, QC.

FQCF (Fédération québécoise des coopératives forestières). 2015. "Historique." *FQCF*. http://www.fqcf.coop/federation-quebecoise-cooperatives-forestieres/histoire/.

FQM (Fédération québécoise des municipalités). 2013. *L'occupation du territoire forestier québécois et les sociétés d'aménagement des forêts*. Mémoire de la Fédération québécoise des municipalités présenté à la Commission de l'économie et du travail, Québec, QC.

Gagnon, R. 2002. *Le program de mise en valeur des ressources du milieu forestier: Bilan quinquennal 1995–2000.* Québec, QC: Ministère des Ressources naturelles, Direction des programmes forestiers Service de l'aménagement forestier, Gouvernement du Québec.

Groupe de travail interministériel sur la forêt habitée. 1996. *La gestion des ressources du milieu forestier habité.* Québec, QC: Ministère des Ressources naturelles, Gouvernement du Québec.

Guertin, C.É. 1997. "Les conditions d'établissement des forêts communautaires au Québec." Master's thesis, Département des sciences du bois et de la forêt, Université Laval, QC.

Hébert-Sherman, D. 2011. "Légitimité politique, droits ancestraux et gestion du territoire forestier: Le cas de la Forêt Habitée de la Doré." Master's thesis, Département d'anthropologie, Université Laval, QC.

Howlett, M., and J. Rayner. 2001. "The business and government nexus: Principal elements and dynamics of the Canadian forest policy regime." In *Canadian forest policy: Adapting to change*, edited by M. Howlett, 23–62. Toronto: University of Toronto Press.

Jacqmain, H. 2008. "Development of a sustainable moose habitat management process that is culturally relevant to the Waswanipi Cree in the boreal black spruce forest of northern Québec." PhD diss., Département des sciences du bois et de la forêt, Université Laval, QC.

Laplante, R. 2010. "Forêt de proximité et nouveau régime forestier: Occasion ratée, rendez-vous reporté." *Revue Vie Économique* 2 (1). http://www.eve.coop/mw-contenu/revues/6/53/RVE_vol2_no1_Laplante.pdf.

Lessard, J. 2012. "Coopératives forestières et communautés pour un développement forestier plus durable." *Revue Vie Économique* 3 (4): 91. http://www.eve.coop/mw-contenu/revues/15/139/RVE_vol3_no4_Lessard.pdf.

Linteau, P.A., R. Durocher, J.C. Robert, and F. Ricard. 1989. *Le Québec depuis 1930.* Vol. 2 of *Histoire du Québec contemporain.* Montreal: Boréal.

Masse, S. 1995. *La foresterie communautaire: Concept, champs d'application et enjeux.* Ste-Foy, QC: Section Planification et Évaluation, Direction du développement forestier, Ressources naturelles Canada, Service canadien des forêts, Centre de foresterie des Laurentides. Fo42–243/1995f.

Mercier, M. 2002. "Forêt habitée et développement durable: Le cas d'un parc expérimental de la nordicité à la ville de La Baie." Master's thesis, Programmes d'études de cycles supérieurs en interventions régionales, Université du Québec à Chicoutimi.

Minville, E. 1944. *La forêt: École des hautes études commerciales.* Montreal: Éditions Fides.

Otis, L. 1989. *Une forêt pour vivre.* Collection Témoignages et Analyses. Groupe de recherche interdisciplinaire en développement de l'Est du Québec. Rimouski: Université du Québec à Rimouski.

Paradis, R. 1980. "La pensée coopérative d'Esdras Minville de 1924 à 1943." *L'Action Nationale* 69 (7): 518–26.

Québec. 2013. *Rendez-vous national de la forêt québécoise, Saint-Félicien, 21–22 novembre 2013.* Québec: Gouvernement du Québec. http://rendezvousdela foret.gouv.qc.ca/pdf/Cadre_financier_presentation.pdf.

–. MRN (Ministère des Ressources naturelles). 1999. *Mise à jour du régime forestier: Synthèse des consultations publiques, automne 1998.* Québec: Ministère des Ressources naturelles.

–. MRNF (Ministère des Ressources naturelles et de la Faune). 2009. *Répertoire des bénéficiaires de CAAF, de CtAF et de CvAF.* 30 June. Québec: Ministère des Ressources naturelles et de la Faune.

–. MTF (Ministère des Terres et Forêts). 1965. *Exposé sur l'administration et la gestion des Terres et Forêts du Québec.* Québec, QC: Ministère des Terres et Forêts.

RESAM (Regroupement des sociétés d'aménagement forestier du Québec). 2012. *Les groupements de propriétaires de lots boisés, présents pour ... faire pousser la forêt.* Québec, QC. http://www.resam.org/wp-content/uploads/ 2012/08/pousser.pdf.

Roy, M.É. 2006. "Des fermes forestières en métayage sur le territoire public québécois: Vers un outil d'évaluation pour les communautés." Master's thesis, Faculté de foresterie, de géographie et de géomatique, Université Laval, Québec, QC.

Saint-Arnaud, M. 2009. "Contribution à la définition d'une foresterie autochtone: Le cas des Anicinapek de Kitcisakik." PhD diss., Institut des sciences de l'environnement, Université du Québec à Montréal.

Summit Partners (Summit Partners on the Future of the Québec Forestry Sector). 2007. "Closing Statement of December 12, 2007." Sommet sur l'avenir du secteur forestier québécois. http://sommetforet.ffg.ulaval.ca/UserFiles/File/ Summit_Closing_Statement.pdf.

Teitelbaum, S. 2009. "An evaluation of the socio-economic outcomes of community forestry in the Canadian context." PhD diss., Faculty of Forestry and Environmental Management, University of New Brunswick, Fredericton.

Teitelbaum, S., T.M. Beckley, and S. Nadeau. 2006. "A national portrait of community forestry on public land in Canada." *Forestry Chronicle* 82 (3): 416–28. http://dx.doi.org/10.5558/tfc82416-3.

UCC (Union catholique des cultivateurs). 1969. *Mémoire de l'Union catholique des cultivateurs et le Comité provincial des offices et syndicats de producteurs de bois aux ministres de l'agriculture et des forêts et au ministre responsable de l'Office de planification du Québec sur l'aménagement de fermes forestières au Québec.* Montréal: Union catholique des cultivateurs.

UCC and FPBQ (Union catholique des cultivateurs et la fédération des producteurs de bois du Québec). 1971. *Réorganization de l'activité forestière rurale.* In collaboration with Louis-Jean Lussier. Québec, QC: UCC and FPBQ.

UMQ (Union des municipalités du Québec). 2000. *Mémoire présenté à la Commission de l'économie et du travail sur le projet de loi 136: Loi modifiant la loi sur les forêts et autres dispositions législatives.* Montréal, QC: Union des municipalités du Québec.

Vachon, B. 1993. *Le développement local: Théorie et pratique.* Boucherville, QC: Gaëtan Morin Éditeur.

–, ed. 1991. *Le Québec rural dans tout ses états.* Montréal: Boréal.

Vincent, O. 1995. *Histoire de l'Abitibi-Témiscamingue.* Vol. 7 of *Collection Les régions du Québec.* Québec, QC: Institut québécois de recherche sur la culture.

Waswanipi Cree Model Forest. 2007. *Ndoho Istchee: An innovative approach to Aboriginal participation in forest management planning.* Waswanipi, QC: Waswanipi Cree Model Forest.

Wyatt, S. 2004. "Co-existence of Atikamekw and industrial forestry paradigms." PhD diss., Université Laval, Québec, QC.

Chapter 4    **Community Forestry on Crown Land in Northern Ontario**
Emerging Paradigm or Localized Anomaly?

*Lynn Palmer, M.A. (Peggy) Smith,
and Chander Shahi*

Ontario has missed multiple opportunities to embrace community forestry as a viable alternative to its predominant industrial system. However, opportunity continues to knock and public support for local decision-making keeps community forestry on the provincial agenda. In this chapter, we adopt a definition of community forestry as a forest-management approach in which "communities play a central role in the decisions" (Teitelbaum and Bullock 2012). More specific principles include participatory governance, rights as they affect the level of authority, local benefits, and increased ecological stewardship related to multiple-use forestry (Duinker et al. 1994; Teitelbaum and Bullock 2012, Teitelbaum this volume).

The roots of forest management in Ontario go back three hundred years to the colonial period. Two main social-ecological systems have developed since that time: one is associated with private property rights and is found predominantly in southern Ontario, and the other is associated with provincially owned Crown land in northern Ontario. This chapter focuses exclusively on the development of community forestry in northern Ontario. While there are community forestry projects in southern Ontario, we deem these to be sufficiently different in terms of property rights regime, forest type, and social context to merit separate analysis (for descriptions, see Teitelbaum, Beckley, and Nadeau [2006] and Teitelbaum and Bullock [2012]).

The Crown land of northern Ontario covers a vast region that contains predominantly boreal forests and a large portion of the Great Lakes–St. Lawrence forest zone. A significant portion of this region was defined as the "area of the undertaking" (AOU) for the purpose of the groundbreaking class environmental assessment (EA) for timber management on Crown lands in Ontario conducted from the late 1980s to 1994 (EAB 1994). The AOU's northern boundary is the commercial limit for logging in northern Ontario and its southern boundary runs from the Mattawa River in the east, across Lake Nipissing, to the French River in the west (Figure 4.1). Beyond the northern boundary is Ontario's Far North region, making up 42 percent of the provincial land base and stretching from Manitoba in the west to the Hudson Bay coast in the north and Quebec in the east (Ontario, NRF 2015b). We have taken the AOU and the Far North region, with its population of just over 800,000 – less than 10 percent of the province's total population (Ontario, Finance 2013), as the geographic focus of this chapter. These two regions of Crown land are both social-ecological subsystems of the main Crown land system, based on their rates and levels of forestry development, which are linked to differences in geographical and social contexts.

The AOU encompasses five cities with over forty thousand inhabitants each: North Bay, Sudbury, Sault Ste. Marie, Thunder Bay, and Timmins. Although they are classified as heavily resource-dependent, these cities are relatively diversified since they function as regional service centres (Southcott 2006). The remainder of the AOU is sparsely populated with smaller, less economically diverse, resource-dependent municipalities that are mostly single-industry forestry or mining towns and First Nation reserves (Southcott 2006; see Figure 4.1).[1] As of 2004, the forest industry employed close to fifty thousand people in the boreal forest throughout Canada, 10 percent of them in logging, 40 percent in the wood industry, and 50 percent in pulp and paper (Bogdanski 2008).

The twenty-four thousand, mainly Aboriginal, inhabitants of the Far North are located in thirty-one First Nation communities – accessible only by air or water and, in some cases, winter roads – and two municipalities, one (Pickle Lake) accessible by all-season road and the other (Moosonee) by rail (Ontario, NRF 2015b). To date, resource development in this part of the province has been limited mainly to hydro-electricity and mineral exploration. Planning for logging in the region has begun only recently through community-based land-use planning. Several provincial parks have been established and the Far North Act, 2010, made a commitment to protect 50 percent of the area.

First Nation communities, which are classified as reserves under federal jurisdiction, have historical and contemporary ties to their traditional territories for subsistence and other purposes. First Nations'

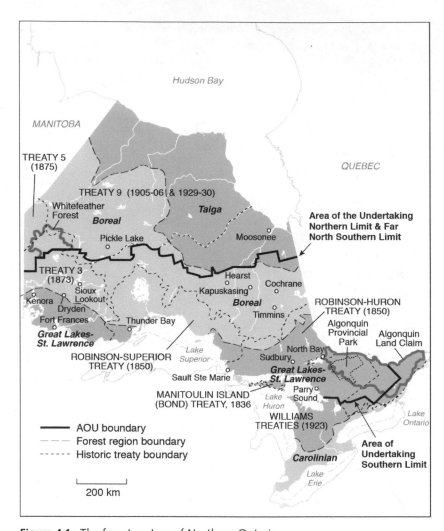

**Figure 4.1**   The forest system of Northern Ontario.
Compiled by Tomislav Sapic in 2015.
*Sources:* Canada, NRCan 2016; Canada, NRCan 2007. Adapted by Eric Leinberger.

traditional territories that are outside reserves encompass large areas of Crown forest lands. Both reserve lands and traditional territories on provincial Crown land are subject to historic treaties (Smith 1998). First Nation communities throughout northern Ontario have historic-ally been largely excluded from the forest-based economy and continue to face much greater economic challenges than municipalities in the same region (Southcott 2006). However, in the wake of recent Supreme

Court of Canada decisions, the protection of both First Nation and Métis rights and their involvement in resource management decisions is becoming a central issue (Gallagher 2012; Coates and Newman 2014).

Ontario's Crown forest-management system was established in the mid-1800s as the province, with constitutional responsibility to manage natural resources within its boundaries, assumed the power to regulate and extract revenues from forests. This centralized command-and-control system, which fit the "staples" model of economic development based on resource extraction and resource commodity export, provided little room for local decision-making (Thorpe and Sandberg 2008). Except for an early proposal for the Nipigon Forest Village in 1944, community forestry was not part of the provincial policy landscape until the establishment of the Algonquin Forest Authority in the 1970s, followed by several community forestry pilot projects in the 1990s.

This chapter draws on several theoretical concepts to characterize the development of community forestry in northern Ontario. Its specific objectives are the following: to describe the historical development of the Crown forest system in Ontario and its relationship to the adaptive cycle (Gunderson and Holling 2002); to draw on complexity theory to help explain community forestry's place within northern Ontario's forest system; and to assess whether Ontario's forest system has truly embraced community forestry throughout the phases of the adaptive cycle, or whether community forestry is simply an anomaly frozen within the dominant industrial forestry paradigm.

## A Complexity Approach to Northern Ontario's Forestry System

In this section, we describe Ontario's forest system and its receptivity to community forestry from the perspective of complex adaptive systems (complexity) theory. A central feature of complexity theory is Gunderson and Holling's (2002) adaptive cycle, a model of systemic change that explains the continuous cycles of disturbance and renewal that occur in a complex adaptive system (CAS). A CAS is a group of systems that exhibit multiple interactions and feedback mechanisms in a non-linear manner to form a complex whole that has the capacity to adapt in a changing environment (Levin 1999; Gunderson and Holling 2002; Holland 2006). A forest system – with its constituent forest ecosystems, social institutions, and actors associated with forest management – is a specific type of CAS composed of linked social and ecological systems (Berkes, Colding, and Folke 2003; Messier, Puettmann, and Coates 2013; Filotas et al. 2014). Complexity theory provides a means to understand the ebbs and flows of both changes and rigidity within a CAS. By viewing the different phases of the Ontario forest system through a complexity lens, it is possible to identify the challenges and opportunities for the transformation of the system that is necessary to support community forestry. Obtaining an understanding from such a complexity perspective can improve the chances for social innovation, such as the implementation of community forests, in which individuals "begin to shift the pattern around us as we ourselves begin to shift" (Westley, Zimmerman, and Patton 2006, 19).

According to Gunderson and Holling (2002), CASs pass through the adaptive cycle's four phases: (1) exploitation (growth) (r) a phase in which the system undergoes a period of rapid growth and its components become routine, dynamics are relatively predictable, and disturbances have a negligible impact on the integrity of the system; (2) conservation (K), a steady-state phase in which the system becomes stabilized but resources are locked up, and there is increasing complexity of system components, rigidity, and vulnerability to shocks that may disturb the system's balance; (3) release ($\Omega$), associated with chaotic collapse of the system, a drastic reduction of structural complexity, and rapid change in the system's properties; and (4) reorganization ($\alpha$), a phase in which innovation is possible through a restructuring of the system but its dynamics are unpredictable. The outcome of reorganization can be a return to a similar state or a transformation – a regime shift – into a new system configuration (Holling 1973, 1986; Walker et al. 2004; Walker and Salt 2006).

The theory of panarchy involves the notion that social-ecological systems function at different but linked scales and that elements of these interacting systems change at different rates, yielding extremely complex interactions (Holling, Gunderson, and Peterson 2002). Viewing systems through a panarchy lens can help explain how they are interconnected and how one system is vulnerable to the effect of other systems going through their own adaptive cycles at various scales.

An understanding of the changes that CASs undergo through adaptive cycles provides insight into how to manage a system's resilience – the amount of disturbance that can be absorbed by a CAS without altering its basic structure and function (Holling 1973, 1986; Walker et al. 2004; Walker and Salt 2006). Features associated with resilience

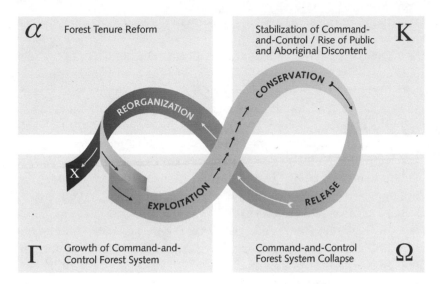

**Figure 4.2**   Phases of the adaptive cycle in northern Ontario's forestry system for the Area of the Undertaking.
*Source:* Adapted from Holling, Gunderson, and Peterson (2002).

in CASs include multiple interactions through web-like interconnectedness; feedback mechanisms and nonlinearity; diversity (of species, knowledge systems, economic options); and the capability for self-organization, learning, and adaptation in the context of change (Berkes, Colding, and Folke 2003; Folke, Colding, and Berkes 2003; Chapin et al. 2004; Armitage 2005; Walker and Salt 2006). In a social-ecological system, adaptive capacity is the collective capacity of the actors in the system to manage resilience by responding to, creating, and shaping variability and change in the state of the system (Folke, Colding, and Berkes 2003; Berkes, Colding, and Folke 2003; Walker et al. 2004).

A resilience approach to natural resource management contrasts with that of command-and-control, which does not address the complexity and uncertainty characteristic of CASs but instead emphasizes optimization and efficiency, top-down control, minimal collaboration among stakeholders, and linear positivistic thinking that attempts to maintain the system in a steady state with predictable outcomes (Walker and Salt 2006; Beratan 2014). Holling and Meffe (1996) describe the "pathology" of the command-and-control approach to natural resource management, which leads to a "cycle of dependency." Pinkerton and Benner (2013) recently demonstrated such an outcome in British Columbia, where forest commodity sawmills exhibited a lack of resilience in the face of the forest economy collapse, exposing the vulnerability of local forest-based economies dependent on these single, large enterprises.

Building system resilience under conditions of change and uncertainty calls instead for an approach that emphasizes flexibility, experiential learning (Holling 1978; Walters 1986; Lee 1993, 1999), collaboration, shared decision-making, and the development of adaptive capacity (Folke, Colding, and Berkes 2003; Folke et al. 2005, Berkes, Colding, and Folke 2003; Lee 1993, 1999). Adaptive capacity fosters the development of innovative solutions in complex social and ecological circumstances such as those that pertain to natural resource management (Walker et al. 2002; Folke, Colding, and Berkes 2003; Gunderson 2003). Adaptive comanagement (Armitage, Berkes, and Doubleday 2007; Armitage, Marschke, and Plummer 2008), collaborative adaptive management (Susskind, Camacho, and Schenk 2012), and adaptive collaborative management (Colfer 2005; Prabhu, McDougall, and Fisher 2007; Ojha, Hall, and Sulaiman 2013) are various terms used to describe natural resource management approaches that address the inherent complexity and uncertainty in CASs by combining adaptive management and collaboration among multiple stakeholders. Such approaches promote power sharing, social learning, and the development of relationships through the building of mutual respect and trust, all aspects that contribute to the development of adaptive capacity and therefore foster resilience in natural resource systems. We suggest that community forestry is aligned with these resilience approaches and, as such, has the potential to promote forest system resilience. With its principles of participatory democracy and multiple use, which necessitate collaboration among a range of actors in the shared management of a diversity of forest products and services, community forestry is a forest governance innovation that recognizes forest system complexity and thereby fosters adaptive capacity.

What follows is an analysis of the development of the forest system in northern Ontario's AOU (from the 1800s to the present), conceptualized through the adaptive cycle framework (see Figure 4.2). Our analysis traces the emergence of community forestry throughout the four phases of this system's adaptive cycle. We then present an analysis of the more recent Far North forest subsystem based on its own adaptive cycle, which is distinct but interconnected with the adaptive cycle of the AOU, as explained by the theory of panarchy.

## Exploitation (Growth) Phase: The Rise of Command-and-Control – 1800s–1930s

The development of northern Ontario's Crown forest system was driven by the desire to exploit the region's timber in order to fuel provincial development following colonization, a major disturbance that caused the collapse of the original social-ecological system configuration based on Aboriginal land use, occupancy, and the fur trade. Following that collapse, the system went through a reorganization phase, emerging as a centralized command-and-control forest system formalized as a policy monopoly governed under the Crown Timber Act of 1849. During the subsequent exploitation (growth) phase, Crown forests were treated as the exclusive domain of the forest industry, their purpose being to generate royalties and employment for the province through sustained-yield timber management (Blouin 1998). Industrial development of Crown forests was the key system driver. Forestry companies focused on the production of high-volume, low-value commodities – pulp and paper and dimensional lumber – for export, primarily to the United States.

Throughout the nineteenth century and up to the 1930s, the system became regularized with the initiation and expansion of licensing of Crown forests to the forest industry, the development of harvesting technologies, and the establishment of municipalities to support local mills. The system dynamics were predictable and were based on a sustained-yield policy. Founded on neoclassical resource economics, this policy was aimed solely at optimization of specific variables – in this case, timber production for export in a staples economy (Innis 1930; Clapp 1998; Howlett and Brownsey 2008). Consideration of the system's social components was restricted largely to ensuring both employment and revenue generation. No attention was given to Aboriginal rights and interests.

## Conservation Phase: Public and Aboriginal Discontent – 1930s–2005

By the 1930s, the industrial forest system was well entrenched and stabilized, thus marking the beginning of the conservation phase. Maximization of timber production was achieved through technological advances in harvesting operations, and benefits were continually returned to those employed in the forestry industry. It was also during this phase that rigidity entered the system. Forest resources became "locked up," with only large forestry companies having licensed access to timber.

At the same time, complexity within the system was increasing as a result of nascent public concern about the management of forest resources. Arthur R.M. Lower (1938) reflected this concern in his book *The North American Assault on the Canadian Forest*, drawing attention to Canada's increasing dependency on US markets. A provincial royal commission on forestry in 1947 addressed concerns about wasteful forestry practices and regeneration, among other matters (Kennedy 1947). This public criticism became a system driver during this phase, leading to pressure for local input into forest management, thereby creating vulnerability in the command-and-control system. The impact of this driver was reflected in changes to provincial forest policy that allowed new initiatives to emerge in the middle of this phase.

The Crown Timber Amendment Act of 1979 established forest management agreements (FMAs), which were implemented in 1980. While these licences maintained the dominant paradigm's goal of sustained timber yield, they shifted responsibility for forest management from government to the forest industry. The key policy change in terms of social objectives was the requirement that FMA holders conduct public meetings during the preparation of management and operating plans. Although the new licensing system was applauded as being "creative and credible" (Fullerton 1984, 66), the Lakehead Social Planning Council raised concerns about the FMAs' exclusive focus on forestry companies and the continued disregard of local and First Nation communities and other stakeholders, which prevented them from participating in decision-making (Lang and Kushnier 1981).

Public criticism continued in this phase. A new era of forest policy that advanced public and Aboriginal participation was ushered in by the class environmental assessment, or EA (EAB 1994). The introduction of the class EA was an acknowledgement of the failure of the sustained-yield policy and of ecosystem management as a preferred approach. The EA Board also laid down groundbreaking terms and conditions governing social aspects, including public input – in particular, through local citizen committees (LCCs) and Aboriginal consultation. Legally binding, these terms and conditions paved the way for significant changes in forest management in Ontario.

The biggest of these changes was a new forestry law, the 1994 Crown Forest Sustainability Act (CFSA). The CFSA enshrined the concept of sustainability as "long-term health and vigour of Crown forests" that would be managed "to meet social, economic and environmental needs of present and future generations." Under the CFSA, the former FMAs were converted to sustainable forest licences (SFLs). These long-term (twenty-year) licences became the mechanism for allocation of Crown forests for harvest by either one (single-entity SFL) forestry company or a group (cooperative SFL) of companies; the licencee would own a processing facility such as a sawmill or pulp mill. A second form of licence, the forest resource licence (FRL), was also created for short-term harvest (up to five years) on an SFL by harvesting companies other than the SFL holder.

Aboriginal organizations, including political territorial organizations and Métis associations (in one case, in partnership with North-watch, a northeastern Ontario environmental NGO), brought to the EA Board their interests in the impacts of forest management on their Aboriginal and treaty rights, their connection to their homelands, and their desire both to share in the benefits of forestry and to participate in decision-making (EAB 1994, 345). These organizations advocated for "co-management" of forests with the Ontario Ministry of Natural Resources, or OMNR (EAB 1994, 366). Such shared decision-making could be considered a form of community forestry, since it would support joint decision-making at the local level, often address revenue sharing, and put traditional land-use activities on the agenda alongside timber harvesting.

The EA Board stated that Aboriginal and treaty rights were outside its mandate but set out several terms and conditions to improve Aboriginal peoples' access to economic benefits from forest management and their participation in forest-management planning. Joint decision-making through comanagement was not among the board's recommendations, and forest comanagement was never implemented. However, several community forestry initiatives were proposed or implemented in response to growing public concern.

### Community Forestry Initiatives

The first documented proposal for community forestry was Auden's (1944) Nipigon Forest Village, which proposed the development of community forestry enterprises based on a multiple-use approach. While the proposal was received warmly in the academic realm, it was disregarded by the provincial government. However, later in the conservation phase, several initiatives did see the light of day. They included the Algonquin Forest Authority, a provincial community forestry pilot program, the Wendaban Stewardship Authority, and

Westwind Forest Management Inc., all of which played a role in showing that alternative forms of tenure were possible.

### Algonquin Forest Authority

Algonquin Provincial Park (see Figure 4.1) was established in 1893 to conserve white pine as a source of timber for future logging. Thus, the park served a dual role of providing recreational activities while also generating economic benefit from commercial logging. However, in the 1960s, with the rise of recreational activities among urbanites in southern Ontario, increasing public concern was expressed about the impact of logging on the park's "wilderness" values. Conflict ensued between loggers and environmentalists, notably the Algonquin Wildlands League (now the Canadian Parks and Wilderness Society – Wildlands League), an environmental NGO that sought an outright ban on logging in the park. The Algonquin Forest Authority (AFA), created by the province in 1974 to address this conflict, became the first local forest governance initiative of its kind (Killan and Warecki 1992).

The AFA is an Ontario Crown corporation enabled by the Algonquin Forestry Authority Act (1990). Its governance structure is composed of a local board of directors, appointed by the province. Additional local public input is provided by an LCC established in 1998 and composed of members with a wide diversity of interests, including First Nations (Callaghan et al. 2008).

Although the AFA was not conceived as a community forestry organization by the province (Usher et al. 1994), it can be classified as one in that it operates under all four community forestry principles. However, the AFA has a restricted set of rights to forest management, covering only a portion of the full range of property rights described by Schlager and Ostrom (1992) as being essential to community-based management.[2] For example, the AFA has rights to timber only, with no authority over non-timber forest resources. Furthermore, while the AFA has some management rights, the province retains full authority over timber allocation, licensing, and approval of management plans. In terms of local benefits, the AFA is able to retain a portion of stumpage revenue to reinvest in its operation although excess profits are taken by the Province.

Research by Bullock and Hanna (2012) points to weaknesses in the AFA governance model: they question whether adequate representation is achieved, given that board members are appointed by the provincial government and receive financial compensation for their work. However, independent audits of the AFA have noted that public participation through the LCC has been significant (KBM 2003; Callaghan et al. 2008).

Another weakness is the lack of sufficient attention to First Nation rights and interests. When the AFA was established, First Nation rights were not addressed, although the Algonquins of Pikwàkanagàn subsequently negotiated access to the park to exercise their hunting rights (APFN 2012). The Algonquin First Nations in the region continue to express concerns about benefits from Algonquin Provincial Park, its significance to their livelihoods and Aboriginal rights, and the complexities relating to their land claim (see Figure 4.1) that includes the park (see Huitema n.d. for a history of the relationship between the Algonquin Nation and the park), although the Algonquins and Ontario have agreed that it will be "preserved for the enjoyment of all" (Ontario 2015).

*Community Forestry Pilot Projects*
In 1991, the Government of Ontario commissioned a forest policy panel to develop a comprehensive framework for forest management in concert with the EA hearings (OFPP 1993). New policy goals in response to the panel's recommendations were developed to address forest sustainability, including community sustainability. One outcome was a pilot community forestry program that was seen by the government as a means to empower communities, in keeping with the new sustainable forest-management paradigm (Smith and Whitmore 1991). In 1991, four community forestry pilot projects were established by the OMNR. Three of these – Geraldton, Elk Lake, and six communities within seventy kilometres of each other along Highway 11 – were on Crown land. A fourth pilot project was established on Wikwemikong Unceded Indian Reserve on Manitoulin Island, an indication of growing provincial involvement in First Nation forest land issues, previously seen as solely a federal responsibility. These pilots were selected out of twenty-two proposals (Harvey and Hillier 1994).

The province provided unprecedented policy and financial support for the program, even if only for its short, three-year lifespan. The program made a valuable contribution to furthering public awareness about community forestry as an alternative forest tenure for Crown land. It also provided several important lessons about factors contributing to the success of community forestry projects – notably, the need for tenure rights that provide security, revenue autonomy, and diversity in governance models (Harvey and Hillier 1994; Harvey 1995; Usher et al. 1994). However, the program involved only a nominal transfer of property rights: pilot projects had limited rights to influence forest-management decisions and garner economic benefits from the forest (Harvey and Hillier 1994; Harvey 1995; Teitelbaum, Beckley, and Nadeau 2006). According to Harvey (1995), because of these limitations, the pilot projects ended up sustaining themselves by providing silvicultural and planning services for the conventional forest industry. Of the four pilot projects, two – the Geraldton and Elk Lake community forestry projects – continue as subcontractors to the forest industry, while the other two are no longer operational. Despite their short duration and limited number, the pilots were an important experiment in alternative governance. However, the limited scale of the pilot program did not allow for "adaptive muddling" (DeYoung and Kaplan 1988) – vigorous experimentation using a diversity of designs and broad-based input that emanates from the bottom up rather than the top down in order to achieve solutions – as called for at the time by several academics (Duinker, Matakala, and Zhang 1991; Matakala and Duinker 1993).

### Wendaban Stewardship Authority

During this phase, a new voice arose to address the use of Crown lands in traditional First Nation territories. The Wendaban Stewardship Authority (WSA), proposed in 1991 in the Temagami region just north of Sudbury and North Bay, was a new type of forest-governance institution intended to promote conflict resolution and cross-cultural collaboration among stakeholders and the estranged First Nation, the Teme-Augama Anishnabai (Shute 1993; Bullock and Hanna 2012).

After struggling since the late 1880s to regain control over their traditional territory, the Teme-Augama mounted several logging road blockades in the mid-1980s. At the same time, environmental organizations were demanding protection of old-growth white-pine forests in the region (Black 1990; Hodgins and Benidickson 1989). The government responded to the escalating conflict and the increasing legal recognition of Aboriginal rights in Canada, including the recognition of these rights in section 35 of the Constitution Act, 1982, by negotiating the WSA with the Teme-Augama.

The WSA was the first mechanism in Ontario for the development of a collaborative approach to the management of forest resources between First Nations and non-First Nations (Laronde 1993). The WSA was intended to create "a regime of co-existence" among key actors for "dialogue, learning and action" (Lane 2006, 391), all practices that support a resilience approach to forest management.[3] The proposed

WSA, similar to the Algonquin Forest Authority in terms of its revenue model and board structure, developed a twenty-year stewardship plan based on ecosystem management, which was by then a cornerstone of the sustainable development paradigm (Benidickson 1996). However, a lack of both government and broad community support meant that the WSA was never legally established. Even though the WSA did not get off the ground, the approach was nonetheless considered a major forest policy breakthrough for its promotion of co-management with First Nations asserting rights over their traditional lands (Benidickson 1996).

### Westwind Forest Stewardship

Westwind Forest Stewardship Inc. has been described as the main example of a large-scale community forest in Ontario (Henschel and McEachern 2002; Clark et al. 2003; Teitelbaum, Beckley, and Nadeau 2006; Bullock and Hanna 2012). In 1998, Westwind received the first sustainable forest licence (SFL) under the Crown Forest Sustainability Act for the French-Severn Forest, which comprised over half a million hectares.

Westwind, based in Parry Sound, is a multi-stakeholder, non-profit community-based forest-management company governed through consensus by a board of directors composed of both community and industry representatives. Westwind is certified by the Forest Stewardship Council, whose standards promote community benefits and environmental responsibility. While some of the companies purchasing wood from the forest management unit are members of Westwind's board, they do not control decision-making but are treated as clients. Profits are returned to support forest management, which can be considered a local benefit. While local and First Nation advisory groups were established from Westwind's inception, a major ongoing challenge has been the lack of participation of local First Nations (Clark et al. 2003; Venne 2007).

The success of this model has been attributed, at least in part, to its geographic and social context (Berry 2006). Westwind is located near the border between northern and southern Ontario, just west of Algonquin Provincial Park, in proximity to large urban centres. Land ownership in this region is equally split between public and private owners, and the region has a diverse economy based on various types of forest use, cottage-based tourism chief among them. This context provides a strong incentive to minimize conflict and create compatibility between the forest and tourism industries (Barron 1998).

Clark et al. (2003) and Barron (1998) assert that Westwind differs significantly from subsequent SFLs because of its joint community-industry board. Like other community-based bodies set up previously during this phase, Westwind has limited forest property rights: the province retains authority over determining the annual allowable harvest, allocating timber, licensing, and approving forest-management plans.

## Resilience in the Conservation Phase

These few community forestry models implemented during the conservation phase, although innovative, did not dramatically alter the entrenched industrial forest system. The system was able to resist pressure for community forestry and maintain its negative resilience through continuing its command-and-control approach. Some initiatives remained at the proposal stage and were never implemented; several community forestry pilot projects did not last, and those that did were not implemented in accordance with all the principles of community forestry. The rigidity of the command-and-control approach meant that the system was able to accommodate only minor variations. Rigidity was evident in the consistent reluctance of provincial authorities to grant enhanced property rights (management and exclusion rights) to communities and in their preference for allocating licences and most timber to large industrial players. As a result, all community forestry attempts, whether enduring or not, amounted to localized experiments only, failing to transform the system. Insufficient devolution of rights by the state to communities has been a common obstacle to community-based forest tenure reform worldwide (Ribot, Agrawal, and Larson 2006; Poteete and Ribot 2011; Cronkleton, Pulhin, and Saigal 2012).

Although the command-and-control forest system maintained substantial negative resilience, it simultaneously became increasingly complex and vulnerable during the conservation phase. Mounting public pressure for a more community-based approach – from First Nations, other local communities, and environmental NGOs – became a new system driver. This rise in public and Aboriginal discontent with the dominant system, along with the interest expressed in community forestry as an alternative, signalled the beginning of a community forestry movement in the early 1990s. However, the policy direction at the end of this phase, which included the discontinuation of some community forestry experiments and an increased focus on industrial tenures (Ontario, Natural Resources 1998), served to stifle the fledgling movement and reinforce the dominant paradigm. Policy momentum to support community forestry therefore stalled during the remainder of this phase.

In spite of the limitations of community forestry attempts during this phase, we suggest that their emergence contributed to the development of policies that promoted a new approach to public participation in forest management. Thus, while the system was not fundamentally transformed at this point, the community forestry initiatives were instrumental in increasing system resilience by laying the basis for further change towards more community-based approaches.

### Release Phase: System Collapse – 2005–9

The forest industry experienced a major downturn in the new millennium due to a combination of changes in global supply and demand, an unfavourable export market in the United States, a rising Canadian currency exchange rate, high energy costs, and competition from lower-cost producers outside Canada (Canada, CFS, NRCAN 2006). The downturn worsened, and the command-and-control system reached a crisis point in 2005, when it lost resilience and was driven by these significant shocks into the release phase.

Over the next few years, the crisis brought severely negative impacts to forest-based communities, with an unprecedented number of mill closures in 2005 and 2006 (OFC, n.d.). Ontario achieved the dubious distinction of leading the pack in mill closures, with fourteen mills mothballed in 2005 (Canada, CFS 2006). The closures led to dramatic declines in forestry employment, out-migration (particularly of youth) from municipalities, the erosion of municipal tax bases, service reductions, a loss of social capital, and a pervasive lack of well-being in affected communities (Bogdanski 2008; Patriquin, Parkins, and Stedman 2009). Permanent layoffs from the forestry industry due to mill closures between 2003 and 2006 were estimated at nearly fifteen thousand in Ontario and Quebec (Canada, CFS 2006).

The release phase is comparable to the third phase of development in a resource staples economy that focuses on the extraction of raw or unfinished bulk commodity products that are sold in export markets after minimal processing (Clapp 1998). The collapse led to widespread recognition in northern Ontario that the tenure system was broken and in need of significant reform. Different actors proposed solutions.

One response was a resurgence of interest in community forestry and a strong push by many communities to implement this approach to foster resilience in forest-dependent communities.

## Community Response

The Task Force on Resource Dependent Communities – set up by the Communications, Energy, and Paperworkers Union of Canada and the United Steelworkers (USW) – recommended that the province reform the tenure system to ensure greater involvement of community stakeholders and workers (Butler, Cheetham, and Power 2007). The USW, whose members had worked at a pulp mill in Kenora (since demolished), demanded the creation of regional timber boards in the northwest of the province (USW, n.d.).

Several new groups sprang up in 2006 and pressed for a fundamental change in the forest system to support community forestry, including Saving the Region of Ontario North Group (STRONG), the Gordon Cosens Survival Group, and the Northern Ontario Sustainable Communities Partnership (NOSCP). STRONG and the Gordon Cosens Survival Group were formed in northeastern Ontario following the shutdown of the Excel sawmill in Opasatika. The latter group submitted for consideration to the OMNR "A Blueprint for Communities' Survival," which proposed a community forestry model for the Gordon Cosens Forest (pers. comm., Marc Guindon, 2006).

The Northern Ontario Sustainable Communities Partnership was formed to advocate for a regional approach to community forestry (NOSCP 2007a). The NOSCP comprises diverse participants that include individual citizens, municipalities, non-governmental organizations, academics, unions, and Aboriginal organizations. Given that it was established as a social network within the dominant forest system, the NOSCP functions as a shadow network – a self-organizing, informal network of people with no official authority, which mobilizes in response to crisis (Gunderson 1999; Olsson et al. 2006; Goldstein 2008). Gunderson (1999, 6) describes shadow networks as groups "where new ideas arise and flourish" and "that explore flexible opportunities for resolving resource issues, devise alternative designs and tests of policy, and create ways to foster social learning."

The NOSCP created space for such possibilities when, in 2007, it developed and distributed for endorsement the Northern Ontario Community Forest Charter (NOSCP 2007b). The charter principles broadly address good governance, shared decision-making, separation of forest management from any one specific user group (i.e., a forestry company), the promotion of a diverse sector through support for best end use of forest resources through value-added production of both timber and non-timber values, less reliance on commodity industries, benefits to local communities from forest development, and the upholding of Aboriginal and treaty rights. The charter commits not only to respect but also to "help resolve" the outstanding issues around implementation of these rights in forest management.

By fostering networking among a wide range of actors through activities like the charter endorsement and an inaugural workshop on community forestry in northern Ontario (NOSCP 2009), the NOSCP also became a bridging organization that furthers vertical and horizontal linkages across multiple organizational levels (municipal to federal) and for geographically dispersed social groups (local to national). Shadow networks that provide these kinds of bridging functions act as a source of resilience by facilitating social learning, building social capital, encouraging trust among actors, and helping create a common vision (Olsson et al. 2006; Berkes 2009).

**Government Response**
The Province's initial response to the forest-sector crisis was to investigate forest-industry concerns through the Minister's Council on Forest Sector Competitiveness (MCFSC 2005). Made up of representatives heavily weighted in favour of the forest industry and focused on "a limited set of forest industry-centred economic factors affecting the efficiency and competiveness of large-scale industrial operations" (Bullock 2010, 99), the council made several recommendations to alleviate the crisis. One of them was to convert single-entity SFLs to cooperative SFLs to improve economies of scale and thereby increase industry competitiveness. The OMNR began an SFL conversion process in the spring of 2007 (Ontario, Natural Resources 2007; Morrow 2007), which continued until the spring of 2010, with the formation of the final cooperative SFL, Miitigoog Limited Partnership. Miitigoog is a 50/50 partnership between three First Nations and several local forestry companies and contractors for the management of the Kenora Forest.

Although cooperative SFLs somewhat increased community involvement in forest management, they were developed primarily as a business model with the goal of increasing the participation of those with a business interest rather than of solving major social issues or addressing treaty rights (Morrow 2007; Ontario, Natural Resources 2007). Given that they did not grant additional forest property rights beyond

those provided by single-entity SFLs, they were a negligible adjust-
ment intended to address only the external economic forces that had
affected the existing system rather than its fundamental restructuring.

As the crisis continued unabated, the Province eventually did take
steps to consider alternative options. In June 2007, the Province appointed an economic facilitator to work with the people of northwestern Ontario to identify initiatives that would build a prosperous
economy. The ensuing Rosehart Report recommended a major forest
tenure reform that would create "quasi-independent" ecosystem-based
authorities managed by boards of directors including First Nations
and forest stakeholders (Rosehart 2008). The recommended approach
would allow both municipalities and First Nations to play a much
greater role in forest management than was possible with either the
cooperative or single-entity SFL models. Rosehart's perspective supported the view, increasingly being expressed by some scholars during
this phase, that the conventional forest system had constrained diversification and innovation, thereby limiting its resilience (Haley and
Nelson 2007; Robinson, 2007, 2009a, 2009b, 2009c).

## Reorganization Phase: Tenure Reform – 2009 and Onward

By the fall of 2009, the Province had recognized that the forestry crisis
could not be resolved by minor adjustments to the existing system.
Amid continued pressure from the forest industry and communities
alike to address the worsening situation, and heeding the recommendations of Rosehart (2008), the Province initiated an unprecedented
forest tenure reform process in September 2009. This policy development moved the system into the reorganization phase of the adaptive
cycle, where it currently remains.

The tenure reform process involved a series of public and Aboriginal
consultations throughout northern Ontario that elicited a widespread
call from communities to accommodate community forestry in a new
forest tenure policy framework (Speers 2010). The consultations contributed to a provincial forest tenure proposal put forward in the
spring of 2010 to create a new forest governance model, the Local
Forest Management Corporation (LFMC) (Ontario, NDMF 2010). The
NOSCP continued with its advocacy work during this time and prepared a commentary challenging the province's tenure proposal,
including recommendations for a forest tenure framework based on
the charter principles (NOSCP 2010). However, extensive lobbying by
the forest industry to slow down and limit the reform had a significant

impact on the outcome. The Ontario Forest Industries Association, which represents a large segment of the forest industry, lobbied hard – with support from the Northwestern Ontario Associated Chambers of Commerce, the Northwestern Ontario Municipal Association, and the Federation of Northern Ontario Municipalities – to maintain existing wood supply commitments and ensure a "measured and moderate" approach to any change in the tenure system (OFIA, NOACC, NOMA, and FONOM 2011, 6). The tenure reform process resulted in the creation of two new forest governance models in the spring of 2011 through the Ontario Forest Tenure Modernization Act, 2011, and new forest tenure policy.

The NOSCP cohosted a second workshop (Palmer, Smith, and Shahi 2012) when the new forest tenure policy framework was announced to promote dialogue about a response. The group subsequently partnered with several organizations to cohost a 2013 conference that expanded networking for community forestry to the national and international levels. One outcome was the NOSCP's commitment to participate in a new national network, Community Forests Canada, spearheaded at the conference to further promote community forestry country wide (Palmer et al. 2013). In keeping with this commitment, the NOSCP was a partner for a 2014 community forestry symposium in Winnipeg to further this new network (Bullock and Lawler 2014).

**New Tenure Models**
The Ontario Forest Tenure Modernization Act outlines how a Local Forest Management Corporation will function and permits the establishment of two such models. A second model created through new policy is the Enhanced SFL, or ESFL (Ontario, NDMF 2011). These new approaches are intended to increase opportunities for local and Aboriginal community involvement and forest-sector competitiveness (Ontario, NDMF 2011). However, neither approach was designed to accommodate the implementation of Aboriginal and treaty rights.

An LFMC is an Ontario Crown corporation comparable to the Algonquin Forest Authority but with a different revenue model. This new type of forest-management company can hold one or more SFLs and has obligations associated with such licences. LFMCs are to have a "predominantly" local board of directors (Gravelle 2011) that includes Aboriginal and other local community representation, with board members appointed by the Province. As with the Algonquin Forest Authority, these corporations retain the base stumpage revenue (which normally goes to the Consolidated Revenue Fund of Ontario) to pay for operating costs and to reinvest in the corporation (Speers 2012). However, if an LFMC makes a profit, the Province has the option of taking a dividend (Ontario, NDMF 2011). An ESFL is an SFL that is "enhanced" so that the licence-holding company of consuming mills and/or harvesters, or a non-profit company, has a shareholder board of directors with minimum representation by Aboriginal and local communities (Ontario, NDMF 2011). The LFMC stumpage pricing model does not apply to ESFLs, which must pay the same Crown timber charges as SFLs. A modified form of forest resource licence known as an enhanced forest resource licence (EFRL) was also created in 2012 as an interim model to allow First Nations and municipalities with established forest-management companies to hold a short-term licence to undertake harvesting and build capacity in forest-management planning activities prior to the establishment of a long-term licence.. This licence is a form of hybrid model between an SFL and FRL (Ontario, NRF 2012).

The Province established the first LFMC and is facilitating the conversion of existing single-entity and cooperative SFLs to ESFLs. This first LFMC, known as the Nawiinginokiima ("working together") Forest Management Corporation (NFMC), became operational in the spring of 2013. It currently holds EFRLs[4] for two forest management units in the vicinity of the municipalities of Marathon, Manitouwadge, White River, and Hornepayne and three First Nations with traditional territories in this area – the Ojibways of Pic River, Pic Mobert, and Hornepayne. The province appointed the board of directors, with local community and Aboriginal representatives who were instrumental in promoting this initiative, as well as two members-at-large with experience in the forest sector. All affected communities are invited to have representatives on the board. The EFRLs are to be converted to SFLs in the near future, and NFMC has applied to obtain an additional two SFLs for nearby forest management units. One of the forest management units currently under an EFRL obtained FSC certification in September 2014.

To contribute to further development of policy relating to ESFLs and their implementation, the Province established a Forest Industry Working Group and – at the request of First Nation organizations, communities, and the NOSCP – a First Nation Working Group in 2010 and a Community Working Group in 2011. The groups became the Joint Working Group, which, in 2012, developed a set of principles to guide ESFL implementation. In 2014, the Joint Working Group was replaced by an Oversight Group, with some new representatives from the various constituencies. Evaluation criteria are currently being developed for a review, to take place in 2016, of both the LFMC and ESFL models, as well as for all other forest-management models (Ontario, NRF 2015a). The inclusion of local communities and First Nations at the negotiating table is a step away from the command-and-control system that historically saw a "business-government nexus" of policy making, with other groups excluded from decision-making (Howlett and Rayner 2001).

**Limitations of the New Tenure System**
While the new tenure system provides some space for the development of community forests, the command-and-control system has not undergone an actual transformation to a new regime that operates on the four community forestry principles of participatory governance, rights, local benefits, and ecological stewardship/multiple use. Nishnawbe Aski Nation, a political territorial organization representing First Nations in the Treaties 5 and 9 areas of northern Ontario, voiced concerns that neither of the two new governance models supports a framework for community-managed forests and that these models are inconsistent with the treaty position that decisions with respect to the land (including forest tenure) are to be community managed. (Palmer, Smith, and Shahi 2012; NAN 2015). NOSCP characterized the new tenure framework as having "a timid beginning with tons of potential" (NOSCP 2011). As in the earlier experiments, the provincial government will retain power over timber allocations, licensing, and approval of forest-management plans, thereby limiting the forest-property rights that are essential for the success of community forestry (Schlager and Ostrom 1992).

The revised system includes only two long-term tenure models. Like the models implemented in the conservation phase, the new system lacks the degree of experimentation, or "adaptive muddling," that is regarded as crucial for success (Duinker, Matakala, and Zhang 1991; Robinson 2009c, 2012) and the level of diversity inherent in resilient systems (Berkes, Colding, and Folke 2003; Folke, Colding, and Berkes 2003; Chapin et al. 2004; Armitage 2005; Walker and Salt 2006). With only two types of long-term tenure models as options, the ability of governance to respond to local needs and conditions is constrained. In addition, both new models continue a singular focus on timber, limiting a broader range of community values; this runs counter to the community forestry principle of multiple use. Public advocacy for community forests has consistently emphasized the need for the development of a broad range of enterprises based not only on timber but also on non-timber products, including tourism and recreation, to support diversification of community economic development. A diversity of enterprises is considered key for the survival of local, forest-based economies in a world dominated by global forces (Orozco-Quintero and Berkes 2010).

Although the new tenure framework suggests that LFMCs will provide "independent, local governance" (Ontario, NDMF 2011, 9), a common concern among many communities is that the appointment, and potential removal, of board members by government will not support true participatory governance, in which local people have meaningful decision-making power through a democratic approach. The imposition by the Province of a governance system also precludes the self-organization principle of resilient systems.

While ESFLs appear to have greater flexibility in governance, with Aboriginal and local communities guaranteed at least minimum representation on ESFL boards, shareholders (i.e., the forest industry) are to have proportional influence over financial decisions. It is therefore questionable how well this model will foster the community forestry principle of local benefits. Concerns have also been raised about the ESFL revenue model, which, unlike that of LFMCs, does not return royalties to local communities. Capistrano and Colfer (2005) and Robinson (2012) point out that for devolution of forest management to be successful, local institutions require revenue and/or taxation powers in order to invest in their people to achieve the continued learning that fosters improvements in forest management.

### Emerging Community Forest Initiatives in the Area of the Undertaking

The tenure reform process has been a driver for the development of several regional community forestry initiatives that involve municipalities and First Nations who share the same forests in a common geographic region. First Nations and settler communities with distinct histories and historically isolated cultures are bridging cross-cultural barriers to collaborate in these initiatives. The concept that "we are all treaty people" provides a powerful foundation for a new relationship among Aboriginal and settler communities based on a respect for Aboriginal rights (Smith, Palmer, and Shahi 2012). In some cases, the initiatives also include forestry companies that are accepting a community-based approach as partners. All of the initiatives involve the Province, which maintains oversight for Crown forests. The processes undertaken to develop these initiatives reflect the start of adaptive and collaborative approaches to forest management that foster resilience.

These emerging initiatives have three objectives: 1) an inclusive, collaborative, and democratic process associated with participatory, regional governance that involves representatives of local stakeholders and First Nation communities; 2) diversified economies based on best end-use of both timber and non-timber forest resources through the development of enterprises that support community and Aboriginal economic development (Boyd and Trosper 2010); and 3) revenue power achieved through resource revenue sharing. While the recognition of Aboriginal and treaty rights is also a key concern, initiatives may adopt the approach used by one BC community forest, the Likely/Xat'súll Community Forest, where First Nation rights, although recognized by the community forest, are promoted through relations between the provincial and federal governments and the First Nation governments (Robinson 2010).

An example of a developing ESFL based on a community forest proposal is found in the Northeast Superior region (Lachance, in preparation). Two Crown forest management units and the 700,000 hectare Chapleau Crown Game Preserve near the municipalities of Chapleau, Wawa, and Dubreuilville are part of the initiative, as are the traditional territories of several First Nations represented by the Northeast Superior Regional Chiefs' Forum (NSRCF). The First Nation participants have proposed the inclusion of portions of several additional adjacent forests that are in their traditional territories. A number of forest companies in the region are also participants. The community forest model is associated with the NSRCF's proposal for a resilient, regional conservation economy that includes value-added timber and non-timber forest products in addition to traditional commodities (Reid-Kuecks et al. 2012). Although the governance structure has not been determined, the province has made a commitment to support and resource the development of the ESFL, and planning is underway for its establishment.

A collaborative community-based process was undertaken to develop the Hearst/Constance Lake First Nation/Mattice-Val Côté community forest model that was proposed during the tenure reform consultations. Although not yet supported by the Province, this model builds on an existing cooperative SFL and cross-cultural relationships developed through an earlier collaborative process involving the municipality of Hearst, Constance Lake First Nation, and local forest industry (Casimirri, in preparation). The spring 2014 closure of a pulp mill in Fort Frances spurred the municipality and First Nations with traditional territories in the region to propose an ESFL with a similar cross-cultural focus for a forest in the area (Hicks 2014). These initiatives also include existing forest industry partners or invite new industry partners who want to participate.

All of these ESFL initiatives display adaptive and collaborative approaches among First Nations and stakeholders, approaches that are transforming the way these different actors are working together in northern Ontario. These and other initiatives that are also developing ESFLs, or are in a transitional state before becoming some form of long-term tenure model, are trying to find their place within the new forest tenure system – notably, the Whitesand First Nation Community Sustainability Initiative, which is a component of the Lake Nipigon ESFL initiative; the Lac Seul, Sapawe, and Kenogami EFRLs, which are to transition to ESFLs; and a recent proposal by three Matawa First Nations for a long-term licence on the Ogoki Forest. With implementation of ESFLs planned over the next several years and the option for an additional LFMC, it is likely that more initiatives will emerge and that established SFLs with a community forestry bent, such as Miitigoog and Westwind, will also become ESFLs.

## Moving Forward in the AOU: Beyond Reorganization

The AOU's forest system remains in the reorganization phase, with its future configuration unpredictable. A new provincial forest-tenure policy framework has created some space for the participation of First Nations and other local communities in new forest-governance structures. At the same time, ongoing resistance from a large segment of the forest industry aimed at maintaining the command-and-control system, despite its negative resilience, has limited the advancement of most community forest models to date and has thus precluded transformative change that would fully support community forestry. This resistance operates as the key driver that maintains vulnerability in the reorganizing system. Such resistance undermines the development of adaptive capacity to foster resilience in the face of additional future shocks that are inevitable in all complex adaptive systems. However, continued pressure for the advancement of community forestry from communities and other organizations is a simultaneous driver that is operating to counteract this negative resilience.

A window of opportunity remains open for revisions to the forest-tenure system, which could provide a future enabling policy environment for community forestry. This opportunity is being pursued in a number of ways. Communities continue to undertake regional, cross-cultural adaptive and collaborative processes that are building adaptive capacity to support the development of community forestry models. First Nation organizations such as Nishnawbe Aski Nation, Matawa First Nations Management, and the Northeast Superior Regional Chiefs' Forum continue to lobby for a tenure system that both promotes community forests and supports protection of First Nation rights and First Nation participation in forest-management decisions. The Northern Ontario Sustainable Communities Partnership continues its activities as a shadow network and bridging organization advocating for community forestry. Emerging co-operative initiatives, as seen in cross-cultural collaboration, and related synergistic effects among numerous actors appear to be driving the system towards a regime shift that could see the creation of a forest-tenure policy that supports community forestry in the future.

## Ontario's Far North: The Whitefeather Forest Initiative

As was the case in the AOU, colonization was a major disturbance in the Far North that caused the collapse of the original social-ecological configuration based on Aboriginal land use and occupancy and the fur trade. However, because of the remoteness of this region, the rate of system change following colonization has been slower, given that industrial timber exploitation did not occur.

Following system collapse due to colonization, the original social-ecological system in the Far North went through a reorganization phase that saw First Nations relocated to reserves, the emergence of a mixed economy based on traditional Aboriginal land use and seasonal wage employment, and a gradual process of change towards the beginning of resource development. The exploitation (growth) phase of the forest system's adaptive cycle was initiated only recently, when one First Nation became involved in planning for what they foresaw as impending forestry development beyond the AOU. The Far North forest system remains in the growth phase of its current adaptive cycle. Forestry development is poised to begin through the Whitefeather Forest Initiative, the first forest-management model that has been developed in the Far North.

First Nations in the Far North are signatories to either Treaty 5 or Treaty 9, signed between 1875 and 1930 (see Figure 4.1). The Ontario Forest Accord – signed by the Ministry of Natural Resources, forest industry representatives, and environmental organizations in 1999 as part of a provincial land-use planning exercise (Ontario, Natural Resources 1999) – stipulated that development in the region was contingent on First Nation consent, environmental assessment, and the establishment of protected areas. This accord led the OMNR to develop the Northern Boreal Initiative, a policy that promoted "community-based land use planning" (Ontario, Natural Resources 2002).

Pikangikum First Nation, whose traditional territory is immediately north of the Far North boundary (see Figure 4.1), was the first community in the Far North to express interest in engaging with the Province to ensure that the community benefited from what seemed inevitable forestry development by being in "the driver's seat" (PFN 2006, 4). The Whitefeather Forest Initiative began in 1999 when Pikangikum approached the OMNR with an "economic renewal" project that led to a joint approach to forest development.

Pikangikum defined the Whitefeather Forest boundaries, encompassing 1.3 million hectares of land north of Red Lake in northwestern Ontario, on the basis of the First Nation's registered traplines. An advisory group made up of community and OMNR representatives provided the mechanism for bridging the gap between the planning approaches of Pikangikum and the OMNR, allowing Pikangikum to work in a "cross-cultural context" (PFN 2006, 12). The planning process covered the development of a land-use strategy, environmental assessment, and, finally, a forest-management plan. The land-use strategy outlined in *Keeping the Land* (PFN 2006) reflected Pikangikum's customary land stewardship traditions. The environmental assessment approval granted in 2009 addressed the unique characteristics of the Whitefeather Forest and the need to respect the customary stewardship practices of Pikangikum, provide continuous habitat for woodland caribou management, and deal with road access issues in order to both provide access to timber and maintain the remote characteristics of the Whitefeather Forest (Ontario, Environment 2009). The forest-management plan was approved in 2012 (Palmer 2012). The OMNR included the unique aspects of this plan in its latest *Forest Management*

*Planning Manual.* In particular, the role of elders in guiding planning and decision-making and the use of traditional knowledge have been recognized (Ontario, Natural Resources 2009). An SFL for the White-feather Forest was issued in 2013 to the Whitefeather Forest Community Resource Management Authority.

The Whitefeather approach is closely aligned with the community forestry principles of participatory governance, rights, local benefits, and ecological stewardship/multiple use. Pikangikum continues to exercise its rights to make decisions about the resources in its territory through its local steering committee and the Whitefeather Forest Management Corporation. The community's approach, outlined in *Keeping the Land* combines modern-day forestry with the continuation of traditional land-use activities based on the community's customary stewardship practices, thus ensuring multiple use and sustainable management. In terms of local benefits, by holding an SFL for the Whitefeather Forest, Pikangikum has the potential to generate employment and revenues.

Other First Nations in the region, like Cat Lake and Slate Falls, have also completed land-use plans. It is likely that additional First Nations in the Far North will pursue provincial forestry licences, under the community-based land-use planning approach captured in the Far North Act, 2010.

While timber exploitation has not been a system driver to date within the Far North forest subsystem, the influence of forestry development in the AOU is nonetheless apparent in the Far North. The northward expansion of forestry activity to the Far North boundary is a driver for First Nations in the Far North to advocate for a community-based approach to forestry development in their traditional territories. Several First Nations, in addition to Pikangikum, are involved in community-based land-use planning that address natural resource development under the Far North Act, 2010. Additional drivers are the legal context that saw the start of a winning streak of Aboriginal cases throughout Canada around the time that interest in forest management was first expressed by Pikangikum (Gallagher 2012) and the lessons learned by the Province in relation to forestry development over a much longer period in the AOU. These drivers contributed to system change by inspiring the political will to develop a supportive policy framework for community forestry in the Far North in an attempt to avoid the negative consequences of the command-and-control approach experienced to the south. The piecemeal approach to forestry development that occurred throughout the AOU will be avoided in the Far North, where forest management is to be implemented from the outset as a component of an overarching community-based land-use planning approach across the landscape to support First Nation values and aspirations. This policy approach to forest management in the Far North fosters resilience in its developing forest system.

The acceptance of community forestry in the forest system of the Far North is affecting change in the AOU forest system by further supporting its ever-intensifying community forestry movement. Viewed through a panarchy lens, these two forest subsystems, which operate at different scales and rates, can be seen to be interconnected such that they are influencing each other to foster greater support for community forestry throughout northern Ontario.

## Conclusions

In this chapter, we have used complexity theory to understand northern Ontario's Crown forest system and the potential for community forestry as an alternative, more resilient form of forest tenure to that of a command-and-control approach. We have characterized Ontario's Crown forest system as a specific type of complex adaptive system, a social-ecological system that has the ability to adapt to a changing environment. We have evaluated community forestry initiatives that have been proposed and implemented in northern Ontario's AOU during the different phases of the forest system's adaptive cycle – the cycle of disturbance and renewal that occurs in complex adaptive systems (CASs) – from the system's inception to the present: the growth phase from the 1800s to the 1930s, the conservation phase from the 1930s to 2005, the release phase from 2005 to 2009, and the reorganization phase from 2009-on. We have differentiated these phases based on features that are characteristic of each phase of a CAS's adaptive cycle. This evaluation explored the resilience of the forest social-ecological system during each phase in terms of its receptivity to community forestry and whether the community forestry principles of participatory governance, rights, local benefits, and ecological stewardship/multiple use were met. We have similarly evaluated the first community forestry initiative in the Far North's developing forest system.

In the AOU, community forestry initiatives have progressed from the formation of the Algonquin Forest Authority in the conservation phase as a reaction to a single issue – a conflict between logging and recreation in Ontario's first provincial park – to proactive regional initiatives in the current reorganization phase. Regional partnerships to develop these initiatives have emerged between historically isolated First Nations and municipalities in northern Ontario. In some cases, the forest industry has become supportive of a more community-based approach. In the Far North, community-based land-use planning fosters control over development in First Nation traditional territories. The context for this direction towards community forestry throughout northern Ontario includes the increasing legal recognition of Aboriginal rights in Canada.

Advocacy for community forestry began as a mere idea that was expressed, but initially disregarded, by the Province of Ontario in the early conservation phase of the AOU forest system's adaptive cycle. This advocacy then increased as the conservation phase progressed, during a period when experimentation with community forestry was supported but stalled in the late conservation phase. The community forestry movement was thus temporarily stifled but subsequently re-emerged during the release phase of the system's adaptive cycle, when the command-and-control approach resulted in a forest sector crisis that led to a major economic downturn in the forest industry. Community forestry advocacy subsequently blossomed to become a well-connected and active community forestry movement in the current reorganization phase of the system's adaptive cycle.

The community forestry initiatives that have emerged in the reorganization phase of the AOU forest system may appear as isolated endeavours that are seemingly mere experiments with only localized impact. Yet when the initiatives are viewed in concert with the broader efforts of organizations such as the NOSCP and the Aboriginal and community working groups that have already influenced provincial forest-tenure policy direction, it can be seen that this combined effort is exerting an effect that Westley et al. (2011, 771) describe as "nibbling at the dominant system." Complexity theory explains how this effect works to reduce the negative resilience of the dominant regime while simultaneously building resilience for innovative alternatives to take hold. This nibbling effect thus makes an important contribution to the process of change in a complex adaptive system and can ultimately drive the dominant system towards a regime shift. Communities and other organizations that are promoting transformational change of the AOU's forest system to embrace community forestry are, in the sense described by Westley et al. (2011), social and institutional entrepreneurs. They are using the window of opportunity provided by the current period of forest-tenure reform to build innovation niches at the local and regional levels in order to link them to the broader (provincial-level) institutional scale. Complexity theory offers an insightful theoretical lens through which to view this process of change.

The Far North forest system is currently in the early growth phase, with forestry development on the verge of implementation under a separate forest-policy framework from that of the AOU. A key difference in this region, where change to the forest system has been slower to develop following colonization, is that community forestry is the accepted forest-policy approach to forest development in the Far North. First Nations, who constitute the largest population in this part of the province, are taking a strong role in forest-management planning that falls under the umbrella of community-based land-use planning.

We have also found it valuable to view the evolution of community forestry within northern Ontario's overall Crown forest system from a complexity perspective because this approach considers system resilience. We argue that community forestry offers the characteristics of resilience within the context of northern Ontario's forest system of operating Crown forests that are shared in a common geographic area by a range of actors with varied perspectives and interests. The community forest movement in the AOU is working to build the forest system's resilience in the face of uncertainty and change. This is achieved through fostering cross-cultural collaboration and social learning in two spheres: the ongoing development of community forestry initiatives that are undertaking adaptive and collaborative management and the local, national, and international networking among communities, supporting organizations, and other stakeholders. The forest-policy direction in the Far North supports a resilience approach to forest management.

Because Ontario remains hesitant about devolving full control of forest-management decision-making to local communities, the forest system in the AOU has not yet undergone a transformative change that supports all four community forest principles of participatory governance, rights, local benefits, and ecological stewardship/multiple use. There is no guarantee that a regime shift to a more resilient system configuration will occur as a result of innovations that arise during the reorganization phase. However, the AOU and Far North forest systems are influencing each other to advance community forestry in the overall Crown forest system of northern Ontario. Given that the community forestry movement continues to build resilience as communities push for this outcome throughout all of northern Ontario, we suggest that the AOU's forest system is in fact being driven towards such a regime shift, which could see a future forest tenure policy framework that supports the implementation of community forests as envisioned by communities. On the basis of this burgeoning movement throughout northern Ontario, we conclude that community forestry is an emerging paradigm rather than a mere localized anomaly frozen within a dominant command-and-control system.

## Notes

1 For statistical purposes, the Métis population is included with municipalities.
2 Property rights as defined by Schlager and Ostrom (1992) comprise a "bundle" of five rights: access, withdrawal, management, exclusion, and alienation. The first two convey rights to enter and obtain resources. The last three are decision-making rights that are particularly significant for forest tenure, since they allow the rights holder to define and adjust rules and standards for exercising other rights. While alienation allows the sale or lease of forest lands, including the other rights, management and exclusion rights convey decision-making power over who has access to a resource and how it is harvested.
3 Lane (2006) describes "co-existence" as "resolving how differing parties can exercise their respective rights in land."
4 Three other ESFLs have been issued since 2012 to First Nation corporations – Obishikokaang Resources Ne-Daa-Kii-Me-Naan and Rainy Lake Tribal Resource management Inc. – owned by First Nations interested in obtaining long-term forest licences.

## References

APFN (Algonquins of Pikwàkanagàn First Nation). 2012. "Moose." *APFN*. http://algonquinsofpikwakanagan.com/harvest_moose_deer.php.

Armitage, D. 2005. "Adaptive capacity and community-based natural resource management." *Environmental Management* 35 (6): 703–15. http://dx.doi.org/10.1007/s00267-004-0076-z.

Armitage, D., F. Berkes, and N. Doubleday. 2007. *Adaptive co-management: Collaboration, learning, and multi-level governance*. Vancouver: UBC Press.

Armitage, D., M. Marschke, and R. Plummer. 2008. "Adaptive co-management and the paradox of learning." *Global Environmental Change* 18 (1): 86–98. http://dx.doi.org/10.1016/j.gloenvcha.2007.07.002.

Auden, A.J. 1944. "Nipigon Forest Village: A prospectus." *Forestry Chronicle* 20 (4): 209–61. http://dx.doi.org/10.5558/tfc20209-4.

Barron, J. 1998. "Community logs: Westwind Forest Stewardship takes a non-profit approach to forest management in Muskoka." *Alternatives Journal* 24 (4): 5.

Benidickson, J. 1996. "Temagami old growth: Pine, politics, and public policy." *Environments* 23 (2): 41–50.

Beratan, K.K. 2014. "Summary: Addressing the interactional challenges of moving collaborative adaptive management from theory to practice." *Ecology and Society* 19 (1): art. 46. http://dx.doi.org/10.5751/ES-06399-190146.

Berkes, F. 2009. "Evolution of co-management: Role of knowledge generation, bridging organizations, and social learning." *Journal of Environmental Management* 90 (5): 1692–702. http://dx.doi.org/10.1016/j.jenvman.2008.12.001.

Berkes, F., J. Colding, and C. Folke, eds. 2003. *Navigating social-ecological systems: Building resilience for complexity and change.* New York: Cambridge University Press.

Berry, A. 2006. *Branching out: Case studies in Canadian forest management.* Bozeman, MT: Property and Environment Research Center. http://www.perc.org/articles/branching-out-case-studies-canadian-forest-management.

Black, B. 1990. "Temagami: An environmentalist's perspective." In *Temagami: A debate on wilderness,* edited by M. Bray and A. Thomson, 141–46. Sudbury/Toronto: Institute of Northern Ontario Research and Development, Laurentian University/Dundurn Press.

Blouin, G. 1998. "Public involvement processes in forest management in Canada." *Forestry Chronicle* 74 (2): 224–26. http://dx.doi.org/10.5558/tfc74224-2.

Bogdanski, B.E.C. 2008. *Canada's boreal forest economy: Economic and socioeconomic issues and research opportunities.* Information Report BC-X-414. Victoria, BC: Natural Resources Canada, Canadian Forest Service, Pacific Forestry Centre.

Boyd, J., and R. Trosper. 2010. "The use of joint ventures to accomplish Aboriginal economic development: Two examples from British Columbia." *International Journal of the Commons* 4 (1): 36–55.

Bullock, R. 2010. "A critical frame analysis of northern Ontario's 'forestry crisis.'" PhD diss., Department of Geography, University of Waterloo, Waterloo, ON.

Bullock, R., and K.S. Hanna. 2012. *Community forestry: Conflict, local values, and forest governance.* Cambridge: Cambridge University Press. http://dx.doi.org/10.1017/CBO9780511978678.

Bullock, R., and J. Lawler. 2014. *Community forests Canada: Bridging practice, research, and advocacy.* Workshop and symposium report, Centre for Forest Interdisciplinary Research and Department of Environmental Studies and Sciences, University of Winnipeg, Winnipeg, MB.

Butler, J., B. Cheetham, and M. Power. 2007. *A solutions agenda for northern Ontario's forest sector.* Communications, Energy, and Paperworkers Union of Canada (CEP) and United Steelworkers (USW) Taskforce on Resource Dependent Communities.

Callaghan, B., T. Clark, C. Howard, M. Leschishin, and P. Shantz. 2008. *Report of an independent audit of forest management of the Algonquin Park Forest for the period 2002 to 2007.* Toronto: Ministry of Natural Resources, Government of Ontario.

Canada. NRCan (Natural Resources Canada). 2016. *Forest classification. Forest regions.* Ottawa: Natural Resources Canada. http://www.nrcan.gc.ca/forests/measuring-reporting/classification/1317.9.

–. NRCan (Natural Resources Canada). 2010. *Historical Indian treaties.* Ottawa: Natural Resources Canada. http://geogratis.gc.ca/api/en/nrcan-rncan/ess -sst/cb216b8f-8893-11e0-8ed0-6cf049291510.html.

–. CFS (Canadian Forest Service). NRCAN (Natural Resources Canada). 2006. *The state of Canada's forests 2005–2006: Forest industry competitiveness.* Ottawa: Canadian Forest Service, Natural Resources Canada.

Capistrano, D., and C.J.P. Colfer. 2005. "Decentralization: Issues, lessons, and re-flections." In *The politics of decentralization: Forests, people, and power,* edited by C.J.P. Colfer and D. Capistrano, 296–313. London, UK: Earthscan.

Casimirri, G. In preparation. "Factors affecting success in a First Nation, govern-ment, and forest industry collaboration process." In *Bridging practice, research and advocacy for communities and their forests,* edited by R. Bullock, P. Smith, L. Palmer, and G. Broad. Winnipeg: University of Manitoba Press.

Chapin, F.S., III, G. Peterson, F. Berkes, T.V. Callaghan, P. Angelstam, M. Apps, C. Beier et al. 2004. "Resilience and vulnerability of northern regions to social and environmental change." *Ambio* 33 (5): 344–49.

Clapp, R.A. 1998. "The resource cycle in forestry and fishing." *The Canadian Geographer* 42 (2): 129–44. http://dx.doi.org/10.1111/j.1541-0064.1998.tb01560.x.

Clark, T., S. Harvey, G. Bruemmer, and J. Walker. 2003. *Large-scale community forestry in Ontario, Canada: A sign of the times.* Paper 0812–C1, submitted to the XII World Forestry Congress, Québec, QC. http://www.fao.org/DOCREP/ARTICLE/WFC/XII/0812-C1.HTM.

Coates, K., and D. Newman. 2014. "Tsilhqot'in ruling brings Canada to the table." *Globe and Mail,* 11 September. http://www.theglobeandmail.com/globe-debate/tsilhqotin-brings-canada-to-the-table/article20521526/.

Colfer, C.J.P. 2005. *The complex forest: Communities, uncertainty, and adaptive col-laborative management.* Washington, DC: Resources for the Future Press.

Cronkleton, P., J.M. Pulhin, and S. Saigal. 2012. "Co-management in community forestry: How the partial devolution of management rights creates challenges for forest communities." *Conservation and Society* 10 (2): 91–102. http://dx.doi.org/10.4103/0972-4923.97481.

De Young, R., and S. Kaplan. 1988. "On averting the tragedy of the commons." *Environmental Management* 12 (3): 273–83. http://dx.doi.org/10.1007/BF01867519.

Duinker, P., P. Matakala, F. Chege, and L. Bouthillier. 1994. "Community forestry in Canada: An overview." *Forestry Chronicle* 70 (6): 711–20. http://dx.doi.org/10.5558/tfc70711-6.

Duinker, P.N., P.W. Matakala, and D. Zhang. 1991. "Community forestry and its implications for northern Ontario." *Forestry Chronicle* 67 (2): 131–35. http://dx.doi.org/10.5558/tfc67131-2.

EAB (Environmental Assessment Board). 1994. *Reasons for decision and decision: Class environmental assessment by the Ministry of Natural Resources for timber management on Crown lands in Ontario.* EA-87–02. Toronto: Ministry of the Environment, Government of Ontario.

Filotas, E., L. Parrott, P.J. Burton, R.L. Chazdon, K.D. Coates, L. Coll, S. Haeussler et al. 2014. "Viewing forests through the lens of complex systems science." *Ecosphere* 5 (1): art. 1. http://dx.doi.org/10.1890/ES13-00182.1.

Folke, C., J. Colding, and F. Berkes. 2003. "Synthesis: Building resilience and adaptive capacity in social-ecological systems." In Berkes, Colding, and Folke, *Navigating social-ecological systems*, 352–87.

Folke, C., T. Hahn, P. Olsson, and J. Norberg. 2005. "Adaptive governance of social-ecological systems." *Annual Review of Environment and Resources* 30 (1): 441–73. http://dx.doi.org/10.1146/annurev.energy.30.050504.144511.

Fullerton, W.K. 1984. "The evolution of Crown land forest policy in Ontario." *Forestry Chronicle* 60 (2): 63–66. http://dx.doi.org/10.5558/tfc60063-2.

Gallagher, B. 2012. *Resource rulers: Fortune and folly on Canada's road to resources.* Waterloo, ON: Bill Gallagher.

Goldstein, E. 2008. "Skunkworks in the embers of the Cedar Fire: Enhancing societal resilience in the aftermath of disaster." *Human Ecology* 36 (1): 15–28. http://dx.doi.org/10.1007/s10745-007-9133-6.

Gravelle, M. 2011. "Modernizing Ontario's forest tenure and pricing system." *Forestry Chronicle* 87 (5): 591–92. http://dx.doi.org/10.5558/tfc2011-064.

Guindon, M. 2006. Personal interview, 3 August 2006.

Gunderson, L. 1999. "Resilience, flexibility, and adaptive management: Antidotes for spurious certitude?" *Conservation Ecology* 3 (1): art. 7.

Gunderson, L.H. 2003. "Adaptive dancing: Interactions between social resilience and ecological crises." In Berkes, Colding, and Folke, *Navigating social-ecological systems*, 33–52.

Gunderson, L.H., and C.S. Holling. 2002. *Panarchy: Understanding transformations in human and natural systems.* Washington, DC: Island Press.

Haley, D., and H. Nelson. 2007. "Has the time come to rethink Canada's Crown forest tenure systems?" *Forestry Chronicle* 83 (5): 630–41. http://dx.doi.org/10.5558/tfc83630-5.

Harvey, S. 1995. *Ontario community forest pilot project: Lessons learned 1991–1994 – Taking stock of Ontario's community forestry experience.* Toronto: Queen's Printer for Ontario.

Harvey, S., and B. Hillier. 1994. "Community forestry in Ontario." *Forestry Chronicle* 70 (6): 725–30. http://dx.doi.org/10.5558/tfc70725-6.

Henschel, C., and G. McEachern. 2002. *A socio-economic feasibility study of the forest tenant model in the Algoma District of Ontario.* Toronto: Wildlands League. http://wildlandsleague.org/attachments/Forest-Tenant-Study.pdf.

Hicks, D. 2014. "District to descend on Queen's Park." *Fort Frances Times*, 3 November. http://www.fftimes.com/node/274476.

Hodgins, B.W., and J. Benidickson. 1989. *The Temagami experience: Recreation, resources, and Aboriginal rights in the northern Ontario wilderness.* Toronto: University of Toronto Press.

Holland, J.H. 2006. "Studying complex adaptive systems." *Journal of Systems Science and Complexity* 19 (1): 1–8. http://dx.doi.org/10.1007/s11424-006-0001-z.

Holling, C.S. 1973. "Resilience and stability of ecological systems." *Annual Review of Ecology and Systematics* 4 (1): 1–23. http://dx.doi.org/10.1146/annurev. es.04.110173.000245.

–, ed. 1978. *Adaptive environmental assessment and management.* New York: John Wiley.

–. 1986. "The resilience of terrestrial ecosystems: Local surprise and global change." In *Sustainable development of the biosphere*, edited by W.C. Clark and R.E. Munn, 292–317. Cambridge: Cambridge University Press.

Holling, C.S., L.H. Gunderson, and G.D. Peterson. 2002. "Sustainability and panarchies." In Gunderson and Holling, *Panarchy*, 63–102.

Holling, C.S., and G.K. Meffe. 1996. "Command and control and the pathology of natural resource management." *Conservation Biology* 10 (2): 328–37. http://dx.doi.org/10.1046/j.1523-1739.1996.10020328.x.

Howlett, M., and K. Brownsey. 2008. "Introduction: Toward a post-staples state?" In *Canada's resource economy in transition: The past, present, and future of Canadian staples industries*, edited by M. Howlett and K. Brownsey, 3–15. Toronto: Emond Montgomery.

Howlett, M., and J. Rayner. 2001. "The business and government nexus: Principal elements and dynamics of the Canadian forest policy regime." In *Canadian forest policy: Adapting to change*, edited by M. Howlett, 23–62. Toronto: University of Toronto Press.

Huitema, M. (n.d.). *Historical Algonquin occupancy Algonquin Park.* Report prepared for Elders without Borders. http://www.lynngehl.com/uploads/5/0/0/4/5004954/huitema_for_swinwood.pdf.

Innis, H.A. 1930. *The fur trade in Canada: An introduction to Canadian economic history.* Toronto: University of Toronto Press.

KBM (KBM Forestry Consultants Inc.). 2003. *Algonquin Park Forest independent forest audit 1997–2002.* Prepared for the Ontario Ministry of Natural Resources. http://www.ontla.on.ca/library/repository/mon/17000/244002.pdf.

Kennedy, H. 1947. *Report of the Ontario Royal Commission on Forestry.* Toronto: Baptist Johnson, Printer to the King's Most Excellent Majesty. https://archive.org/stream/reportontforest00onta#page/n3/mode/2up.

Killan, G., and G. Warecki. 1992. "The Algonquin Wildlands League and the emergence of environmental politics in Ontario, 1965–1974." *Environmental History Review* 16 (4): 1–27. http://dx.doi.org/10.2307/3984947.

Lachance, C. In preparation. "Northeast Superior Regional Chiefs' Forum community forestry framework development process." In *Bridging practice, research and advocacy for communities and their forests*, edited by R. Bullock, P. Smith, L. Palmer, and G. Broad. Winnipeg: University of Manitoba.

Lane, M.B. 2006. "The role of planning in achieving Indigenous land justice and community goals." *Land Use Policy* 23 (4): 385–94.

Lang, P., and J. Kushnier. 1981. *The forest industry in northwestern Ontario: A socio-economic study from a social planning perspective.* Thunder Bay, ON: Lakehead Social Planning Council.

Laronde, M. 1993. "Co-management of lands and resource in n'Daki Menan." In *Rebirth: Political, economic, and social development in First Nations*, edited by A.-M. Mawhiney, 93–106. Sudbury, ON: Institute of Northern Ontario Research and Development, Laurentian University.

Lee, K.N. 1993. *Compass and gyroscope: Integrating science and politics for the environment.* Washington, DC: Island Press.

–. 1999. "Appraising adaptive management." *Ecology and Society* 3 (2): art. 3.

Levin, S.A. 1999. *Fragile domain: Complexity and the commons.* Reading, UK: Perseus Books.

Lower, A.R.M. 1938. *The North American assault on the Canadian forest: A history of the lumber trade between Canada and the United States.* Toronto: Ryerson Press.

Matakala, P., and P. Duinker. 1993. "Community forestry as a forest-land management option in Ontario." In *Forest dependent communities: Challenges and opportunities*, edited by D. Bruce and M. Whitla, 26–59. Sackville, NB: Rural and Small Town Research and Studies Programme, Mount Allison University.

MCFSC (Minister's Council on Forest Sector Competitiveness). 2005. *Final report: May 2005.* Ottawa: Minister of Natural Resources, Government of Canada. https://dr6j45jk9xcmk.cloudfront.net/documents/2849/mnr-e000248.pdf.

Messier, C., K.J. Puettmann, and K.D. Coates. 2013. *Managing forests as complex adaptive systems: Building resilience to the challenge of global change.* New York: Earthscan.

Morrow, L. 2007. *Introduction to the shareholder SFL Conversion Initiative.* Class presentation, Forest Policy. Thunder Bay ON: Lakehead University.

NAN (Nishnawbe Aski Nation). 2015. Ontario's Forest Tenure Modernization Act. http://www.nan.on.ca/article/ontarios-forest-tenure-modernization-act--465.asp.

NOSCP (Northern Ontario Sustainable Communities Partnership.) 2007a. "Community-based forest management for northern Ontario: A discussion and background paper." *NOSCP.* http://noscp.ca/?page_id=145.

–. 2007b. "Charter." *NOSCP.* http://noscp.ca/?page_id=37/.

–. 2009. "Press Release." Press release, 6 March. *NOSCP.* http://noscp.ca/wp-content/uploads/2012/09/Press-Release-6mar091.pdf.

–. 2010. *Response to Ontario's proposed framework to modernize Ontario's forest tenure and pricing system.* Submitted to Ontario Ministry of Northern Development, Mines and Forestry, 18 May.

–. 2011. "Ontario's Forest Tenure Modernization Act: A timid beginning with tons of potential." Press release, 19 May. *NOSCP.* http://noscp.ca/wp-content/uploads/2012/09/NOSCP_Press_Release_19may111.pdf.

OFC (Ontario's Forestry Coalition). N.d. "Sawmill closures (temporary and permanent)." *OFC.* http://www.forestrycoalition.com/closures.html.

OFIA, NOACC, NOMA, and FONOM (Ontario Forest Industries Association, Northwestern Ontario Association of Chambers of Commerce, Northwestern Ontario Municipal Association, and Federation of Northern Ontario Municipalities). 2011. Letter addressed to party leader, 5 July. http://www.ofia.com/

files/Pre-election%20template%20letter_OFIA-NOMA-NOACC-FONOM_%20
to%20Provincal%20Party%20Leaders%20of%20the%20Liberal%
20NDP%20and%20PC%20parties%20July%205%202011.pdf.

OFPP (Ontario Forest Policy Panel). 1993. *Diversity: Forests, people, communities
– Proposed comprehensive forest policy framework for Ontario*. Toronto: Government of Ontario.

Ojha, H.R., A. Hall, and R. Sulaiman V. 2013. "Adaptive collaborative approaches in
natural resource governance: an introduction." In *Adaptive collaborative approaches in natural resource governance: Rethinking participation, learning
and innovation*, edited by H.R. Ojha, A. Hall, and R. Sulaiman V, 1–19. London
and New York: Routledge.

Olsson, P., L. Gunderson, S. Carpenter, P. Ryan, L. Lebel, C. Folke, and C.S. Holling.
2006. "Shooting the rapids: Navigating transitions to adaptive governance of
social-ecological systems." *Ecology and Society* 11 (1): art. 18.

Ontario. 2015. "The Algonquin land claim." *Government of Ontario*. https://www.
ontario.ca/aboriginal/algonquin-land-claim.

Ontario. Environment. 2009. "MNR declaration order for forest management on
the Whitefeather Forest (MNR-74)." *Government of Ontario*. https://www.
ontario.ca/environment-and-energy/mnr-declaration-order-forest
-management-whitefeather-forest-mnr-74.

Ontario. Finance. 2013. *Ontario population projections update, 2012–2036*. Toronto:
Ministry of Finance, Government of Ontario. http://www.fin.gov.on.ca/en/
economy/demographics/projections/projections2012-2036.pdf.

Ontario. Natural Resources. 1998. "Toward the development of resource tenure
principles in Ontario: A discussion paper on natural resource tenure."
Toronto: Ministry of Natural Resources, Government of Ontario.

–. 2002. *Community-based land use planning: Northern Boreal Initiative – Developing new, sustainable forest management opportunities with First Nation
communities in the Far North of Ontario*. Thunder Bay, ON: Northern Boreal
Initiative, MNR Field Services Division. http://www.web2.mnr.gov.on.ca/mnr/
ebr/cat-slate/C-LUP.pdf.

–. 2007. *Sustainable forest licence conversion initiative: Implementation strategy*.

–. 2009. *Forest management planning manual for Ontario's Crown forests*. Toronto: Queen's Printer for Ontario. https://www.ontario.ca/document/forest
-management-crown-forests.

Ontario. NDMF (Northern Development, Mines, and Forestry). 2010. *Ontario's
forests, Ontario's future: A proposed framework to modernize Ontario's forest
tenure and pricing system*. Toronto: Ministry of Northern Development,
Mines, and Forestry, Government of Ontario. http://www.ontla.on.ca/library/
repository/mon/24005/301179.pdf.

–. 2011. *Strengthening forestry's future: Forest tenure modernization in Ontario*.
Toronto: Ministry of Northern Development, Mines, and Forestry.

Ontario. NRF (Natural Resources and Forestry). 2012. "More local say in forest
management." News release, 31 July. http://news.ontario.ca/mnr/en/2012/
07/more-local-say-in-forest-management.html.

–. 2015a. "Forest tenure modernization." *Government of Ontario.* https://www. ontario.ca/environment-and-energy/forest-tenure-modernization.

–. 2015b. "Far North of Ontario." *Government of Ontario.* https://www.ontario.ca/ rural-and-north/far-north-ontario.

Ontario Forest Accord Advisory Board. 2001. State of the Ontario Forest Accord: An interim report of the Ontario Forest Accord Advisory Board. http://www. ontla.on.ca/library/repository/mon/1000/10294267.pdf.

Orozco-Quintero, A., and F. Berkes. 2010. "Role of linkages and diversity of partnerships in a Mexican community-based forest enterprise." *Journal of Enterprising Communities: People and Places in the Global Economy* 4 (2): 148–61. http://dx.doi.org/10.1108/17506201011048059.

Palmer, A. 2012. *2012–2022 Forest management plan for the Whitefeather Forest— final.* Red Lake, ON: Red Lake District, Ontario Ministry of Natural Resources and Whitefeather Forest Management Co. Ltd. http://www.efmp.lrc.gov.on.ca/ eFMP/viewFmuPlan.do?fmu=994&fid=100106&type=CURRENT&pid=100106& sid=11582&pn=FP&ppyf=2012&ppyt=2022&ptyf=2012&ptyt=2017&phase=P1.

Palmer, L., P. Smith, and R. Bullock. 2013. "Community Forests Canada: A new national network." *Forestry Chronicle* 89 (2): 133–42. http://dx.doi.org/ 10.5558/tfc2013-028.

Palmer, L., P. Smith, and C. Shahi. 2012. *Building resilient northern Ontario communities through community-based forest management.* Report from SSHRC public outreach workshop, Lakehead University, Thunder Bay, ON, 17 May 2011. http://noscp.ca/wp-content/uploads/2012/09/Workshop_Report_Final_ 27Jan2012.pdf.

Patriquin, M.N., J.R. Parkins, and R.C. Stedman. 2009. "Bringing home the bacon: Industry, employment, and income in boreal Canada." *Forestry Chronicle* 85 (1): 65–74. http://dx.doi.org/10.5558/tfc85065-1.

PFN (Pikangikum First Nation). 2006. *Keeping the land: A land use strategy for the Whitefeather Forest and adjacent areas.* In cooperation with Ontario Ministry of Natural Resources. https://whitefeatherforest.ca/wp-content/uploads/ 2008/06/land-use-strategy.pdf.

Pinkerton, E., and J. Benner. 2013. "Small sawmills persevere while the majors close: Evaluating resilience and desirable timber allocation in British Columbia, Canada." *Ecology and Society* 18 (2): art. 34. http://dx.doi.org/ 10.5751/ES-05515-180234.

Poteete, A.R., and J.C. Ribot. 2011. "Repertoires of domination: Decentralization as process in Botswana and Senegal." *World Development* 39 (3): 439–49. http://dx.doi.org/10.1016/j.worlddev.2010.09.013.

Prabhu, R., C. McDougall, and R. Fisher. 2007. "Adaptive collaborative management: A conceptual model." In *Adaptive collaborative management of community forests in Asia: Experiences from Nepal, Indonesia, and the Philippines*, edited by R. Fisher, R. Prabhu, and C. McDougall, 16–49. Bogor, Indonesia: Center for International Forestry Research.

Reid-Kuecks, B., N. Hughes, S. Manhas, and E. Enns. 2012. *Building a conservation economy in the Northeast Superior Region of Ontario: A blueprint.* Victoria, BC: Ecotrust Canada.

Ribot, J.C., A. Agrawal, and A.M. Larson. 2006. "Recentralizing while decentralizing: How national governments reappropriate forest resources." *World Development* 34 (11): 1864–86. http://dx.doi.org/10.1016/j.worlddev.2005.11.020.

Robinson, D. 2007. "When Canadian forest policy fails." *Northern Ontario Business,* 5 May. http://www.northernontariobusiness.com/DisplayArticle.aspx?id=15752andLangType=1033andterms=when%20canadian%20forest%20policy%20fails.

–. 2009a. "The science of community forests. Part I: Approaching regime change systematically." Working paper #2–08. Sudbury, ON: Institute for Northern Ontario Research and Development, Laurentian University. http://noscp.ca/wp-content/uploads/2012/09/Science_of_Part_I.pdf.

–. 2009b. "The science of community forests. Part II: The simple theory of forests with joint products." Working paper #4–08. Sudbury, ON: Institute for Northern Ontario Research and Development, Laurentian University. http://noscp.ca/wp-content/uploads/2012/09/Science_of_Part_II.pdf.

–. 2009c. *Forest tenure systems for development and underdevelopment.* Sudbury, ON: Institute for Northern Ontario Research and Development, Laurentian University.

–. 2012. "Where is the science at with forest tenure reform? The new literature." In Palmer, Smith, and Shahi, *Building resilient northern Ontario communities,* 11–15.

Robinson, E. L. 2010. "The cross-cultural collaboration of the community forest." *Anthropologica* 52 (2): 345–56.

Rosehart, R. 2008. *Northwestern Ontario: Preparing for change.* Northwestern Ontario economic facilitator report. Prepared for Government of Ontario. http://www.nodn.com/upload/documents/ecdev-net/1nw-regional/reports/rosehart-report-feb-08.pdf.

Schlager, E., and E. Ostrom. 1992. "Property-rights regimes and natural resources: A conceptual analysis." *Land Economics* 68 (3): 249–62. http://dx.doi.org/10.2307/3146375.

Shute, J.J. 1993. "Co-management under the Wendaban Stewardship Authority: An inquiry into cross-cultural environmental values." Master's thesis, Department of Geography, Carleton University, Ottawa. https://curve.carleton.ca/system/files/theses/23806.pdf.

Smith, P. 1998. "Aboriginal and treaty rights and Aboriginal participation: Essential elements of sustainable forest management." *Forestry Chronicle* 74 (3): 327–33. http://dx.doi.org/10.5558/tfc74327-3.

Smith, P., L. Palmer, and C. Shahi. 2012. "We are all treaty people: The foundation for community forestry in northern Ontario." In *Pulp friction: Communities and the forest industry in a global perspective,* edited by R.N. Harpelle and M.S. Beaulieu, 100–20. Thunder Bay, ON: Northern Studies Press.

Smith, P., and G. Whitmore. 1991. *Community forestry.* Proceedings of the Lakehead University Forestry Association 23rd Annual Symposium, 25–26 January. Occasional Paper 8. Centre for Northern Studies, Lakehead University, Thunder Bay, ON.

Southcott, C. 2006. *The north in numbers: A demographic analysis of social and economic change in northern Ontario.* Thunder Bay, ON: Northern Studies Press.

Speers, M. 2010. "Ontario Ministry of Northern Development, Mines, and Forestry tenure and pricing review." Presentation at A New Approach: Tenure and Reform in Ontario's Forests, 42nd Annual Lakehead University Forestry Symposium, Lakehead University, Thunder Bay, ON, 15 January.

–. 2012. "Where are we at with forest tenure reform?" In Palmer, Smith and Shahi, *Building resilient northern Ontario communities,* 3–10.

Susskind, L., A.E. Camacho, and T. Schenk. 2012. "A critical asessment of collaborative adaptive management in practice." *Journal of Applied Ecology* 49 (1): 47–51. http://dx.doi.org/10.1111/j.1365-2664.2011.02070.x.

Teitelbaum, S., T. Beckley, and S. Nadeau. 2006. "A national portrait of community forestry on public land in Canada." *Forestry Chronicle* 82 (3): 416–28. http://dx.doi.org/10.5558/tfc82416-3.

Teitelbaum, S., and R. Bullock. 2012. "Are community forest principles at work in Ontario's county, municipal, and conservation authority forests?" *Forestry Chronicle* 88 (6): 697–707. http://dx.doi.org/10.5558/tfc2012-136.

Thorpe, J., and A. Sandberg. 2008. "Knotty tales: Forest policy narratives in an era of transition." In *Canada's resource economy in transition: The past, present, and future of Canadian staples industries,* edited by M. Howlett and K. Brownsey, 189–207. Toronto: Emond Montgomery.

Usher, A., N. Richardson, M. Michalski, and P. Usher. 1994. *Partnerships for community involvement in forestry: A comparative analysis of community involvement in natural resource management.* Prepared for Community Forestry Project, Ontario Ministry of Natural Resources. Toronto: Queen's Printer for Ontario.

USW (United Steel Workers). N.d. "Wood Council." *USW.* http://www.usw.ca/districts/wood?id=0003.

Venne, M, 2007. *An Analysis of Social Aspects of Forest Stewardship Council Forest Certification in Three Ontario Case Studies.* Waterloo, ON: Master of Environmental Studies, Wilfrid Laurier University. http://scholars.wlu.ca/cgi/viewcontent.cgi?article=1876&context=etd.

Walker, B., S. Carpenter, J. Anderies, N. Abel, G. Cumming, M. Jannssen, L. Lebel, J. Norerg, G.D. Peterson, and R. Pritchard. 2002. "Resilience management in social-ecological systems: A working hypothesis for a participatory approach." *Conservation Ecology* 6 (1): art. 14.

Walker, B., C.S. Holling, S.R. Carpenter, and A. Kinzig. 2004. "Resilience, adaptability, and transformability in social-ecological Systems." *Ecology and Society* 9 (2): art. 5.

Walker, B., and D. Salt. 2006. *Resilience thinking: Sustaining ecosystems and people in a changing world.* Washington, DC: Island Press.

Walters, C.J. 1986. *Adaptive management of renewable resources.* New York: Collier Macmillan.

Westley, F., P. Olsson, C. Folke, T. Homer-Dixon, H. Vredenburg, D. Loorbach, J. Thompson et al. 2011. "Tipping toward sustainability: Emerging pathways of transformation." *Ambio* 40 (7): 762–80. http://dx.doi.org/10.1007/s13280 -011-0186-9.

Westley, F., B. Zimmerman, and M.Q. Patton. 2006. *Getting to maybe: How the world is changed*. Toronto: Vintage Canada.

Chapter 5     **Forests and Communities on the Fringe**
An Overview of Community Forestry in
Alberta, Saskatchewan, and Manitoba

*John R. Parkins, Ryan Bullock, Bram Noble,*
*and Maureen G. Reed*

This chapter offers a regional perspective on the forest sector and community forestry initiatives within the three Canadian Prairie provinces. Despite being classified as prairie, these three provinces contain almost 67 million hectares of forest land, which is approximately 22 percent of total forest area in Canada (Canadian Forest Service 2011). Not only does this region contain vast quantities of forest, but it also contains large and diverse human settlements that are embedded within the boreal forest (see Figure 5.1). According to Canadian census data (Statistics Canada 2011), there are 498 individual human settlements (census subdivisions) within the boreal regions of Alberta, Saskatchewan, and Manitoba, hosting approximately 940,000 people. Some of these communities are quite large, such as Fort McMurray (Wood Buffalo Municipality), Alberta (pop. 66,896); Prince Albert, Saskatchewan (pop. 35,129); and Thompson, Manitoba (pop. 12,829). Others are smaller communities such as Fisher River Indian Reserve, Manitoba (pop. 1,129), and the Village of Arran, Saskatchewan (pop. 40). On average, the size of communities in these forested regions is approximately 1,900 residents. These statistics suggest that forests are large and communities are numerous across the Prairie provinces, and such material conditions raise the possibility that models of community forestry could be prevalent within these settings in ways that are similar to other parts of Canada.

Yet there has been limited uptake of community forestry initiatives within this region, a phenomenon that is explored in some detail below. With large forest regions and numerous forest communities, why has

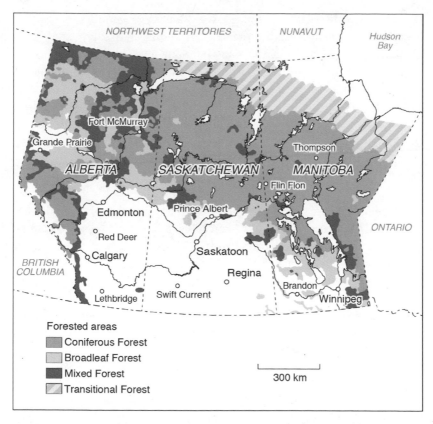

**Figure 5.1**   Forest areas and major population centres in Alberta, Saskatchewan, and Manitoba.

*Source:* "The Atlas of Canada." *Natural Resources Canada.* Canada, NRCan 2016. http://www.nrcan.gc.ca/earth-sciences/geography/atlas-canada. Adapted by Eric Leinberger.

community forestry been slow to develop? What are the barriers and challenges to community forestry in this region?

We address these questions by exploring three short case studies of forestry initiatives: the Pasquia-Porcupine Forest Management Agreement, in east-central Saskatchewan; Northwest Communities Forest Products, in northwest Saskatchewan; and the Weberville Community Forest Association, in northwest Alberta. A suitable case in Manitoba was not identified during the research for this chapter. Although only one of the case studies calls itself a community forest, all of them offer insights into recent efforts and emerging models of community involvement in forest management in this region of Canada.

After reviewing these examples, we identify a number of factors associated with forestry in the region, along with the social, geographic, and economic conditions that can help explain the limited extent of community forest development. Finally, the chapter concludes with some thoughts on the future of alternative forest management in the region. Before proceeding to these case studies, however, we provide some context through a short overview of the forest industry in Alberta, Saskatchewan, and Manitoba.

## Forest Industry and Management in the Prairie Provinces

The forest industry in Alberta is considerably larger than in Saskatchewan and Manitoba combined (see Table 5.1), yet the recent history of forest sector industrialization in the Prairie provinces is somewhat similar across these three jurisdictions. In the early 1980s, provincial leaders were looking for ways to diversify the economies of their provinces, create new sources of employment for stagnating western economies, and avoid the pitfalls of excessive dependence on single industries like energy, in Alberta, and agriculture, in Saskatchewan and Manitoba. Coming relatively late to forest industrialization (compared to British Columbia or Ontario) and having large tracts of public forest land available for development, the provincial governments set up large-scale industrial partnerships with international players such as Weyerhaeuser and Mitsubishi to kick-start a global-scale industry. The rapid pace of forest industrialization in Alberta has been documented and critically analyzed by Pratt and Urquhart (1994).

In return for tax breaks and loan guarantees, these industry players provided high-paying jobs and royalties to the Crown, which offered considerable benefits to host provinces for a number of years. By the

TABLE 5.1  Forest industry statistics for the Prairie provinces

| Province | Direct jobs (2010) | Hectares harvested (2009) | Provincial ownership (%) | Domestic exports, C$ (2010) |
|---|---|---|---|---|
| Alberta | 16,000 | 71,249 | 89 | 2.24 billion |
| Saskatchewan | 3,600 | 7,920 | 90 | 224 million |
| Manitoba | 5,000 | 13,648 | 95 | 285 million |

Source: Canadian Forest Service 2011.

late 1990s, however, the Canadian forest sector had moved into sharp decline, with significant restructuring of the industry that continued for more than a decade. Between 2003 and 2009, mill closures and curtailments caused job losses that amounted to 1,722 in Alberta, 1,668 in Saskatchewan, and 190 in Manitoba (Canadian Forest Service 2009). Given total forest sector employment numbers in each province, these losses were most significant in Saskatchewan. Current forest industry conditions for the Prairie provinces are summarized in Table 5.1. In ways that are similar to other provinces, public forests are managed in the Prairie provinces according to conditions set out in renewable long-term leases. As an example, every ten years in Saskatchewan, following a state of the forests report, a forest accord is completed that sets out the province's goals for forest management. The forest accord is operationalized regionally through integrated forest land-use plans, which determine the most appropriate mix of land uses in a forest-management area. Forest companies wishing to harvest in a forest-management area must enter into a forest management agreement (FMA) – a formal agreement between the province and a company concerning licensing, harvesting, and management – and prepare a twenty-year forest-management plan (Gachechiladze, Noble, and Bitter 2009).

One of the strategies that forest companies and governments often use to facilitate a more community-based approach to forest planning and management is through processes of public participation that include open houses, public advisory committees, and website-based portals to solicit ideas and feedback on local forest-planning activities. The advisory committee is a particularly common tool for public participation in Canada and is popular in all three Prairie provinces. In a survey conducted in 2004, Parkins and colleagues identified 101 such committees from coast to coast, with 14 in Alberta, 3 in Saskatchewan and 4 in Manitoba (Parkins et al. 2006). Research indicates that many participants of these committees are satisfied with their opportunity to contribute to forest-management decision-making and feel good about the diverse perspectives that are represented in group discussions. Other research, however, is more critical of various aspects of these participatory processes, including inadequate gender representation (Richardson et al. 2011) and limited opportunities for group learning (McGurk, Sinclair, and Diduck 2006). Given this general background to the conventional industrial model of forestry in these three provinces, the next section delves more deeply into the context of several community forestry initiatives and their potential as an alternative to the industrial model.

## Case Studies of Community-Based Forestry Initiatives in the Prairie Provinces

This book identifies four principles at the heart of community forestry in Canada and in other countries: participatory governance, rights, local benefits, and ecological sustainability. These four principles represent major themes in the community forestry literature, and they reflect what it means to undertake community forestry and how community forestry might be distinct from other models of forest management. Community forestry, however, is not a static concept: it tends to evolve in ways that respond to the social and material context in which it emerges. For instance, the principles of participatory governance and local benefits exist on a continuum, with limited community control or benefit at one end and complete control and benefit at the other (Krogman and Beckley 2002; Davis 2008). With this basic concept of community forestry as a backdrop, the following three case studies are described in relation to these principles. Do these cases represent alternatives to conventional, large-scale industrial forestry, reflecting some enhanced elements of public engagement and thus a type of forestry initiative that is more indicative of community forestry?

### Pasquia-Porcupine FMA Area, Saskatchewan

The Pasquia-Porcupine FMA) area is located near the Saskatchewan-Manitoba border, surrounding the communities of Hudson Bay and Cumberland House, east of the town of Carrot River. The FMA area encompasses approximately two million hectares, of which more than 50 percent is commercially viable forest (SMLP 1997). In addition to forestry, other land uses in the FMA area include agriculture (cattle grazing); mineral exploration; tourism and recreation; and traditional uses including hunting, trapping, fishing, and use of timber for fuel. The Pasquia-Porcupine FMA is currently held jointly by Weyerhaeuser and Edgewood Forest Products. Because of a recent transfer of ownership, among other reasons, renewal of the FMA was delayed until 2015; however, consultation plans in preparation for renewal have already been developed by Weyerhaeuser and submitted to the province.

Two attributes of the Pasquia-Porcupine FMA are of particular interest in looking for attributes of the community forestry model in

Saskatchewan. First, with regard to local benefits and local management, the *Pasquia-Porcupine Integrated Forest Land Use Plan* (1998) established "administrative" and "partnership" agreements between the province and local communities. Administrative agreements were established to facilitate ongoing collaboration with local communities and rural municipalities for the purpose of land-use planning in the area, including forest resources management. Four Aboriginal partnerships, or comanagement agreements, were also established to involve groups in community-based, consensus decision-making. There is no sharing of jurisdiction or control; rather, the partnerships focus on facilitating Aboriginal engagement in resource management and participation in resource inventories in the FMA area. Of the four Aboriginal partnerships, the Cumberland House Development Agreement, negotiated in 1989, was the only one implemented to ensure improved socio-economic opportunities for the community. It involved, among other things, the transfer of $13 million in annual grants over ten years and up to fifty thousand acres of land (SERM 1998).

A second attribute of the Pasquia-Porcupine FMA concerns forest community engagement and participatory governance. Section 39(2) of the Forest Resources Management Act requires that licensees consult with Aboriginal and other affected land users in the licensing area during the preparation of a forest management plan (FMP). One of the objectives of the Pasquia-Porcupine FMP was to ensure that communities affected by the implementation of the FMA's operational plans have an opportunity to provide input to each plan before it is approved and during its implementation (SMLP 1997). A Human and Community Development Committee was established to engage representatives from communities, government agencies, and First Nations in communication about matters concerning the FMP and impacts on and benefits to local communities. This strategy was carried forward in Weyerhaeuser's current FMA renewal process, in the form of a "public advisory group" for the FMA review and renewal, representing, among others, Aboriginal communities and townships. This approach is not unique to the Pasquia-Porcupine FMP or to the forestry sector. However, in a sparsely populated region where communities and affected interests have tended to lack the organization to make their concerns known on a large scale (Rayner and Needham 2009), the use of community or public advisory committees has proven effective in organizing and communicating local concerns about such matters as local employment, training, community needs, and forest resource monitoring.

### Northwest Communities Forest Products Ltd., Saskatchewan

Saskatchewan's first community-based timber licence was formally announced in 2002 following almost two decades of Aboriginal-led efforts to strengthen community involvement in northern forest economic development and management. Located within the north-western portion of the provincial forest, this licence provided a 770,000 hectare operating area for community-owned Northwest Communities Forest Products Ltd. Northwest is owned by seven Métis communities (Beauval, Buffalo Narrows, Green Lake, Île-à-la-Crosse, La Loche, Patuanak, and Pinehouse); it is one of very few Métis-held Crown forest tenures in Canada (Saskatchewan 2002; NAFA, n.d.).

Northwest emerged out of existing working relationships between an Aboriginal-owned forest company and two industrial licence-holders. The earlier expansion of large-scale commercial forestry operations in the region during the 1980s had stirred social conflict with traditional forest users that led to prolonged protests and blockades (Chambers 1999). As a result, in 1990, the mills joined together to form a non-profit management company – Mistik Management – mandated to plan, harvest, and reforest the land base in direct consultation with local communities (Beckley and Korber 1996). Consequently, Mistik initiated several co-management boards with local Aboriginal communities to elevate local involvement in forest planning. Such efforts were formally recognized in a 1993 memorandum of understanding (MOU) with the province to continue developing co-management arrangements at the local and regional levels. The MOU gave forest industry players the power to create co-management boards. Initially, the province took a hands-off approach to let grassroots processes evolve.

The boards were small scale, since they were formed on the basis of individual communities and corresponding fur conservation areas.[1] Boards reviewed plans and directly advised on how much logging and reforestation occurred, as well as where and how it took place (Beckley and Korber 1996). However, the boards and the relations with industry players advanced without direct provincial oversight, and by the mid-1990s, the Province was growing uncomfortable with the high level of control vested in local groups and industry partners (Chambers 1999). For example, the community of Beauval, which had the most active and vocal co-management board, negotiated a separate MOU with one industrial partner in January 1998 to expand co-management control into a neighbouring industrial licence area that included the other half of their fur conservation area. The Province repeatedly, yet unsuccessfully, tried to rein in Beauval through more formal comanagement agreements that would have limited their decision-making authority (Chambers 1999).

Perhaps not by coincidence, a consortium of communities, including Beauval and several neighbouring communities with active comanagement boards, was soon offered a conditional allocation of surplus timber that had been relinquished by a major forest company. In March 1999, the Province signed a letter of intent with Northwest to commence negotiation for an FMA for a portion of forest land adjacent to Beauval (SERM 2000). This arrangement would confer long-term timber rights to manage a specific area of land and to secure fibre for use by existing local companies (e.g., an oriented strand board facility and sawmill). The community-owned Northwest would have to meet the same licensing and planning requirements as any other industrial licence-holder, including the development of a twenty-year forest management plan. An interim timber supply licence was proposed in 2000 and later granted, in 2002, to help the company commence operations (Saskatchewan 2002; SERM 2000). The 2002 agreement also outlined steps for Northwest to eventually take over another portion of forest in northern reserve areas that had been relinquished by Mistik (Saskatchewan 2002).

**Weberville Community Forest Association, Alberta**
The Weberville Community Forest Association (WCFA) is located in the Peace Region of northwest Alberta. Initiated in 2007 and formally organized in 2009, this association exists across approximately thirty-three thousand hectares of forest land to "promote woodlot stewardship and sustainable landscape management of private forest land in Alberta" (WCMF 2010). Founders of the association included novaNAIT, a branch of the Northern Alberta Institute of Technology in Peace River, Alberta, and FPInnovations, a forest research and product development company in Canada. Both organizations recognized an opportunity and a need to work with woodlot owners in the region to realize common goals related to forest management. The association focuses its activities on a region encompassing primarily private forest land, as well as a small section of public forest land. The area has approximately three hundred residents dispersed over the County of Northern Lights. The town of Peace River is about twenty-five kilometres south of the designated community forest area. Current membership on the board of governors includes representatives from

Alberta Sustainable Resource Development (renamed Agriculture and Forestry in 2015), the federal government, the Prairie Farm Rehabilitation Administration, the local watershed council, and local land owners. In 2010, the WCFA joined the Canadian Model Forest Network to elevate its profile and improve access to potential resources for research and educational purposes.

The mission of the WCFA is "to enhance woodlot management on private land by providing support for landowners, land managers and others who influence land use practices in Alberta" (WCMF 2010). The association seeks to facilitate and coordinate landscape level responses to issues of mutual interest among landowners in the region. These issues include the development of multiuse trail systems for recreation; the coordination of logging activities; tree planting; the sharing of equipment such as mowers and chain saws; access to technical staff who conduct forest inventories and land assessments; and ongoing projects such as carbon credit opportunities, conservation credit opportunities, and other forest-management activities for which organizing at a landscape level is advantageous.

Education is the primary direct community-level benefit of WCFA activities, one example being the trail system, which includes voluntary membership and interpretive elements. Educational opportunities are advertised broadly. Residents of the town of Peace River also enjoy enhanced access to infrastructure and educational opportunities on the WCFA land base. The association maintains an ongoing half-time position of general manager with funds from a provincial government grant, but long-term funding to support staff and ongoing activities remains a challenge for this group.

### Locating "Community" in Community Forestry

Cases of community forestry in the Prairie provinces are few and far between, and those that do exist often do not fit well with the four key principles of community forestry outlined earlier. Scholars usually reserve the terms *community forest* and *community forestry* to refer to specific institutional arrangements in which locally owned and controlled decision-making entities hold direct legal authority and are explicitly mandated to manage a particular forest land base for collective local benefit (Krogman and Beckley 2002; Danks 2008). These institutional conditions are not observed in the cases discussed above. Instead, it may be more accurate to describe these cases as being characterized by community-based forestry, collaborative woodlot management, or enhanced public input rather than calling them "community forestry." The term *community-based forestry* is particularly fitting in the context of at least two of the cases described above (Pasquia-Porcupine and Weberville), because it describes joint efforts by communities and their allies (i.e., nongovernment organizations) that are intended to shape management decision-making and outcomes for community benefit on lands over which these groups do not hold exclusive tenure.

The Pasquia-Porcupine case is far from the normative model of community forestry outlined in the introductory chapter of this collection. Aboriginal communities and townships in the Pasquia-Porcupine FMA area have no harvesting or forest land rights. Rather, administrative and partnership agreements concern such broad matters as collaboration and the assurance of ongoing communication in forest resource management. The *Pasquia-Porcupine Integrated Forest Land Use Plan* explicitly notes that these agreements and partnerships do not deal with matters of jurisdiction or shared power with communities, and they confer no exclusive use, control, or rights over a partnership area (SERM 1998). The agreements do not establish forest or land tenure arrangements. Weyerhaeuser and Edgewood's recent consultation plan for FMA renewal is perhaps most telling of the extent of community forestry: although consultation with forest communities is a required part of the FMA renewal process, the communities – including Carrot River, Cumberland House, and Hudson Bay – are referred to in the consultation plan as "forest fringe communities" (Weyerhaeuser and EFP 2013, 7), yet all three of these communities are technically within the Boreal Region of Canada.

The degree and success of community forestry relies, in large part, on the tenure system and the extent of local control (Davis 2008). In the Pasquia-Porcupine case, most of the communities are outside the planning and FMA area. There is collaboration – perhaps even a degree of collaboration that is better than the "norm" in forest management – as evidenced by formal agreements and partnership arrangements; however, local community control over forest lands and harvesting rights is non-existent. The Pasquia-Porcupine case is a typical example of the role of communities in Saskatchewan's forestry sector: they sit on the fringe of community forestry.

In the Northwest Communities Forest Products case, the industrial community forest model does achieve a high level of local control through the timber licence held by a community-owned company. This model strives to achieve broader participatory governance through a board of directors that represent communities located directly within and adjacent to the FMA area, as well as through a public advisory committee. This arrangement also helps to satisfy provincial interest in having more regional representation of communities and eco-systems rather than a narrow and concentrated set of interests from single communities overseeing smaller, isolated land bases (Chambers 1999). With a unique tenure arrangement, a community-owned company, and a co-management agreement, the Northwest case demonstrates stronger principles of community forestry, especially in terms of participatory governance, rights, and local benefits. These elements map closely onto the four principles outlined earlier, yet the initiative as a whole does not define itself explicitly in terms of community forestry and has no formal linkages to broader community forest initiatives in Canada or at international levels.

Through its name, the Weberville Community Forest Association is most clearly self-identified as a community forestry initiative, and it exhibits strong elements of community forestry in terms of local benefits and multiple-use forestry. Yet the association has limited connection to broader principles of community forestry in several key areas. First, the concept of community is neither clearly defined nor centrally placed within the association's mandate. Rather, the Weberville community includes mostly landowners who are finding ways to work together and open up their land for coordinated education, recreation, and commercial benefits, with some of these benefits extending to communities beyond the defined land base (such as Peace River). There is no change in land ownership or tenure structure in relation to this community forest, and in some sense, the community forest might be more accurately described as a woodlot association, which is more common in parts of eastern Canada where private forest land is prevalent (Dansereau and deMarsh 2003). The WCFA can also be seen to play a role in extension forestry and servicing small-scale service provision arrangements for enhancing forestry practices on smaller private woodlots. Table 5.2 summarizes the community forestry characteristics across all three case studies.

### Potential Barriers to Community Forestry Development in the Prairie Provinces

Through examining the lack of development in community forestry across the Prairie provinces, we have identified a number of potential barriers, with a focus on unique industrial, political, and geographic

TABLE 5.2   Summary of key elements of three community forestry
initiatives in the Prairie provinces

| Community forestry characteristic | Pasquia-Porcupine FMA, Saskatchewan | Northwest Communities Forest Products, Saskatchewan | Weberville Community Forest Association, Alberta |
|---|---|---|---|
| Participatory governance | Consultation with Aboriginal people and other affected land users (required by the Province) | Comanagement boards with Métis replaced with community-owned management company | Board of governors representing local landowners and interest groups |
| Local management | Formal comanagement agreement with local municipalities and Aboriginal communities | Community-managed timber lands with advisory committees | Informal agreement to coordinate and promote specific woodlot management practices |
| Rights | Allocated on long-term lease | Allocated to seven Métis communities | No allocation of rights; all rights held by member landowners |
| Local benefits and use | Integrated planning, joint decision-making | Economic development, preservation of cultural values and land use, respect for ecological values related to fur management | Education, recreation, coordinated forest harvesting and conservation efforts |
| Ecological sustainability | Public advisory group recommendations on forest management and land and resource use | Fur conservation | Tree planting, carbon credit, and conservation credit initiatives |

factors. First, the development of forestry in this part of Canada has a
relatively short and somewhat unique trajectory. While some parts of
the boreal forest were cleared for agriculture in the mid-twentieth century, large-scale industrial forestry developed in the Prairie provinces
much later than in other parts of the country, expanding rapidly in
Alberta in the 1980s and in Saskatchewan in the 1990s. The overall

expansion of forestry in the Boreal Region was probably made possible by a general decline in the North American timber supply, leading the industry to explore new areas for forest development in what were previously considered less productive regions that were also further away from markets (Bouman and Kulshreshtha 1998). In addition, products using hardwoods from the boreal forest became significant, including pulp and paper, oriented strand board, and waferboard. Alberta was the most aggressive of the three Prairie provinces in expanding the industry. Between 1986 and 1995, the Government of Alberta provided about $4 billion in financial assistance to international partners to "develop" the boreal forestry economy, primarily through expansion of the pulp and paper sector (Urquhart 2001). In the late 1990s, the Government of Saskatchewan announced seven new major industrial forestry projects. Forestry also became subject to environmental impact assessment procedures with provisions for public involvement. In addition, forest companies were explicitly required to include Aboriginal people and local communities in these new operations, with the idea that forestry would bring opportunity to northern communities and boost the provincial economy. This combination of large companies and community involvement, then, appeared to offer government the best of both worlds: that is, the potential benefits of larger, international companies (such as access to capital and markets) while ensuring that the interests of local, northern, and Aboriginal communities were also included in decisions. Given this policy thrust, governments and communities were much less interested in smaller-scale, community-oriented development. From the outset, the focus was on large-scale industrial development under long-term lease from the provincial government.

Second, in the Prairie region, experience with public participation in land and resource management is more recent and diffuse, and less vocal, than in other provinces. British Columbia's War in the Woods, a phrase coined during massive public campaigns against provincial land-management practices during the 1990s (Hayter 2003), was preceded by a series of smaller-scale "skirmishes" dating back to the 1970s. These ongoing "battles" placed an international spotlight on British Columbia and its government, and particularly on its charismatic old-growth coastal rainforests and land-tenure system (Haley and Nelson 2007). By contrast, in the Prairie region, environmental organizations were not as prominent in debates related to the protection of boreal forest ecosystems. Across the Prairie provinces, forestry was not an iconic activity. Unlike in British Columbia, forestry did not

define the Prairie region. Furthermore, the boreal forest is not a charismatic landscape like the old-growth temperate rainforest of the West
Coast. The emergence of both the forest industry and the environmental organizations that seek to protect forest landscapes are much
more recent in the Prairie region than in other parts of the country.
More specifically, in Alberta, there were several attempts in the early
1990s and 2000s to undertake integrated approaches to resource
management with extensive involvement from local residents and
key stakeholders. Despite repeated disappointment in the application
of initiatives such as Special Places 2000 (Alberta Environment, 1992),
public discontent and public debate was muted (Hanson 2013). Similarly, in Manitoba, ministerial discretion was used effectively to reduce
opposition to proposed expansion and upgrading of mill capacity
(Urquhart 2001). At the same time, in Saskatchewan, environmental
organizations lacked capacity for sophisticated analyses demanded of
public participation opportunities, and more general public knowledge
about forestry activities was deemed to be low (Urquhart 2001). Given
the lack of explicit demand for community engagement and alternative
forestry models among environmental organizations and other community organizations in the Prairie provinces, it is not surprising that
community forestry has not yet emerged as a significant player in the
forestry landscape.

Third, different types of community relations have emerged in
relation to forestry practices. As mentioned previously, in Saskatchewan, governments assumed that public involvement was addressed
through the then-new provisions to environmental legislation and in
the directive by government to ensure that Aboriginal peoples and/or
northern communities be directly engaged in any new forestry projects (Urquhart 2001). Community forestry became encoded in law
and planning as a way to address some of the principles of community forestry without embracing the model in a more holistic sense.
However, given the novelty of the forest industry itself, and of formal
public participation processes such as Pasquia-Porcupine and local
control-oriented grassroots initiatives such as Northwest Communities,
it could be argued that the region has adopted a different form of
industry-community relations than in other Canadian provinces.

Fourth, in addition to factors related to industrial structure and
regulation discussed above, aspects of the physical and human geography may also constrain capacity for community forestry in the region. The forest industry has not developed as fully in the Prairie
region as in other parts of Canada, in part because its climate does

not provide conditions for a sufficiently large and fast-growing forest cover. In addition, while the northern regions of the Prairie provinces share hinterland characteristics with other provinces, the small size of communities and the geographic distance between them are more pronounced than elsewhere, limiting forestry development. While community forestry is not necessarily an outgrowth of industrial forestry, the absence of infrastructure may be a dual hindrance to the establishment and flourishing of both industrial and community forestry.

Community forestry may also be constrained by human capacity at site. For example, Leake, Adamowicz, and Boxall (2006) reported that forest dependence in Canadian communities was positively and significantly related to unemployment and the incidence of poverty for people in private households over the period 1986–96, while Patriquin, Parkins, and Stedman (2007) found that socio-economic status as measured by employment, poverty, and income was lower in boreal communities than in non-boreal communities, with a gap that appears to be growing rather than declining. Different explanations have been offered, including the "underinvestment in human skills, power structures dominated by large scale corporate interests, moral exclusion through social constructs of nature, [and] mobility and availability of resources fixed to a land base" (Patriquin, Parkins, and Stedman 2007, 282). These factors suggest that a lack of investment in human capacity rather than social conflict or ecological considerations have been at play, possibly limiting community-based initiatives in the forest sector.

Furthermore, within this broad definition of forest-based community, Aboriginal people – both male and female – are the most marginal in the structure of forestry employment. Despite composing a large proportion of the northern population (provincial government websites report 20 percent in northern Alberta, 84 percent in northern Saskatchewan, and 65 percent in northern Manitoba), Aboriginal people are concentrated in lower-paying and less stable occupations (Mills 2006). And despite rising legal recognition of the rights of Aboriginal people to land and to decisions about how environmental resources are used, they still experience economic, social, and political exclusion at all geographic scales, which will take years to address (Reed and Davidson 2011). These observations suggest that it may take considerable time, investment, and learning to build the capacity within boreal communities to match the requirements for a vibrant community forestry sector.

## Conclusion

Community forest initiatives seek to improve management through facilitating stakeholder cooperation and coordination of, for example, local knowledge and technical inputs rather than direct management authority tied to a specific land base (Bullock and Hanna 2012). As such, community forest proponents often seek to build public awareness and coordinate projects on small private forest holdings, local government-owned forests, Crown land, or some combination thereof. In light of these conceptual and practical distinctions – and with the presence of large forest regions and numerous forest communities – why has community forestry been slow to develop in northern Alberta, Saskatchewan, and Manitoba? What are the barriers and challenges to community forestry in these settings? Providing more detailed answers to these questions is challenging, but answers are likely to touch on the history of forestry in these regions, as well as on politics, policy, culture, geography, and a range of other factors.

As this chapter describes, interacting factors contribute to limited uptake of the community forestry model – factors associated with both the timing and character of the industry and the social and geographic characteristics of the region. With these questions in mind, this chapter has examined the emerging forest industry in Alberta, Saskatchewan, and Manitoba, along with case studies of three community-based forest initiatives. These cases do not fit the strict criteria of community forestry, yet they represent some limited experiments with alternative forestry in the region. Innovation in community forestry in this region is likely to remain limited as publics and governments remain focused on other concerns such as mining and energy development, issues that often consume the interests of these boreal communities and those that regulate them.

### Notes

1 Fur conservation areas were management units established during the mid-1940s to help recover beaver populations while maintaining commercial fur harvesting operations.

### References

Alberta Environment. 1992. *Special places 2000: Alberta's natural heritage.* Edmonton, AB: Government of Alberta.

Beckley, T.M., and O. Korber. 1996. *Clear cuts, conflict, and co-management: Experiments in consensus forest management in northwest Saskatchewan.* Report

NOR-X-349. Edmonton, AB: Natural Resources Canada, Canadian Forest Service, Northern Forestry Centre.

Bouman, O.T., and S.N. Kulshreshtha. 1998. "A case of integrated development in the boreal forest of Saskatchewan, Canada." *Commonwealth Forestry Review* 77 (4): 254–61.

Bullock, R., and K. Hanna. 2012. *Community forestry: Local values, conflict, and forest governance.* Cambridge: Cambridge University Press. http://dx.doi.org/10.1017/CBO9780511978678.

Canadian Forest Service. 2009. "Mill and machine curtailments in the Canadian forest industry, 2003–2009." Ottawa, ON: Natural Resources Canada.

–. 2011. "The state of Canada's forests: Annual report." Ottawa, ON: Natural Resource Canada.

Chambers, F. 1999. *The future of grass-roots co-management in Saskatchewan.* Working paper 1999–15. Edmonton: Sustainable Forest Management Network.

Danks, C. 2008. "Institutional arrangements in community-based forestry." In *Forest community connections: Implications for research, management, and governance*, edited by E. Donoghue and V. Sturtevant, 185–204. Washington, DC: Resources for the Future.

Dansereau, J.P., and P. deMarsh. 2003. "A portrait of Canadian woodlot owners in 2003." *Forestry Chronicle* 79 (4): 774–78. http://dx.doi.org/10.5558/tfc79774-4.

Davis, E.J. 2008. "New promises, new possibilities? Comparing community forestry in Canada and Mexico." *BC Journal of Ecosystems and Management* 9 (2): 11–25.

Gachechiladze, M., B.F. Noble, and B.W. Bitter. 2009. "Following-up in strategic environmental assessment: A case study of 20-year forest management planning in Saskatchewan, Canada." *Impact Assessment and Project Appraisal* 27 (1): 45–56. http://dx.doi.org/10.3152/146155109X430362.

Haley, D., and H. Nelson. 2007. "Has the time come to rethink Canada's Crown forest tenure systems?" *Forestry Chronicle* 83 (5): 630–41. http://dx.doi.org/10.5558/tfc83630-5.

Hanson, L. 2013. "Changes in the imaginings of the landscape: The management of Alberta's rural public lands." In *Social transformation in rural Canada: Community, cultures, and collective action*, edited by J.R. Parkins and M.G. Reed, 148–68. Vancouver: UBC Press.

Hayter, R. 2003. "'The War in the Woods': Post-Fordist restructuring, globalization, and the contested remapping of British Columbia's forest economy." *Annals of the Association of American Geographers* 93 (3): 706–29. http://dx.doi.org/10.1111/1467-8306.9303010.

Krogman, N., and T. Beckley. 2002. "Corporate 'bail-outs' and local 'buyouts': Pathways to community forestry?" *Society and Natural Resources* 15 (2): 109–27. http://dx.doi.org/10.1080/089419202753403300.

Leake, N., W. Adamowicz, and P. Boxall. 2006. "An econometric analysis of the effect of forest dependence on the economic well-being of Canadian communities." *Forest Science* 52: 595–604.

McGurk, B., A.J. Sinclair, and A. Diduck. 2006. "An assessment of stakeholder advisory committees in forest management: Case studies from Manitoba, Canada." *Society and Natural Resources* 19 (9): 809–26. http://dx.doi.org/10.1080/08941920600835569.

Mills, S. 2006. "Segregation of women and Aboriginal people within Canada's forest sector by industry and occupation." *Canadian Journal of Native Studies* 26: 147–71.

NAFA (National Aboriginal Forestry Association). N.d. "Metis rights and participation in the forest sector." *Forest home.* http://nafaforestry.org/forest_home/metis.html.

Parkins, J.R., S. Nadeau, L. Hunt, J. Sinclair, M. Reed, and S. Wallace. 2006. *Public participation in forest management: Results from a national survey of advisory committees.* Information Report NOR-X-409. Edmonton, AB: Natural Resources Canada, Canadian Forest Service, Northern Forestry Centre.

Patriquin, M.N., J.R. Parkins, and R. Stedman. 2007. "Socio-economic status of boreal communities in Canada." *Forestry* 80 (3): 279–91. http://dx.doi.org/10.1093/forestry/cpm014.

Pratt, L., and I. Urquhart. 1994. *The last great forest: Japanese multinationals and Alberta's northern forests.* Edmonton, AB: NeWest Press.

Rayner, J., and F. Needham. 2009. "Saskatchewan: Change without direction." *Policy and Society* 28 (2): 139–50. http://dx.doi.org/10.1016/j.polsoc.2009.05.003.

Reed, M.G., and D. Davidson. 2011. "Terms of engagement: The involvement of Canadian rural communities in sustainable forest management." In *Reshaping gender and class in rural spaces*, edited by B. Pini and B. Leach, 199–220. Aldershot, UK: Ashgate.

Richardson, K., A.J. Sinclair, M.G. Reed, and J.R. Parkins. 2011. "Constraints to participation in Canadian forestry advisory committees: A gendered perspective." *Canadian Journal of Forest Research* 41 (3): 524–32. http://dx.doi.org/10.1139/X10-220.

Saskatchewan. 2002. "Northwest partnership receives 5-year timber licence." News release, 2 November. http://www.saskatchewan.ca/government/news-and-media/2001/november/02/communitybased-forestry-partnership.

SERM (Saskatchewan Environment and Resource Management). 1998. *Pasquia-Porcupine integrated forest land use plan: Background document.* Regina: Saskatchewan Environment and Resource Management.

–. 2000. *Pinehouse-Dipper integrated forest land use plan: Draft background information document.* Regina: Saskatchewan Environment and Resource Management

SMLP (Saskfor-MacMillan Limited Partnership). 1997. *Twenty-year forest management plan and environmental impact statement for the Pasquia-Porcupine Forest Management Area: Main document.* Saskatchewan: SMLP.

Statistics Canada. 2011. Census of Canada. Data documentation for profile series part A and part B. Ottawa, ON: Statistics Canada.

Urquhart, I. 2001. "New players, same game? Managing the boreal forest on Canada's prairies." In *Canadian forest policy: Adapting to change*, edited by M. Howlett, 316–47. Toronto: University of Toronto Press.

WCMF (Weberville Community Model Forest). 2010. "About Us." *WCMF*. http://www.wcmf.ca/Aboutus.html.

Weyerhaeuser and EFP (Edgewood Forest Products). 2013. *Public consultation plan for Pasquia Porcupine 2015–2035 forest management plan*. http://www.environment.gov.sk.ca/2014-019PublicConsultationPlan.

Chapter 6    **Community Forestry in British Columbia**
From a Movement to an Institution

*Lisa Ambus*

Community forestry in British Columbia emerged, in its current institutional form, in the 1990s (Haley 2003). This was a decade of heightened social conflict focused on the forests of British Columbia, dubbed in the media as "the War in the Woods." The forest industry, dominated by a handful of large corporations with close linkages to provincial government, was under fire from labour, environmentalists, communities, and First Nations (Wilson 1998). In the midst of this conflict, a variety of actors began looking to community forestry as a source of common ground (Pinkerton 1993).

Different groups supported community forestry for different reasons. For environmentalists, community forestry was seen as inherently more environmentally benign than industrial-scale logging (Hammond 1991; Dunster 1994; M'Gonigle 1998). For the labour movement, community forestry was a means to preserve local jobs, while shifting power away from corporations and government. With most of British Columbia subject to unresolved questions of Aboriginal title, some First Nations saw the community-based model as an opportunity to take greater control of their traditional territories (Nathan 1993; Booth 1998; Curran and M'Gonigle 1999). Communities saw it as a way to "repatriate" benefits from the forest and create more stability through a diversified local economy (Allan and Frank 1994; Weir and Pearse 1995). Thus, the mid-1990s saw the emergence of a community forestry movement consisting of a loose coalition of communities and other groups, which aimed to capitalize on potential opportunities for alternative and community-based approaches to forestry as government sought out policy solutions to the War in the Woods (Wilson 1998).

In 1997, the provincial government announced the Community Forest Pilot Project, and within the year, it unveiled a new forest tenure – the Community Forest Agreement (CFA). Since 1999, when the first "pilot" CFAs were awarded, there have been significant shifts in the provincial forestry sector – in policy and legislation, global markets, and the forested landscape itself due to the mountain pine beetle epidemic. (See a detailed description of these changes in Chapter 7.) Despite these changes and challenges, as of 2015, some eighteen years later, the CFA is a permanent part of the forest tenure regime in British Columbia, growing to fifty-three  communities managing a total of 1.42 million hectares, approximately 1.4 percent of the total provincial land base.[1] While to date there has been no comprehensive accounting of the total economic benefit generated by CFA licensees, the significant growth of the provincial program and the continued demand for CFAs among communities are indicators of its perceived success. Currently, British Columbia has one of the largest networks of community forests in Canada.

Community forestry in British Columbia, in many respects, remains an ongoing experiment in constrained devolution of forest management to a local level (Ambus and Hoberg 2011). With the institutionalization of community forestry in the CFA, the vision of the community forest movement became bound within a set of rules determined, ultimately, by the provincial government. As a result, CFA holders struggle to work within the limitations of their management mandates and strive to balance the high, and sometimes contradictory, expectations of diverse local stakeholders (McIlveen and Bradshaw 2005–6; Ambus, Davis-Case, and Tyler 2007).

In this chapter, we explore how the interests and interactions of a range of actors influenced the evolution of community forestry in British Columbia: from a concept and a movement to a policy initiative, and finally to its establishment in the institution of the CFA. We also briefly examine some of the outcomes of CFAs observed to date and review those outcomes against the early vision of community forestry expressed by advocates.

This chapter draws from primary research conducted in 2005 in the form of semi-structured interviews with twenty-seven individuals – community forest managers, advisory committee members, government staff, and BC Community Forest Association representatives. A survey was also administered to twenty-four managers of community forests to gather data about the on-the-ground outcomes of the operations of CFA holders. Document analysis served as a complementary research method, including review of provincial forestry legislation, government reports, and management plans and reports prepared by individual CFA licensees.

## The Context for Community Forestry in British Columbia

### Regulatory Context

Some 95 percent of British Columbia is made up of Crown land, held by the provincial government.[2] The provincial forest supports a range of public values, including drinking watersheds, recreation, and scenic viewscapes, and with the exception of areas set aside for protection, much of the land base that supports merchantable timber is actively managed for commercial forestry operations.

Timber rights are allocated through tenures issued under the provincial Forest Act. Tenures fall into two broad categories. Area-based tenures, including Tree Farm Licences, community forest agreements, and woodlot licences, assign exclusive timber rights and management responsibilities within a defined area. Volume-based tenures, including forest licences, assign rights to harvest a specified volume of timber, within broad, common-access timber supply areas. The government sets the total harvestable volume, or annual allowable cut (AAC), for both area-based tenures and timber supply areas on sustained-yield principles, taking into account land-use zones such as protected areas and other management constraints on the land base.

Forest harvesting and management in British Columbia are governed by the Forest and Range Practices Act, which sets out management objectives for forest values, including biodiversity, soils, riparian areas, and cultural heritage resources. The act is a results-based regulatory framework under which tenure holders must achieve prescribed outcomes through harvesting and silviculture activities.[3]

### Socio-economic Context

Industrial forestry is woven into the social and economic fabric of British Columbia. Most timber harvesting rights are allocated to a handful of large, vertically integrated forest product companies. The dominance of these companies in the forest sector, and in the provincial economy as a whole, is reflected in the large number of communities that have a long history of economic dependence on lumber or pulp mills.

The indelible link between forests and communities in British Columbia has been recognized in a series of Royal Commissions on provincial forest policy going back several decades (Haley 2002). A number of these commissions set out proposed policies to enable community-based forestry. For example, in 1991, a Forest Resources Commission recommended that 50 percent of the timber volume apportioned to forest licences be reallocated to small area-based tenures managed by communities, woodlot operators, and First Nations (Pearse 1992).[4] The commissions' recommendations regarding community-based forestry were not taken up as government policy, possibly because of the cost of compensating existing tenure holders and the lack of available land (Mitchell-Banks 1999; Haley 2002). Another plausible reason is that British Columbia's tenure regime has been shaped by continuous adaptation to the interests of large corporate tenure holders (Pearse 1992; Haley and Nelson 2007).

Despite the orientation of the policy framework towards industrial interests, among the ranks of forest licensees are a handful of communities that obtained conventional forestry tenures prior to the introduction of the CFA in 1998. For example, the Municipalities of Mission and Revelstoke manage Tree Farm Licences near their communities, and the Villages of Creston and Kaslo were awarded volume-based forest licences in 1995, which they endeavoured to manage as community forests.[5] While these community forests generated some local benefits in the form of direct revenues and employment and were leading examples of community forestry, the prevailing view was that existing forms of tenure limited communities' ability to realize the more holistic approach that community forestry was thought to entail and that there was limited interest and capacity among communities to manage industrial forest tenures (Haley and Mitchell-Banks 1997; Burda et al. 1997; Anderson and Horter 2002).

## The Conditions for Change

It is against this backdrop that British Columbia's nascent environmental movement began gaining ground in the 1980s and 1990s. Early campaigns played out in relatively isolated areas of the province, including the Carmanah Valley, Haida Gwaii, and the Central Coast, with a focus primarily on winning formal protection for old-growth forests and wilderness areas. These campaigns were met with some success, as government moved forward with park designations for some areas in an effort to defuse the conflicts. However, at the provincial scale, these early compromises were relatively small punctuations in what was business-as-usual for the forest industry.

In 1993, a watershed moment in the decade-long War in the Woods took shape in the wake of the 1992 UN Conference on Environment and Development in Rio de Janeiro. The growing sophistication of environmental organizations, relative proximity to urban centres, emergent public environmental awareness, and celebrity interest converged in the Clayoquot Sound campaign, on the west coast of Vancouver Island. A summer of blockades of logging activity led to court injunctions and the largest mass arrests in Canadian history, and drew prolonged national and international attention to forest management in British Columbia (Wilson 1998).

Under intense political pressure, the provincial government responded with a series of policy initiatives aimed at deflating the conflict.[6] In 1991, it established a Commission on Resources and Environment (CORE) to undertake public regional land-use planning processes. These planning processes involved multistakeholder roundtables represented by the forest industry, environmental groups, labour, local government, and other actors, which were charged with presenting consensus recommendations to government.[7] In 1993, the government launched a Protected Areas Strategy, which was intended to address its commitment to attempt to meet the Rio Conference's land protection targets. In 1995, it introduced the Forest Practices Code, a regulatory framework that established environmental standards for forestry planning and practices.

The participatory planning processes led by CORE were intended to set high-level priorities and objectives for regional land use, taking into account both timber and other values, such as recreation and biodiversity. However, the mandates given to the planning tables by government included a proviso that objectives set for other values must avoid unduly impacting timber supply. The plans were meant to guide rather than to restructure how forests were managed through existing tenures.

The government's policy response remade the face of forest management in British Columbia, thereby bringing some measure of closure to the War in the Woods. However, not all actors considered these measures to be adequate. Forest licensees complained about the impacts of new regulations on operating costs. Unions opposed the creation of parks, which removed land from the "working forest" and which they feared would result in job losses. The new regulatory and policy framework for forestry did not go far enough for many environmentalists, who objected to the government's constraints on land-use planning (Cashore et al. 2001).

## Coalescence of a Movement

While the government's measures achieved some containment of the debate over forest management in British Columbia, some actors continued to focus on the failure of those measures to address forest governance. As one critic put it, the fundamental issue was "not just an issue of means, but of the ends for our forests – and whose interests should determine these ends" (M'Gonigle 1998, 103). In particular, critics focused on the failure of government to reshape and localize forest governance. Under the Forest Practices Code, major licensees continued logging new territory, and mill-dependent communities continued to be either sustained or shuttered by strategic investment decisions made by large corporations.

Localized land-use conflicts continued as timber-harvesting interests appeared to overshadow other forest values. These local conflicts were exacerbated by a sense that licensees had little accountability to communities, and when conflicts attracted political attention, a common perception was that government came out in favour of licensees. The continued frustrations shared by environmentalists, labour, First Nations, and local communities prompted a search for alternative forest governance models (M'Gonigle and Parfitt 1994; Burda et al. 1997; M'Gonigle 1998; Luckert 1999; Pinkerton et al. 2008).

The concept of community forestry began to take shape as a desirable alternative to the industrial forestry model and a necessary complement to the policy solutions proffered by government. In the early 1990s, community forestry was featured as a panel topic at municipal conventions, and a series of conferences on the topic were held. At around the same time, an unconventional alliance of environmental, labour, and First Nation organizations dubbed the Tin Wis Coalition began advocating for legislative changes that would devolve control over forests to local community forest boards (Pinkerton 1993).[8]

The vision and the movement for community forestry in British Columbia coalesced around a number of themes.[9] Forests should be managed for multiple values, not just timber. Communities should have control over what happens in their "backyards" and should have choices about land use in those areas, unconstrained by government directives. A resilient, diversified, value-added market economy should be developed to extract the maximum value from each log harvested. Jobs and revenues should stay in local communities, protected from the flux of global markets and distant, profit-focused corporate decisions.

## Creation of the Community Forest Agreement

### The Community Forest Advisory Committee (CFAC)

With growing momentum behind the idea of community forestry, the provincial government announced its Community Forest Pilot Project in 1997.[10] The pilot aimed to "increase the direct participation of communities and First Nations in the management of local forests and to create sustainable jobs" and was seen as "the first step towards giving communities the flexibility to manage local forests for local benefits" (British Columbia, Forests 1998). Interviewees for this study suggested that the introduction of the pilot project appears to have been partly motivated by criticism from communities involved in land-use planning processes led by CORE. Another motivating factor may have been the opportunity for community forests to harvest timber in "socially constrained" areas such as drinking watersheds, thereby offsetting the impact on timber supply resulting from the Protected Areas Strategy and the Forest Practices Code (Haley and Mitchell-Banks 1997).

Exploration of policy options began in earnest when the minister of forests appointed a Community Forest Advisory Committee to provide recommendations on a new institutional mechanism for community forestry. The membership of the committee represented a cross-section of actors, including local municipal government, environmentalists, academia, industry, labour, and First Nations.

The vision among many interested communities and other stakeholders, including members of the CFAC, was to create something new and unique that represented a significant departure from, and counterpoint to, the conventional industrial forestry model (Pinkerton 1993; Burda et al. 1997; Cortex Consultants 1997; Haley 2003).

However, the government's appetite for wholesale tenure reform, or for radical new models that departed from the established tenure system, proved to be limited. The committee's mandate was to develop a model that would fit within the same legislation and regulations governing other types of tenures – such that the community forest tenure would exist alongside, rather than replace, existing industrial tenure types.

Within these bounds, the committee's underlying goal was to design a tenure model that would provide as much flexibility as possible to communities, recognizing that community goals and objectives would probably differ significantly from those of major licensees and would vary among communities themselves. As one committee member recalled in an interview, "We tried to create a situation where communities could make community forestry into what they thought it should be." A core principle adhered to by the committee was the desire to give communities as much power as possible, while staying within its mandate.

The CFAC advocated for a tenure that afforded comprehensive rights and management responsibilities for all resources within the tenured area – such as water, recreation, botanical forest products, and sand and gravel – while respecting the government's ultimate jurisdiction to oversee the management of Crown land (CFAC 1998). The committee suggested that the government award the tenure in perpetuity to provide communities with greater security and that the new tenure include fewer operational constraints than existing industrial tenure types to allow communities more flexibility to manage for a variety of local objectives.

The committee also envisioned communities having autonomy over key forestry decisions, including setting their AAC and harvesting practices, provided environmental standards were met. A former committee member recalled, "The general opinion of the committee was that communities – if we were going to give them autonomy, if we were going to empower them – should have the freedom to decide how they were going to manage their forest, provided that there was the safety net of the Forest Practices Code Act."

**Institutional Design**

After some deliberation on the committee's recommendations, the provincial government introduced the CFA into its forest tenure legislation.[11] The CFA reflected a number of themes from the committee's recommendations and the community forest movement's vision for a new approach to forest management and governance, while also bearing many of the features of existing industrial tenure models.[12]

The CFA is an area-based tenure, like a Tree Farm Licence, giving the tenure holder both the exclusive right to harvest timber in a defined area on the land base and the stewardship responsibilities for that area. CFA holders must meet the same environmental standards as other tenure holders and must pay timber royalties – called "stumpage" – to the government for the volumes they harvest. Unlike other tenures, the CFA provides some rights – though not exclusive rights – to manage for botanical non-timber forest resources.

CFAs can only be held by a community-based organization, though no particular institutional model is prescribed; the enabling legislation allows for CFAs to be held by municipalities, cooperatives, societies, corporations, First Nation bands, or other legally constituted bodies. CFA areas are generally located in close proximity to municipalities and often include the prominent viewscapes, drinking watersheds, local recreation features, and interface zones. However, CFA licensees are expected to manage for these values while still fulfilling harvesting obligations under the Forest Act.

A notable, if arguably symbolic difference between CFAs and other tenures is that the original legislation that created the CFA allowed these tenures to be awarded for up to ninety-nine years, a significantly longer duration than the twenty-five-year Tree Farm Licence. However, during the early years of the CFA program, communities were only eligible for limited five-year "probationary" licences. These probationary CFAs were evaluated by government, and if the tenure holders demonstrated success – evidenced primarily by their financial viability – they were eligible for a long-term CFA. As the program evolved, the decision was made to eliminate both the probationary period and the ninety-nine-year duration and to move all of the long-term CFAs to a twenty-five-year renewable licence.

## Implementation

To implement the CFA pilot project, the government broadcast a request for proposals, inviting communities throughout British Columbia to apply to be granted one of three pilot CFAs. A total of eighty-eight communities expressed interest in participating, and twenty-seven developed full proposals. Of these full proposals, three communities were offered probationary CFAs in 1999. There was significant demand for CFAs, and in 2000, the government expanded the pilot project to include ten communities. In 2015, fifty-three communities held CFAs, managing a total of 1.41 million cubic metres (approximately 1.9 percent of the total provincial AAC), and a further three communities were at various stages in the application process.

During the pilot phase, the CFAC was heavily involved in reviewing and evaluating applications submitted by communities seeking a CFA. As the CFA program was expanded, the selection process was changed to one by direct invitation from the minister, which led to criticism. According to interviewees for this study, some people perceived the invitations to be politically motivated – a "good news" story used by the government to offset pressure from forestry-dependent communities negatively impacted by the downturn in the forest sector.

In rolling out the community forest program, the government heralded the CFA as a model of "local control for local benefit" (British Columbia, Forests 1998) that would see decision-making vested in communities. However, analysis of the CFA indicates that while the tenure gives communities the power to make some operational decisions, such as when and how to harvest timber, they have limited authority to make more strategic "collective choice" decisions such as those affecting the rate of harvest or the rules used to determine their AAC (Ambus and Hoberg 2011). Viewed in this light, British Columbia's experience with devolution of rights and responsibilities through the CFA echoes lessons from other jurisdictions, in which the government states its commitment to empowering communities but devolves only limited substantive authority to the local level (Ribot 2002; Shackleton et al. 2002).

Determining whether the CFA devolves *enough* authority to the local level depends on what outcomes the community aims to achieve. Vesting greater authority over "collective choice" decisions at the local level might provide CFA licensees with a broader suite of tools to manage for their desired outcomes. For example, a community that prioritizes water quality over timber values may seek to reduce the level of timber harvesting and lower its AAC. Currently, communities have limited influence over this key management decision within the terms and conditions of the CFA.

There have been a number of adjustments to the CFA tenure since its inception. These adjustments – for example, the elimination of the five-year probationary licence and stumpage relief – reflect operational tweaks to the CFA program. There was limited appetite within government for fundamentally restructuring the tenure system at the time the CFA pilot was introduced, and there has not been a substantial restructuring of the CFA tenure since.

While the institutional form of the CFA has largely remained a constant, the past decade and a half have seen significant shifts in provincial forest policy and legislation, including the introduction of a "results-based" Forest and Range Practices Act and the Forestry Revitalization Plan.[13] The mountain pine beetle epidemic, affecting massive swaths of forest throughout the interior, has catalyzed changes to the landscape and community economies through temporary "uplifts" in AAC and a surge of additional dividends, followed by a sharp decline in harvestable volume.

**Advocacy for Community Forests**

The community forestry movement in British Columbia has benefited from the support of a broad and diverse suite of actors. Its early support by a loose coalition of environmentalist, labour, and First Nation organizations was complemented by academics who dedicated substantial scholarly effort to articulating its theoretical and policy basis (Pinkerton 1993; Dunster 1994; M'Gonigle and Parfitt 1994; Burda et al. 1997; Curran and M'Gonigle 1999; Booth 1998). Over the past decade, as the CFA has become an established part of the provincial tenure regime, scholarly interest in community forestry in British Columbia has kept pace (McIlveen and Bradshaw 2005–6; McCarthy 2006; Reed and McIlveen 2006; Vernon 2007; Pinkerton et al. 2008; Bullock and Hanna 2007; Usborne 2010; Ambus and Hoberg 2011).

In 2002, the BC Community Forest Association (BCCFA) was formed to provide a collective voice for all operational CFAs and aspiring community forest organizations throughout the province. The BCCFA has played a key role in providing support to CFA holders at all stages of development (Gunter 2004; Mulkey and Day 2012). Each year the BCCFA holds an annual conference focused on topics of interest or key challenges facing CFA licensees. One of the association's major achievements was the successful negotiation of a change to the stumpage regime for CFAs, vastly reducing the royalties paid by CFA communities to government.[14] In doing so, the BCCFA laid out in clear terms the demonstrable financial challenges to the viability of community forests under CFA tenure based on competitive pressures from major industrial licensees. The BCCFA has remained an active and vocal advocate for community forestry and has maintained regular engagement with successive government ministries. Currently, the association is a unique example of an organization in Canada that provides a common voice for community forests.

## Community Forestry Outcomes

There has yet to be a comprehensive accounting of whether community forestry in British Columbia is living up to the expectations of the movement. Several preliminary reviews of aspects of the CFA program have indicated that the program has achieved "mixed results with respect to the achievement of its broad range of program objectives" (British Columbia, Finance 2004, 1; see also McIlveen and Bradshaw

2005–6; Meyers Norris Penny and Enfor Consultants 2006; and Ambus 2008). What follows is a short description of socio-economic and ecological outcomes of community forestry, drawn from the literature and research interviews.

### Local Challenges, Innovations, and Benefits

Since the inception of the program in the late 1990s, communities holding CFAs have faced a difficult economic context, stemming from the general downturn in the forest sector and, in the interior of the province, the onset of the fast-spreading mountain pine beetle epidemic. On the coast, forest companies saw their operating costs rise as the industry was forced to shift its focus from old-growth to second-growth forests and as opportunities to harvest high-value timber close to processing facilities dwindled. On the north and central coasts, land-use planning premised on an ecosystem-based management approach resulted in heightened environmental standards and operating costs.

Within this context, CFA licensees face unique challenges related to the small size of their operations. Their relatively small timber-harvest volume provides little competitive advantage on the marketplace and results in higher costs vis-à-vis stewardship planning and government administration as compared to large licensees. They must fulfill the same environmental obligations as large licensees, while harvesting substantially smaller volumes. As one informant commented, "All tenures have added responsibilities, but these have a disproportionate effect on the small guys. There are extra costs to community forests because of economies of scale." The concentration of public values near communities can also bring additional costs and challenges, as expectations for community consultation and participatory management planning are high. Moreover, many CFA holders operate in difficult terrain (Pinkerton et al. 2008).

As new start-up enterprises, community forests also face acute financial challenges in the first several years after being issued a CFA. Because many of them operate on a not-for-profit basis, they have had difficulty attracting private investment or start-up capital, particularly when margins for forest companies generally are slim. As one manager stated in an interview, CFA licensees "start out broke or deep in debt, with no expertise, no capital. They are starting out fresh when they are already up to their ears in mud."[15]

Community forests are also nested within and dependent on a broader industrial forestry sector. Most CFA holders operate as "market loggers," with raw logs as their primary product (Ambus 2008). These are sold to primary breakdown facilities (lumber mills), which are often owned by major licensees. Through their influence on market prices and their control of purchasing policies, major licensees have some control over the financial fortunes of CFA licensees. As a nested part of the broader forest industry, these small operators are subject to the market influences on the industry as a whole, including US housing starts and state-controlled Asian economies. Because they are small, place-based enterprises often operating with small margins, they lack the flexibility of the major licensees to shift capital and investment among broader holdings and can be more susceptible to market trends.

Despite the challenges, many CFA communities have managed to establish viable commercial forestry enterprises. In some circumstances, these successes have been due to temporary increases in harvest levels related to the mountain pine beetle epidemic, which is also spurring innovation and diversification. (See Chapter 7 for a case study of Burns Lake Community Forest.) For example, some CFAs in the interior of the province are planning for "life after the beetle" through investing profits in trust-like arrangements to stretch the economic benefits of the pine beetle uplift. Others are initiating dialogue within their communities to explore options in anticipation of drastically reduced harvest levels and a reduced flow of local benefits.

There are other examples of innovation in community forests. The Bamfield Huu-ay-aht Community Forest for instance, which has the smallest annual harvest of all CFAs at only one thousand cubic metres, identified research, not timber harvesting, as its primary management goal (Morgan 2002). Interviewees suggested that many CFA licensees are seeking ways to extract more value from their timber profile. Some have focused on harvesting to supply niche markets, such as log home building. Other communities are actively seeking to expand their tenured area in order to increase their annual harvest of timber and improve economies of scale.

The inclusion of some rights to botanical forest resources in the CFA tenure has been looked to as an opportunity for economic diversification (Cocksedge 2006; Ambus, Davis-Case, Mitchell, and Tyler 2007). However, these rights are not exclusive and such resources are treated as open access for public use. Markets for non-timber forest products are elusive, usually seasonal, and often marginal at a commercial scale. Consequently, to date, it appears that CFA holders have not realized any notable commercial benefits from their rights to non-timber forest products (Ambus 2008; Pinkerton et al. 2008).

Most community forests that operate under a CFA use the dividends earned from timber sales for projects and programs within the community. One informant offered the Cheslatta Carrier Nation as an example: it used proceeds from its CFA to finance a hot lunch program for all students at its local elementary school. Some CFA community forests have developed grant programs, while others have established partnerships with local service or economic development organizations. Still others have established local-first contracting policies to help keep jobs within communities.

Thus, community forests under CFA tenure can provide targeted financial support to worthy causes that may otherwise struggle to find municipal or provincial government funding. For example, in Likely – a small, rural unincorporated village in the interior of British Columbia – the Likely Xat'súll Community Forest purchased an ambulance for the volunteer paramedic service and pays for its servicing and fuel. The McBride Community Forest contributed significant funds to help pay for the construction of a new community centre. Some CFA holders donate loads of free firewood to elders or single parents. While major industrial licensees also support community causes, many community forests under CFA tenure see these benefits and services as part of their core mandate.

**Environmental Stewardship**

A central principle of the community forestry movement in British Columbia is the presumption that vesting management control at the local level will result in "greener" environmental outcomes than the conventional industrial model. As David Haley, a member of the CFAC stated, "Management practices within community forests are generally more innovative, diverse and labour intensive than on [sic] other forms of tenure and provisions are made for a broader spectrum of forest values" (Haley 2002, 61).

The capacity of community forests to implement this principle has been influenced by a range of factors, including timber profile, expertise in alternative harvesting methods, and the local community's environmental sensibilities. Furthermore, the paramount objective of a CFA-tenured forest is financial viability. Immediate economic pressures often factor into decision-making about harvest methods, in turn influencing environmental outcomes. In a recent study, informants described harvesting methods employed by CFA holders as not unlike those used by industrial operations working in the same region (Ambus 2008). Similarly, in a case from Vancouver Island, British Columbia, the primary objective of the community forest was the enhancement of the local economy, although environmental practices were also an important consideration, principally through conformance to legal obligations (Davis 2008).

Some CFA licensees have gone to great lengths – often at great expense – to adopt "alternative" forest-management approaches (Elias 2000; Pinnell and Elias 2002). As described in Chapter 10, research at the Harrop-Procter and Creston Valley Community Forests has revealed substantial efforts on the part of these communities to prioritize watershed protection over economic values. However, these may be the exception rather than the rule. A recent study concerning climate change by Furness and Nelson (2012) reveals that just under one-third of thirty-eight surveyed community forests were already integrating adaptation techniques into their work.

Some respondents described CFA holders as more responsive to local environmental concerns because of their roots in the community. One informant noted:

> With the kind of governing body that we have, we can be more sensitive and responsive to what the community sees is important. [The community forest] has five owners – an environmental group is 20 percent owner in a logging company! Community forests have to be accountable to neighbours because directors live in town. You don't get that kind of accountability from a large company. For example, logging practices ... have changed because there were mountain lady slippers growing in the area. We changed because of flowers.

## Governance Challenges

The governance of community forests under CFA tenure can be complex and challenging. This is due in part to the context in which community forestry operates, with competing interests on the land base and/or controversies over the acceptability of forestry practices. Indeed, there are several examples of governance challenges within existing CFA forests. For example, in one community, the community forest's board of directors was ousted for being too "green." In another case, local environmentalists blockaded harvesting operations taking place in forests that were within the community's drinking watershed. In yet another instance, a board of directors was replaced following a controversy over the board's decisions to invest in community start-ups that proved to be economically unviable. Thus, it appears that some community forests are constrained by pre-existing conflicts in values and land use, while others lack the capacity to properly navigate the challenging task of forest governance.

Reed and McIlveen (2006) explored the tension between recruiting board members with experience versus building in broad participation by the community. Their case study of the Burns Lake Community Forest reveals that despite early efforts at inclusivity and consensus-based decision-making, the board composition eventually reflected a business-driven model, incorporating only those community members with tools of bargaining power and influence.

There are also examples of CFA licensees who have been innovative in creating a governance structure that reflects the diversity of their communities, have strong mechanisms for downward accountability, and enjoy community support (Pinkerton et al. 2008; Leslie, in this volume). Still other community forests have succeeded in diffusing conflict in situations where there are competing visions among local forest interests (Bullock 2012).

## Conclusion

While community forests are now an established part of the forestry landscape in British Columbia, the CFA program is still a relatively small initiative at a provincial scale, accounting for approximately 1.5 percent of the total provincial forested land base. Nonetheless, the program remains popular among communities, with a steady stream of requests to government for new or expanded CFAs. It remains to be seen whether the provincial government will consider any major changes to the scale of the program or the institutional design of the CFA.

Over the past decade, community forestry in British Columbia has evolved from an alternative ideal to an institution in practice. The laudable, if arguably romantic, vision for community forestry in British Columbia – forests managed for multiple use, community control, diversified value-added market economies, the retention of jobs and revenues in local communities – has been realized, in various ways, through the implementation of the CFA.

The BCCFA, which replaced earlier coalitions as the main advocate for community forestry, continues to lobby for community forests, with some notable successes. The causes taken up by the BCCFA reflect many of the practicalities of community forestry, such as royalties, marketing, and tenure administration. The association also strives to keep alive the vision of community forestry as a unique, community-based forest governance model (Mulkey and Day 2012).

When asked to recount the chronology of events leading to the development of the CFA, some informants lamented the perceived dilution or co-optation of the "essence" of community forestry that underpinned the movement during the early 1990s. There is a feeling among some CFA managers that despite government's stated intentions for the program, CFA holders are viewed and treated the same as other forest licensees. While these sentiments speak to bureaucratic attitudes, they may also reflect some of the sense of loss that accompanies transitions from visionary ideas to more mundane realities of day-to-day operations.

Balancing the practicalities of running viable forestry businesses against local expectations and the community forestry movement in order to realize a radical departure from the industrial forestry model is, perhaps, the quintessential challenge facing community forests in British Columbia. That balance is struck, and restruck, every day in communities across the province.

### Notes

1  Updates on the program's status – including the number of CFAs awarded, area, and harvest volume – are posted online by the Province at http://www.for.gov. bc.ca/hth/timber-tenures/community/reports.htm.
2  By the late 1800s, historic treaties between the Crown and First Nations had been concluded across much of Canada. However, the vast majority of First Nations in British Columbia did not conclude treaties, and many First Nations dispute the Crown's jurisdiction and assert Aboriginal title to their traditional territories.
3  In November 2002, the Forest and Range Practices Act replaced the Forest Practices Code Act, under which specific management techniques were prescribed by the government. The transition to a regime in which the ends, rather than the means, of forestry are regulated shifted management accountabilities from government to licensees.
4  Recommendations for community forestry also appeared in the 1945 and 1957 commissions led by Gordon Sloan, the 1976 Pearse Royal Commission, and the 1991 Peel Forest Resources Commission.
5  For a history of the Mission Municipal Forest, see Allan and Frank (1994), and see Weir and Pearse (1995) for background on the Revelstoke Tree Farm Licence. The communities of Kaslo and Creston, which both held forest licences, were later awarded CFAs (see Armstrong 2000; Gunter 2000).

6  For detailed background, see Wilson (1998) and Cashore et al. (2001).
7  First Nations were invited to participate in land-use planning as stakeholders; while some participated, many boycotted the processes, asserting that the unique nature of their interests and rights required a unique response from government.
8  The Tin Wis Coalition brought together the David Suzuki Foundation, the Canadian Pulpworkers Union, the Sierra Club, and the Nuu-chah-nulth Tribal Council (Nathan 1993).
9  For an elaboration of these themes and a detailed vision of an alternative and community-based model of forestry in British Columbia, see M'Gonigle and Parfitt (1994) and Burda et al. (1997).
10  The Community Forest Pilot Project was introduced as a minor component of the 1997 Jobs and Timber Accord, in which the government committed to new jobs in the forestry sector.
11  The CFA was established through Bill 34 of the Forest Statutes Amendment Act, 1998, which added new sections to the Forest Act,1996.
12  See Ambus and Hoberg (2011) for a detailed comparison of the CFA to other tenure models.
13  Through the Forest Revitalization Plan, approximately 20 percent of timber volume allocated to volume-based licences across the province was taken back from existing (mainly major industrial) licensees. The plan was intended to create a market-based pricing system based on a fair representation of the overall timber supply (British Columbia, Forests and Range 2004). Of the 20 percent tenure take back, approximately 1 percent was reallocated to create new CFAs, woodlot licences, and tenure opportunities for First Nations.
14  As a result, CFAs in the interior region of the province pay only 15 percent of the tabular rates calculated for other licensees, and on the coast, CFAs pay 20 percent.
15  Interview with community forest manager, March 2015.

### References

Allan, K., and D. Frank. 1994. "Community forests in British Columbia: Models that work." *Forestry Chronicle* 70 (6): 721–24. http://dx.doi.org/10.5558/tfc70721-6.

Ambus, L. 2008. "The evolution of devolution." Master's thesis, Department of Forest Resources Management, University of British Columbia, Vancouver.

Ambus, L., D. Davis-Case, D. Mitchell, and S. Tyler. 2007. "Strength in diversity: Market opportunities and benefits from small forest tenures." *BC Journal of Ecosystems and Management* 8 (2): 88–99.

Ambus, L., D. Davis-Case, and S. Tyler. 2007. "Big expectations for small forest tenures in British Columbia." *British Columbia Journal of Environment and Management* 8 (2): 46–57.

Ambus, L., and G. Hoberg. 2011. "The evolution of devolution: A critical analysis of the community forest agreement in British Columbia." *Society and Natural Resources* 24 (9): 933–50. http://dx.doi.org/10.1080/08941920.2010.520078.

Anderson, N., and W. Horter. 2002. *Connecting lands and people: Community forests in British Columbia*. Victoria, BC: Dogwood Initiative.

Armstrong, L. 2000. "Just who will log? And how? 'Driving the bus....'" *Ecoforestry* 15 (3): 22–25.

Booth, A. 1998. "Putting 'forestry' and 'community' into First Nations' resource management." *Forestry Chronicle* 74 (3): 347–52. http://dx.doi.org/10.5558/ tfc74347-3.

British Columbia. Finance. 2004. *Final report on the Community Forest Pilot Project*. File no. 050067. Victoria, BC: Internal Audit and Advisory Services, Office of the Comptroller General.

British Columbia. Forests. 1998. "Legislation enables new community forest agreements." Press release, 16 June.

British Columbia. Forests and Range. 2004. *Report on the Community Forest Agreement Program 2002–2004.* Victoria, BC: Ministry of Forests and Range, Government of British Columbia.

Bullock, R. 2012. "Reframing forest-based development as First Nation-municipal collaboration: Lessons from Lake Superior's north shore." *Journal of Aboriginal Economic Development* 7 (2): 78-89.

Bullock, R., and K. Hanna. 2007. "Community forestry: Mitigating or creating conflict in British Columbia?" *Society and Natural Resources* 21 (1): 77–85. http://dx.doi.org/10.1080/08941920701561007.

Burda, C., D. Curran, F. Gale, and M. M'Gonigle. 1997. *Forests in trust: Reforming British Columbia's forest tenure system for ecosystem and community health.* Victoria, BC: Eco-research Chair on Environmental Law and Policy, University of Victoria.

Cashore, B.W., G. Hoberg, M. Howlett, J. Rayner, and J. Wilson. 2001. *In search of sustainability: British Columbia forest policy in the 1990s.* Vancouver: UBC Press.

CFAC (Community Forest Advisory Committee). 1998. *Final recommendations on attributes of a community forest tenure.* Victoria, BC: Community Forest Pilot Project, Ministry of Forests, Government of British Columbia.

Cocksedge, W. 2006. *Incorporating non-timber forest products into sustainable forest management: An overview for forest managers.* Victoria, BC: Royal Roads University.

Cortex Consultants. 1997. *Community Forest Pilot Project background discussion paper 1: Designing a community forestry tenure for British Columbia.* Victoria, BC: Cortex Consultants.

Curran, D., and M. M'Gonigle. 1999. "Aboriginal forestry: Community management as opportunity and imperative." *Osgoode Hall Law Journal* 37 (4): 711–74.

Davis, E.J. 2008. "New promises, new possibilities? Comparing community forestry in Canada and Mexico." *BC Journal of Ecosystems and Management* 9 (2): 11-25.

Dunster, J. 1994. "Managing forests for forest communities: A new way to do forestry." *International Journal of Ecoforestry* 10 (1): 43–46.

Elias, H., 2000. How a Persistent Community Fashioned Its Own Forestry Future. *Ecoforestry,* 15 (2): 18-26.

Furness, E., and H. Nelson. 2012. "Community forest organizations and adaptation to climate change in British Columbia." *Forestry Chronicle* 88 (5): 519–24. http://dx.doi.org/10.5558/tfc2012-099.

Gunter, J. 2000. "Creating the conditions for sustainable community forestry in BC: A case study of the Kaslo and District Community Forest." Master's thesis, School of Resource and Environmental Management, Simon Fraser University, Burnaby, BC.

–, ed. 2004. The community forestry guidebook: Tools and techniques for communities in British Columbia. FORREX Series 15. Kamloops/Kaslo, BC: Forest Research Extension Partnership (FORREX)/BC Community Forest Association.

Haley, D. 2002. "Community forests in British Columbia: The past is prologue." *Forests, Trees, and People Newsletter* 46: 54–61.

–. 2003. "Community forests: An old concept for a new era in British Columbia." *FORUM Magazine* 10 (2): 18-19.

Haley, D., and P. Mitchell-Banks. 1997. *Community forestry in BC: Opportunities and constraints.* Vancouver: Faculty of Forestry, University of British Columbia.

Haley, D., and H. Nelson. 2007. "Has the time come to rethink Canada's Crown forest tenure systems?" *Forestry Chronicle* 83 (5): 630–41. http://dx.doi.org/10.5558/tfc83630-5.

Hammond, H. 1991. *Seeing the forest among the trees: The case for wholistic forest use.* Winlaw, BC: Polestar Press.

Luckert, M.K. 1999. "Are community forests the key to sustainable forest management? Some economic considerations." *Forestry Chronicle* 75 (5): 789–92. http://dx.doi.org/10.5558/tfc75789-5.

M'Gonigle, M. 1998. "Structural instruments and sustainable forests: A political ecology approach." In *The wealth of forests: Markets, regulation and sustainable forestry,* edited by C. Tollefson, 102–19. Vancouver: UBC Press.

M'Gonigle, M., and B. Parfitt. 1994. *Forestopia: A practical guide to the new forest economy.* Madeira Park, BC: Harbour.

McCarthy, J. 2006. "Neoliberalism and the politics of alternatives: Community forestry in British Columbia and the United States." *Annals of the Association of American Geographers* 96 (1): 84–104. http://dx.doi.org/10.1111/j.1467 -8306.2006.00500.x.

McIlveen, K., and B. Bradshaw. 2005–6. "A preliminary review of British Columbia's community forest pilot project." *Western Geography* 15–16: 68–84.

Meyers Norris Penny and Enfor Consultants. 2006. *Community forest program review.* Victoria, BC: Ministry of Forests and Range, Government of British Columbia.

Mitchell-Banks, P. 1999. "Tenure arrangements for facilitating community forestry in British Columbia." PhD diss., Forest Resources Management, University of British Columbia, Vancouver.

Morgan, D. 2002. "Community forestry Canadian west-coast style: The Bamfield Huu-ay-aht Community Forest." *Forests, Trees, and People Newsletter* 46: 62–65.

Mulkey, S., and J.K. Day, eds. 2012. *The community forestry guidebook II: Effective governance and forest management.* FORREX Series 30. Kamloops and Kaslo,

BC: Forum for Research and Extension in Natural Resources (FORREX) and BC Community Forest Association.

Nathan, H. 1993. "Aboriginal forestry: The role of the First Nations." In *Touch wood: BC forests at the crossroads*, edited by K. Drushka, B. Nixon, and R. Travers, 137–70. Madeira Park, BC: Harbour.

Pearse, P. 1992. *Evolution of the forest tenure system in British Columbia*. Victoria, BC: Ministry of Forests, Government of British Columbia.

Pinkerton, E. 1993. "Co-management efforts as social movements: The Tin Wis Coalition and the drive for forest practices legislation in British Columbia." *Alternatives* 19 (3): 32–38.

Pinkerton, E., R. Heaslip, J. Silver, and K. Furman. 2008. "Finding 'space' for co-management of forests within the neo-liberal paradigm: Rights, strategies, and tools for asserting a local agenda." *Human Ecology* 36 (3): 343–55. http://dx.doi.org/10.1007/s10745-008-9167-4.

Pinnell, H., and H. Elias. 2002. "How the Harrop-Procter community is harvesting its forest." *Ecoforestry* 17 (4): 8–16.

Reed, M. G, and K. McIlveen. 2006. "Toward a pluralistic civic science? Assessing community forestry." *Society and Natural Resources* 19 (7): 591–607. http://dx.doi.org/10.1080/08941920600742344.

Ribot, J., 2002. *Democratic Decentralization of Natural Resources: Institutionalizing Popular Participation*. Washington, DC: World Resources Institute, 30.

Shackleton, S., B. Campbell, E. Wollenberg, and D. Edmunds. 2002. "Devolution and community-based natural resource management: Creating space for local people to participate and benefit?" *ODI Natural Resource Perspective* 76: 1-6.

Usborne, A. 2010. *Planning a sustainable approach to community forest management with the Katzie First Nation at Blue Mountain and Douglas Provincial Forests*. Burnaby, BC: Simon Fraser University.

Vernon, C. 2007. "A political ecology of British Columbia's community forests." *Capitalism, Nature, Socialism* 18 (4): 54–74. http://dx.doi.org/10.1080/104 55750701705088.

Weir, D., and C. Pearse. 1995. "Revelstoke Community Forest Corporation: A community venture repatriates benefits from local public forests." *Making Waves* 6 (4): 4–13.

Wilson, J. 1998. *Talk and log: Wilderness politics in British Columbia*. Vancouver: UBC Press.

# PART 2
## Case Studies: Connecting Principle and Practice

## Chapter 7

**Community Forestry in an Age of Crisis**
Structural Change, the Mountain
Pine Beetle, and the Evolution of
the Burns Lake Community Forest

*Kirsten McIlveen and Michelle Rhodes*

Structural changes within British Columbia's forest sector have led to a "crisis" in the forest industry and forest-dependent communities. More recently, these changes have been exacerbated by the dramatic loss of wood fibre as a result of the mountain pine beetle (MPB) epidemic in BC's interior. Smaller community-based operators face a web of challenges as they manoeuvre the complex environment of forest health challenges, global competition, and structural change in the forest industry.

One such operator is the Burns Lake Community Forest (BLCF), located in the heart of British Columbia's timber-dependent north-central interior – and the pine beetle epidemic. The BLCF, born with great community expectations (McIlveen 2004), was an early model of community forestry developed under the provincial Community Forest Pilot Project, implemented in 1998. Indeed, it has experienced success in reaching many of its goals and has generated significant revenue from its forestry operations. Of late, however, the BLCF has faced difficulties in dealing with external forces beyond its control – most notably, an infestation by the MPB. This endemic insect threatens to annihilate the forest base that underpins the community forest licence.

In this chapter, we identify five stages of development, and the key factors that constrained or enabled the forest's operations, during the first half-decade of the Burns Lake Community Forest Pilot (CFP).[1] Enablers and constraints were first identified in part from survey and interview data collected by McIlveen (2004) from key stakeholders in the BLCF during a two-and-a-half-year period (2000–2002). More recent follow-up research – in the form of telephone interviews, email interviews, and informal person-to-person interviews (all with those directly and indirectly involved with the BLCF), as well as reading reports produced within the ministry responsible for forests – provides a longitudinal perspective on these same issues.

The community of Burns Lake (pop. 2,390) is located in the northern interior of the province, halfway between Prince George and Terrace in the Bulkley-Nechako Regional District. Over the last two decades, Burns Lake has faced a series of economic challenges, including structural changes in the BC forest sector, the pine beetle crisis, and, most recently, the loss of one of the region's largest mills to fire. Despite its scenic setting and related tourism potential and the development of mining opportunities, downturns in the forest sector have resulted in cumulative job losses and out-migration. Burns Lake experienced a 3.7 percent decline in population between 2006 and 2011, reversing a previous growth trend (BC Statistics 2014), and forestry, wood manufacturing, and paper-products manufacturing constitute a shrinking percentage of the community's economic activity (BC Statistics 2014).

The forest sector still accounts for nearly half of the region's employment. However, changes brought on by a move to flexible production in the industry over the last three decades, as well as increasing overall economic volatility (Hayter 2000; Edenhoffer and Hayter 2013), have highlighted the need for new strategies for forest tenure and production in order to provide some community stability. The BLCF developed within this context. This community forest is one of the most successful of the community forests established as part of BC's pilot project, and as a result the BLCF was granted a twenty-five-year licence in 2004. Despite being significantly smaller in terms of operations and revenues than other forest product companies in the region, the BLCF has emerged as one of the largest employers in the Burns Lake region (McIlveen 2004).

In response to the crisis in the forest sector, the Village of Burns Lake began developing the Burns Lake Recovery Strategy and a community transition Plan in early 2012, including rebuilding the Babine Forest Products Mill, which burnt down in January 2012. These plans are also designed to address a more vexing problem: timber supply. Like other timber-dependent communities and regions in the province, the Burns Lake area faces a dramatic reduction in timber supply in the wake of the MPB outbreak and the resulting reductions in labour and difficulties in long-term economic planning. The outbreak has had a much greater effect on small, area-based tenures like the BLCF than on larger timber companies. When community forest operations are unable to secure fibre supply locally, they have limited ability to move further afield to find more resources.

These ecological and economic changes mean that the success – or lack thereof – of provincial and community-driven forestry initiatives needs to be measured in part by how operations such as the BLCF respond to crisis. The objective, then, of the following case study is to illustrate not only the operational difficulties that smaller, community-based operations such as the BLCF face but also how these organizations deal with simultaneous challenges occurring at multiple scales. The study begins with an introduction to the development of community forest initiatives in the province in the 1990s and an in-depth examination of the growth of the BLCF, itself situated deep in the heart of the pine beetle epidemic. The BLCF did many things "right" on the road towards gaining long-term tenure over its forest lands, but it still must contend with the same forest health challenges, global competition, and structural change in the forest industry that affect larger timber firms (e.g., Canfor, Tolko, West Fraser, etc.) operating in the province.

The pine beetle infestation raises concerns over the ongoing viability of community forest projects in heavily affected areas, given the loss of fibre in concert with economic shocks. It also raises questions about whether a community forest, with its small size and limited authority, can manage its resources and harvest logs at a scale needed to supply local mills with competitively priced inputs. Can or should a community forest be the driver of timber-dependent community recovery, particularly in light of compromised forest health and supply? More significantly, the mounting challenges faced by Burns Lake and other community forests raises the questions of the degree to which government responsibilities for forest management should be devolved to local entities and whether, in spite of these immense challenges, community forestry may prove to be more appealing and resilient.

## Postwar Expansion and Structural Change in BC's Forest Sector

Commercial logging in British Columbia dates back to the mid-nineteenth century, but large-scale, highly mechanized production did not begin until nearly eighty years later. Based on a Fordist model of production, BC mills met a global demand – and, in particular, an insatiable American appetite – for wood products and pulp and paper. The geography of the mills had also changed by the early 1900s, as coastal forests began experiencing fall-down and as expanding infrastructure and population opened up interior forests.[2]

Demand for BC wood and paper products grew as the economies of the largest consumers of these products – the US, Europe, and Japan – expanded in the years following the Second World War. In order to meet the demand for fibre in both the near and long term, the Province granted renewable long-term leases of twenty-five years that covered large areas of forests to large firms on the condition that they invested in "forest production complexes" (Edenhoffer and Hayter 2013) and with the hope that this would result in the maintenance of steady timber supplies through practices supporting "sustained yield" (i.e., a continuing supply of timber). The number of sawmills increased from the 1930s to the 1960s, because of the expansion of road and rail systems, efficiencies in production gained through Fordist production, and appurtenancy regulations that required firms to invest in mills proximate to sites of timber extraction (Hayter 2000; Edenhoffer and Hayter 2013). The Province even invested in the creation of communities that would support the timber industry (as well as other resource activities) through the Instant Towns Act of 1965 – a measure that would, literally, carve communities out of the wilderness with the goal of integrating the hinterland and reducing regional disparities (Markey, Halseth, and Manson 2008; McGillivray 2010).

During the late 1960s and the 1970s, investments became spatially diffused while becoming industrially concentrated. Most new investments were focused in the interior, where some new "instant towns" were created, and other communities expanded in response to ramped-up investment. Soon the volume of wood cut in the interior surpassed that of the coast (McGillivray 2010; Edenhoffer and Hayter 2013). Meanwhile, multinational corporations used mergers and buyouts on the path to becoming highly integrated – both vertically and horizontally. By the end of the 1970s, a small number of firms had gained control over extraction and wood-processing in most of the province. Control over supply, in other words, allowed these firms (most notably, MacMillan Bloedel) to dominate production and ensure the supply needed to maintain high levels of output (Hayter 1976; Edenhoffer and Hayter 2013). The old mills on the coast were now under serious pressure to compete, and new investments were required to keep them competitive (Hayter 1976; McGillivray 2010). By the late 1970s, total production had increased, but output was generated from fewer, more efficient mill complexes.

During this Fordist period, the Ministry of Forests, timber corporations, and unions reached agreements on tenure, production targets, mill location, and the like behind closed doors (Hayter 2000). Ostensibly, the goal was to provide stability in an industry plagued in the past by labour action and shortages (especially important in British Columbia, a province with a history of labour action in its forests) and, in doing so, to allow for efficient production of timber on public lands. The relative cohesion among these groups, the importance of the forest-products industry to the province's economy, decades of timber-based economic growth, industrial integration, and the perception of inexhaustible forests all contributed to industrial stability (Boyd 2003; McCarthy 2006). Indeed, the industry provided wealth to many British Columbians and stable employment for many workers (M'Gonigle 1997; Marchak, Aycock, and Herbert 1999; Hayter 2000).

As new mills sprang up in remote coastal and interior communities in the province, a core-periphery pattern was reinforced, in which the greatest benefits were concentrated in the urban-industrial-political cores of southwestern British Columbia and southern Vancouver Island. The immense wealth generated by the forest sector flowed to the core, which was also the centre for decision-making authority related to forest planning. More remote communities were left with far fewer opportunities and incentives for diversification and value-added production. Thus, interior communities were particularly prone to the staples trap, which occurs when a commodity that dominates an economy's exports no longer generates an adequate income, and to Dutch disease, which occurs when resource exports push out other activities, like manufacturing. The high cost of unionized labour, "expensive public infrastructure that is dedicated to resource exports, and supportive government policies and rhetoric," gave the appearance of permanent stability while discouraging alternative industrial development (Edenhoffer and Hayter 2013, 141). As a result, more remote forest-dependent communities experienced significant social and economic upheaval when the industry began to undergo substantial changes in the late 1970s.

This Fordist "compromise" was soon challenged, and the result was massive social and economic upheaval taking place over decades for resource-dependent communities. Following a series of oil shocks, increasing global competition from low-cost producers, and, most significantly, a deep global recession in the early 1980s, BC's forest sector underwent transformational change (Hayter 2000; Markey, Halseth, and Manson 2008; Edenhoffer and Hayter 2013). The recession also provided the conditions for the rise of neoliberalism (Hayter and

Barnes 2012), although Fordist institutions (e.g., unions) were strongly resistant to changes brought on by neoliberal philosophies. By this time, BC's model of production had become inefficient and costly, and as a result, the province had become one of the world's highest-cost forest regions (Marchak, Aycock, and Herbert 1999; Edenhoffer and Hayter 2013).

In addition, the industry faced a number of challenges: fall-down; growing conflicts with the United States over softwood timber imports, since forest-dependent communities in the United States were also dealing with many of the same structural changes; the MPB infestation, which spanned almost twenty years; even greater reliance on favourable exchange rates to drive production; and changes in consumption patterns, as consumers of wood products began demanding more specialized outputs that could not be effectively produced on such a large scale. In addition, the industry moved towards increasing technological efficiencies gained through modernization. This reduced the need for labour, both in the forest and in the mill (Hayter 2000; Marchak, Aycock, and Herbert 1999; Edenhoffer and Hayter 2013).

BC's remote communities faced long-term job losses and were ever more vulnerable to the fluctuations of commodity markets. The industry entered a "new normal" in forest production: repeated periods of boom and bust (Robson 1996; Marchak, Aycock, and Herbert 1999; Hayter 2000; Stiven 2000; Edenhoffer and Hayter 2013). Forestry and mill jobs were cut in half over the next three decades, with greater losses experienced in pulp and paper (Edenhoffer and Hayter 2013). More remote mills faced closure after the end of appurtenancy regulations in 2013, effectively providing forest firms greater licence to spatially amalgamate holdings. Following the housing market crash in the United States in 2007, the losses in the forest sector were particularly severe – more than 116,000 jobs were lost in 2008 and 2009 alone, although whether most of these jobs would have been lost eventually because of further industrial restructuring is not entirely known. Plant closures, job losses, and poor returns on investment led many to declare a state of crisis in the industry (e.g., Beckley 1998; Marchak, Aycock, and Herbert 1999; Hayter, Barnes, Bradshaw, 2003).

Tensions have escalated over access to supply and markets, exacerbated by ongoing concerns over low stumpage fees, continued negotiations and conflicts resulting from the land claims process, increasing pressure from environmental groups, and forest health concerns (Clapp 1998; Hayter 2000; Markey, Halseth, and Manson 2008; Edenhoffer and Hayter 2013). While there are differing estimates concerning an appropriate annual allowable cut (AAC), critics of the industry asserted that forest resources in most of British Columbia were mismanaged and that the AAC in many areas was set higher than what is ecologically sustainable (M'Gonigle 1997; Burda 1999; Marchak, Aycock, and Herbert 1999; Gunter 2000). However, Hayter (2000) recognizes that in the 1990s, not all of the AAC was consumed, for market and cost reasons, despite fears of wood-fibre shortage.

In British Columbia, a push for greater public involvement in forest planning coincided with these larger shift changes in the industry. It had become clear to many in rural British Columbia that traditional models of forestry were not addressing community interests. In the postwar years, the forest-tenure arrangement gave large, integrated firms extensive control over public forests, and the leases granted to forestry companies did not mandate them to maximize employment and income for communities, nor to protect ecosystems (M'Gonigle and Parfitt 1994; Booth 1998; Burda 1999; Hayter 2000).

## The Mountain Pine Beetle: Making a Bad Situation Worse

By the time the MPB epidemic arrived in the 1990s, BC's forest industry was already in a state of extended crisis. The MPB had the effect of worsening many of the challenges already faced by communities, including the search for markets, questions over long-term supply, and pulse harvesting for timber recovery. Pulse harvesting refers to the depletion of the resource often to the point of exhaustion, often followed by abandonment of the resource on a regional scale (Perlin 1991).

Burns Lake sits in the heart of "the most extensive mountain pine beetle epidemic in recorded history" (Parfitt 2005, 13). The provincial height of the beetle epidemic occurred in 2005 in the Lakes District, which includes Burns Lake. Burns Lake and its surrounding region are pine dominant. Approximately 77 percent of the Lakes Timber Supply Area (TSA), of which the BLCF is a part, consists of pine and spruce, with minor amounts of balsam. In the BLCF, pine accounts for 85 percent of the forested land base (British Columbia Mid-Term Timber Supply 2012). Most of this forest base is older (over forty years) second- and third-growth forest and is thus relatively evenly aged. The vast majority of production from these forests was for lumber.

As of 2015, the Ministry of Forests, Lands, and Natural Resource Operations estimated that the MPB had killed a cumulative total of 723 million cubic metres of timber since the most recent infestation began (British Columbia, FLNRO 2015). The most recent outbreak has been linked to a changing climate and past forest-management policies. Early and/or extended cold spells typically kill off many beetle pupae and eggs, reducing the impact of the insect. Such cold spells have become increasingly uncommon, even in the northern half of British Columbia, a region known for long, cold winters. This, along with the abundance of even-aged stands resulting from the replanting of pine through commercial operations and fire suppression, has contributed to the most recent outbreak. Some of the most extensive losses have been recorded in the central interior and around Burns Lake.

Following an MPB outbreak, affected trees are often harvested before wood quality is compromised, at least for use as lumber. Most affected pine can be harvested as sawlogs, with the percentage of viable sawlog timber steeply decreasing beyond two years after death and becoming "uneconomical" to harvest between four and eight years after death. These figures are somewhat variable, since much depends on the moisture content of the trees and related factors (British Columbia, Forests and Range 2007). Past that point, the most common alternative uses for the dead timber are pulp, wood pellets, and bioenergy. These end uses have lower value than lumber, and harvest may not provide sufficient returns.

Provincially, the response to the outbreak has been to dramatically escalate logging rates through "salvage" licences issued to companies to log beetle-attacked wood. AACs have been increased to account for these operations (British Columbia, Forests and Range 2007). According to the timber supply review for the Lakes, Prince George, and Quesnel TSAs, the increased logging rates approved for 2005 were 3.17 million cubic metres, while the projected logging rates for 2015 are 1.4 million cubic metres. This dramatic decline in logging rates represents a decline of 55.8 percent (British Columbia Mid-Term Timber Supply 2012). To encourage investment, British Columbia has allowed companies to harvest diseased timber at the cut-rate price of twenty-five cents per cubic metre, compared to fourteen dollars or more for higher-grade timber (McKenna 2010). The BC government's decision to raise the AAC in order to harvest greater volumes of MPB-infected timber has been opposed by environmental organizations as compromising long-term ecological sustainability (Hughes and Dreyer 2001). Indeed, this decision may have had mixed economic benefits as well.

Ironically, the increased supply of fibre met a bottleneck in terms of processing, since many mill operations had already been shut down;

conversely, harvesting of the MPB-killed timber was a factor in the creation of "supermills" that embody the very qualities of Fordist-style production (Hayter and Barnes 2012). Outside of Burns Lake, Canfor reopened its sawmill in nearby Houston. With its $26.4 million upgrade, Canfor laid claim to the largest sawmill in the world. Lumber production increased from an already formidable 450 million board feet per year to 600 million. Within a year, Canfor posted a record net income of $420.9 million and announced that it would invest $104 million to build a second mill in Vanderhoof, similar in size to its Houston operation. The result is that there are now two large mills near Burns Lake (Parfitt 2012).

This increased lumber production has caused some concern over whether the forests can actually supply the required milling horsepower (Parfitt 2012). Parfitt sees BC's forest industry as increasingly vulnerable due to a diminished pool of commercially harvestable trees. Around Burns Lake, local mills can consume 1.9 million cubic metres, but in the near future, forests will only be able to provide for 26 percent of mill needs (Parfitt 2012). Furthermore, as Edenhoffer and Hayter (2013, 144) note, "the government's decision to increase AACs in response to a pine beetle epidemic that quickly threatened large swaths of BC forests first fed the boom in the early 2000s and then deepened the bust starting in 2007." In short, the accelerated harvesting of MPB-killed timber has only exacerbated existing trends.

## The Emergence of Community Forestry in British Columbia

Support for developing a community forestry tenure accelerated in the 1990s, especially from within forest-dependent communities (Beckley 1998). In British Columbia and elsewhere, community forestry has been seen as a forest tenure option that provides community-scale benefits, including increased local decision-making over forest use and management, greater economic stability through consumptive and non-consumptive forest industries, and better and more ecologically and culturally sustainable forest management (Bullock and Hanna 2012). Given the importance of the resource in British Columbia, the growing dissatisfaction with forest management, the erosion of forest industry jobs, and increasing concerns about the environmental impact of industrial forestry practices, community forests provide a competing vision for how local supplies could be managed and for what purposes. Furthermore, interior communities, in particular, face a lack of alternatives to forestry for economic development due to factors related to distance, infrastructure, knowledge of and familiarity with alternatives, and competition between small communities.

Community-managed forests are not new to the province, but their numbers increased substantially beginning in the 1990s. In south-western British Columbia, the Mission Municipal Forest (MMF) has been in operation since the 1940s and has held a municipal Tree Farm Licence since 1958, the only municipality to do so until the 1990s (District of Mission, 2005). Bullock and Hanna (2012) have profiled other provincial community forestry initiatives, including the Creston Valley Forest Corporation (CVFC), a group formed by community residents and businesses concerned about wood supply and ecological health, and the Denman Community Forest Cooperative, an organization that evolved from conflicts over clear-cut logging on Denman Island. In 1998, as a result of growing public support for community forestry across the province, the provincial government established the Community Forest Pilot Project (CFPP) and created a new form of community forest tenure – the community forest pilot agreement, or CFPA. The  pilot project program was unique in that it began the establishment of a comprehensive network of community-managed forests on public land (Community Forestry Forum 2000; McIlveen and Bradshaw 2005–6; Teitelbaum, Beckley, and Nadeau 2006).

Under the CFPP, the government allocated forest land and managerial autonomy to ten communities. The aims of the program were to increase community involvement in local forest land, "to provide opportunities at the community level to test some new and innovative forest management models," and to maintain "forest-related community lifestyles and values, while providing jobs and revenue that contribute to community stability" (British Columbia, Forests 1999, 1).

The initial ten community forest pilots, or CFPs, were geographically dispersed, varied in size, and reflected a diversity of interests and community types – from newly formed organizations, to First Nations, to villages and districts. While all of the CFPs had a commercial agenda – some, for instance, saw community forestry as an opportunity to create employment and revenue for the community – only some wanted to employ conventional industrial forestry practices (i.e., to operate as small-scale industrial foresters). Others chose to focus on watershed protection, non-timber forest products such as botanicals, and eco-tourism. Still others, given their small size, focused on education and recreational opportunities. Finally, some of the CFPs saw community forestry as an opportunity to enhance community cohesion and build bridges with various stakeholders in the community (British Columbia, Forests 1998).

The progress of the ten community forest pilots varied significantly. By 2002, only three of them had progressed to the stage of harvesting and selling logs, five had made no progress beyond their initial selection of pilot sites, and two had reached stages between these two extremes. This variation can be explained by identifying the various constraints and enablers affecting the initial implementation and functioning of the CFPP. In spite of the difficulties faced by the CFPs, the BC government renewed its commitment to community forestry again in 2003 as part of the Forest Revitalization Act, which, among other changes, sought to reallocate 20 percent of current tenure holdings from larger firms to a variety of smaller holders (including Aboriginal groups and community forests) and auctions (Hayter and Barnes 2012). As of January 2015, approximately 1.42 million hectares were being managed as community forests; there were 52 active community forests in the province, and another three communities were in the application process (British Columbia, FLNRO 2015).

The Burns Lake Community Forest (see Figure 7.1) is considered one of the CFPP "success stories." The Burns Lake CFP had, in its proposal, identified several key objectives: generating a source of revenue and employment for the community, testing innovative forest practices such as labour-intensive harvesting, developing trail systems, working with local educational institutions for training opportunities, encouraging stakeholder cooperation in the community forest, and encouraging First Nations to develop and market traditional botanicals (British Columbia, Forests 1999). In 2004, having achieved a sufficient level of success, the Burns Lake Community Forest Ltd. became the first organization in British Columbia to be offered a long-term community forest agreement. This twenty-five-year renewable forest tenure replaced the original community forest pilot agreement and provided the BLCF with the long-term operational certainty needed to expand its operations. It also ensured that the community would have a say in how the forest resources surrounding Burns Lake would be managed for at least the next quarter century.

### The Burns Lake CFP: Constraints and Enablers

The progress of the Burns Lake CFP stood apart, even at an early stage, from that of the other CFPP applicants. Of the ten CFPs, Burns Lake was

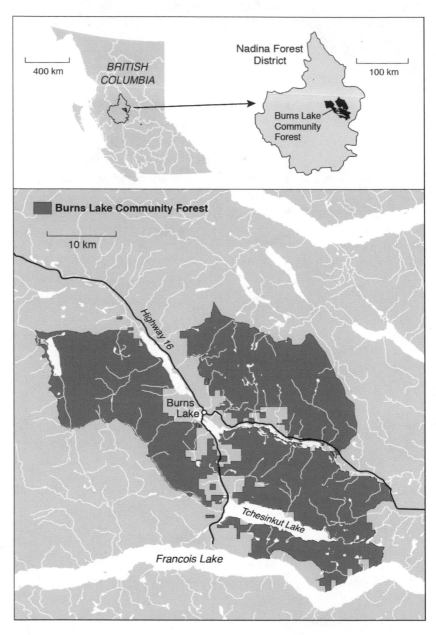

**Figure 7.1** Location of Burns Lake Community Forest Ltd.

*Source:* Burns Lake Community Forest Management Plan #2, 2010–15. Adapted by Eric Leinberger.

one of three to be at the stage of selling logs by the end of the pilot project period. What were the factors that contributed to Burns Lake's success, and what were the limits that constrained the community forest's expansion?

The following discussion identifies the five stages of development of the Burns Lake CFP and the key factors that constrained or enabled its operations during its first half-decade (see Table 7.1). These enablers and constraints were first identified from survey and interview data collected by McIlveen (2004) from key stakeholders in the Burns Lake

TABLE 7.1  Stages, Enablers, and Constraints in the Burns Lake CFP's Evolution

| Stage | Enablers | Constraints |
|---|---|---|
| Secure a forest land base | • Lack of competition for adjacent lands<br>• Sub-boreal forest, with mature stands<br>• Streamlined process for approval due to MPB<br>• Expanding AAC to accommodate more harvesting of MPB-killed timber | • High scenic and recreational value (makes harvest more difficult)<br>• Mountain pine beetle infestation (affects quality, mode of extraction)<br>• Reduction in AAC (supply is less available after 2013) |
| Draw on community skills | • Dependence on industry<br>• Common vision<br>• High level of volunteerism<br>• Effective leadership | • Lack of representativeness in leadership structure |
| Comply with provincial regulatory system | • Expertise in complying with provincial regulations | • Revenue appraisal system established for large-scale forestry, disincentives for smaller-scale production and unique product development |
| Secure markets and exist within a global environment | • Interest in developing diversified product profile using consumptive and non-consumptive forest products | • Distance from markets<br>• High instability in BC forest sector<br>• Traditional reliance in BC on forest exports to a narrow range of markets<br>• Global competition and neoliberal policies |

CFP during a two-and-a-half year period (2000–2). The results of this earlier investigation largely supported prior research assumptions about community forest sustainability, assumptions that were informed by the literature on community development, property rights and resource control, and globalization.

In order to survive and ultimately succeed, British Columbia's CFPs needed to secure a forest land base, draw on community skills, comply with the regulatory system, and, finally, secure markets and exist within a complex global environment. While each stage contributes to success, the order of the stages is important. Furthermore, the more enabling factors there are within each stage, the more likely it is that the CFP will progress to the next stage.

As identified in Table 7.1 and discussed below, each of the four stages of the Burns Lake CFP's development represents significant variability based on current and projected economic and ecological conditions. In other words, for a community forest to move through each of these stages requires that it anticipate environmental change (stage 1), changing demographics (stage 2), and shifting regulatory and economic policies (stages 3 and 4). The more dynamic the external influences, the greater the difficulty in trying to create long-term sustainability. Conversely, however, these very challenges, and the responses of the community and the province, may increase the long-term resiliency of the organization.

### Stage 1: Secure a Forest Land Base

A vital first stage for all community forests is to secure a forest land base (e.g., M'Gonigle and Parfitt 1994; Gunter 2000). In the case of the Burns Lake CFP, the lack of competition for the forest land base allowed the CFP to secure an area adjacent to the community. While the provincial ministry responsible for forests held the forest land base, the area had not been harvested for several reasons. It was a politically contentious area, since it surrounds several large lakes that are used for recreation, its proximity to the community presents visual quality constraints, and it is in the traditional territory of the Burns Lake Band. Given the potential for conflict, the provincial government may have decided that the community was better suited to manage this area.

The forest land base for a community forest must have adequate merchantable timber (i.e., sufficient quantity) in order to support a successful timber business (Gunter 2000). While the Burns Lake CFP had one of the largest areas of the ten CFPs, many wanted a larger forest land base and AAC, since they felt that the small size constrained their capacity to generate revenue for the community, become competitive, and achieve economies of scale. Increases to the AAC would allow for more flexible marketing opportunities. However, although the initial AAC size (approximately twenty-three thousand cubic metres) was small compared to most industrial tenures, the relatively limited size did not constrain the Burns Lake CFP: it has grown to employ more than three dozen people and generate revenue through harvesting and selling logs to the local mills.

The quality of the Burns Lake CFP's timber profile was well suited to enhanced management. The forest is situated in the sub-boreal spruce zone, and while it supports several tree species of varying age classes, with pine and spruce dominating, it is in the dry interior, with a less varied ecosystem than that of the coast. Nevertheless, the Burns Lake CFP's forest land base has had adequate reserves of mature timber to generate revenue.

A key limitation of the quality of Burns Lake CFP's forest land base has been the MPB outbreak. One of the first operational challenges that the Burns Lake CFP, and later the BCLF, faced was the management of an area-based tenure hit hard by the MPB. In a 2001 survey of community forest participants, completed as part of a review of the Burns Lake CFP experience (McIlveen 2004), respondents felt that the MPB outbreak was both an enabler and a constraint. The managers had planned to horse-log 25 percent of the AAC in order to create more employment through labour-intensive harvesting. The need to manage the forest in response to infestation resulted in having to abandon, at least for the time being, this form of niche production. In order to quickly remove infested pine, the labour-intensive and slower-paced horse-logging plans were replaced by the faster mechanized-logging system. Increasing the short-term AAC has resulted in fewer future jobs provided by the Burns Lake CFP. It has also made the operation vulnerable to lower prices for infested wood.

More recent evidence suggests that the MPB epidemic is both a short-term enabler and a potentially longer-term constraint and that the quality of the forest land base is indeed an important factor in influencing success. The BLCF sought and received approval to log more than twice the amount called for in its initial management plan (BLCF 2011). Initially, in 2000, an AAC of 23,677 cubic metres was awarded to the BLCF. The provincial government has reconfigured the BLCF boundaries to approximately four times its original size, thus attempting to ensure a reasonable AAC for the future (see Table 7.2).[3] According to a former manager, the management team was able to market this

TABLE 7.2   Burns Lake Community Forest Ltd. Activity Summary, 2001–11

| Year | Area harvested (gross ha) | Area harvested (NAR ha) | Volume harvested (m³) | Stumpage paid (C$) | Hectares planted (NAR ha) | Total trees planted | Season |
|------|------|------|------|------|------|------|------|
| 2001 | 98.3 | 97.7 | 47,653.100 | 261,655.64 | 33.6 | 71,490 | Spring |
| 2002 | 160.7 | 149.1 | 63,986.600 | 223,321.84 | 66.6 | 51,255 | Spring |
| 2003 | 189.9 | 175.8 | 69,016.131 | 297,031.08 | 188.5 | 306,115 | Spring |
| 2004 | 293.0 | 263.5 | 89,804.791 | 425,318.60 | 88.7 | 200,000 | Spring |
| 2005 | 1168.7 | 1065.3 | 309,163.791 | 611,859.08 | 252.9 | 404,560 | Spring |
| 2006 | 2883.7 | 2292.4 | 516,658.520 | 1,509,533.11 | 375.3 | 522,000 | Spring |
| 2007 | 1272.4 | 1043.2 | 479,898.216 | 1,273,027.47 | 1125.2 | 1,660,080 | Spring |
| 2008 | 928.3 | 751.6 | 300,297.994 | 520,422.83 | 1674.0 | 1,498,945 | Spring |
|  |  |  |  |  |  | 1,006,940 | Summer |
| 2009 | 1014.5 | 811.2 | 191,529.922 | 127,054.05 | 1689.8 | 2,372,290 | Spring |
| 2010 | 805.4 | 664.3 | 178,968.190 | 183,024.19 | 735.7 | 1,363,776 | Spring |
| 2011 | 116.4 | 108.5 | 102,615.277 | 58,805.41 | 432.0 | 607,760 | Spring |
|  | 8931.3 | 7422.6 | 2,349,592.532 | 5,491,053.30 | 6662.3 | 10,065,211 |  |

Notes: NAR ha = Net Area to be harvested; ha = hectares.
Source: Burns Lake Community Forest, n.d.

wood almost exclusively to local mills at a rate that was profitable. In an interview on 2 February 2012, he stated, "It has been a priority for our group to salvage this dead pine, with the community benefiting through local employment, hundreds of thousands of man-hours of local employment, and contributions to the community – donations totalling over $3.2 million to date."[4]

While this represents considerable profit, it is profit that resulted from an increased AAC due to the epidemic and that may shorten the lifespan of the BLCF. As the former manager stated, one of the biggest hurdles that the BLCF faces is the decrease in the AAC, which was reduced to 100,000 cubic metres annually in 2013. MPB infestation results in dead pine trees and in decaying trees, and less merchantable volume.

As the pine beetle has affected area-based tenures across the province, licence holders – including woodlots, First Nations woodland operations, and other community forests – have received expanded forest access. The province has also recommended a "targeted fertilization program" for the Lakes Timber Supply Area, in which the BLCF is located. The question remains as to whether expanded forest access in the short term will provide enough fibre to sustain area-based operations until supply can be regenerated through fertilization.

## Stage 2: Draw on Community Skills

Previous experience with community involvement in resource management contributed to the community support for the Burns Lake CFP. Given an adequate forest land base, a community's strengths will have a large influence on the success of any community forest initiative, and on British Columbia's CFPs in particular. The literature recognizes that community support is an essential factor (Skutsch 2000; Klooster 2000). The experience of the Burns Lake CFP largely supports this view.

Community support for the Burns Lake CFP was facilitated by the community's general sense of dependence on the forest industry and a belief that through the Burns Lake CFP, the community could gain greater control of the local forest resource. This sense of dependence is reflected in the following comments from a local forester from a 2001 survey (McIlveen 2004):

> Just because it affects so many of us, we understand what happens when there are control issues, when corporations decide what they want to do and then an entire community is affected regardless if we have any feet in the forest industry ... The city slickers down in Victoria don't understand what it is like to be so dependent on the forest resource.

Respondents in the 2001 survey also felt that the Burns Lake CFP process was unique because the community was able to unite behind the common vision of generating jobs and revenue for the community (McIlveen 2004).

The level of support for the Burns Lake CFP was reflected in the degree of volunteerism. The steering committee devoted two years of volunteer work, and approximately $100,000 worth of volunteer time went into the Burns Lake CFP proposal in order to generate public support and input (McIlveen 2004). One respondent to the 2001 survey felt that the tremendous effort put into the proposal created a greater sense of ownership of the process: "We had to go through an incredible process to get to where we are now, which has strengthened us. The other CFPs have been more or less given to the communities and they haven't had to go through the process that we have; there isn't as much ownership of the community forest by other communities"(McIlveen 2004).

Indeed, initial community support and the lack of stakeholder conflict were key factors in developing the proposal and in gaining the confidence of the Ministry of Forests in order to secure the land base. However, once the BLCF was established (i.e., the pilot status was converted to a community forest agreement and the long-term community forest tenure was awarded), community support became less important. Rather, it was the level of forestry and business expertise and capable leadership that contributed to its ongoing success.

The experience of leadership within the Burns Lake CFP, particularly of the elected manager and the board, was vital for the ongoing success of the project (McIlveen and Bradshaw 2009). This is consistent with other studies that have pointed to the role that expertise and leadership play in successful community ventures (Aspen Institute 1996; Litke and Day 1998; Markey and Roseland 1999). Generally, governing boards must be representative of the community in order to be effective. In the case of the Burns Lake CFP, however, this was not so, and it appears that it need not be so for the success of the community forest. While community participation and representation was apparent in the design stage of the initiative, it was not so for the operational stage. The Burns Lake CFP owed much of its initial success to a small group of members who formed the project steering committee and, later, sat on an even smaller board. The board included a former manager of the local forest district; two forestry consultants; a manager from a regional forest company, Babine Forest Products; representatives from two local Aboriginal groups; a representative of the Village of Burns Lake; and a business consultant. While some effort was given to ensure adequate representation, it appeared that the primary aim in establishing the board was to secure individuals with forestry experience and/or business expertise. It was more important that the board be "effective" than "representative" and that the community forest be a self-sufficient business that generated some jobs and profits for the community rather than a project that brought together and involved community members. As one respondent in the 2001 study suggested, "Community forestry tends to be idealized ... This isn't a feel good thing; realistically people are more concerned that it not cost them money" (McIlveen and Bradshaw 2009).[5] The community's support and the level of expertise allowed the board to gain the confidence and support of the Ministry of Forests, a requirement for securing the community forest tenure. Appropriate expertise enabled the Burns Lake CFP to comply with aspects of the provincial regulatory system and to generate some employment and profit. Through the CFP process, the community's awareness of the forest industry and the forest itself increased (McIlveen 2004).

**Stage 3: Comply with the Provincial Regulatory System**
After the land base has been secured and the workforce has been established, the CFP is ready to operate under the Forest Act. As part of the provincial regulatory system, the CFPs must operate under the provincial revenue appraisal system, the system that determines royalty payments to the Crown for the use of timber. Under this system, the Burns Lake CFP, and the subsequent BLCF, have been required to generate a certain level of royalties for the Crown (Bill 34, s. 43.3[d]).

The CFPs, and the BLCF, are subject to woodlot regulations and are bound by the same revenue appraisal system that is applied to larger forest licences. The revenue appraisal system that is applied to larger forest licences is geared towards conventional harvesting of larger clear-cuts and does not reflect the operating costs of smaller-scale forestry. Respondents to the 2001 survey felt that the costs of labour-intensive forestry such as horse-logging, small clear-cuts, and partial cutting were greater than the cost of conventional harvesting such as larger clear-cuts. This system was seen as a disadvantage to a community forest seeking to create unique product identity around smaller-scale methods of harvest. Respondents also felt that the costs of lower-impact forest management – such as narrower roads, the use of skid trails, and intensive stream assessment – were not accounted for in the stumpage calculation. One comment of a survey respondent reflects this concern:

> The stumpage system doesn't reflect the type of harvesting that most of the population wants – selective harvesting, alternative harvesting systems with the least impact on the forest. The community forests are forced to clear cut, which is against the proposal ... It is the economics that determine the management plan and not the other way around. We need more flexibility.

While the Burns Lake CFP has been successful in generating some profit, generating more profit within niche markets and testing innovative forestry management appears to be constrained by the need to supply royalties based on high-volume harvesting. This provincially imposed constraint on innovation simply compounds the problems posed by the severe forest-health concerns in the Burns Lake region and the BLCF's associated obligation to the Ministry of Forests to manage for the MPB.

More recently, there is some renewed optimism for the future of forestry in British Columbia, despite the MPB infestation, which has reduced fibre supply for many years into the future and which co-incided with a collapse in the US housing market, sharply cutting demand for BC lumber products. Optimism stems from the fact that negative forest-health conditions will eventually recede and that the industry has demonstrated resilience in its ability to respond to shifting markets. China accounts for 18.4 percent of British Columbia's commodity exports – a total value of more than $6.2 billion, and exports have increased by an average of almost 20 percent annually from 2005 to 2014. British Columbia has emerged as a leading supplier of softwood lumber to China with 43 percent of total softwood lumber imports representing over $1.4 billion in 2014 (British Columbia, Canada 2015). Furthermore, proximity to the United States continues to be a competitive advantage; it remains the most important market for Canada's forest sector despite the impact of the housing market collapse and economic downturn.  In 2014, softwood lumber exports totaled $8.3 billion, a 16.3% increase over 2013. Further, in 2014, the value of Canada's forest product exports increased by 9.8% over 2013, rising to $30.8 billion from $28.4 billion (Natural Resource Canada 2015).

### Stage 4: Secure Markets and Exist within a Complex Global Environment

CFPs that have secured a forest land base, drawn on the strengths of their community, and complied with the provincial regulatory system will not necessarily be successful unless they can secure adequate markets and exist within a complex global environment. BC's forest economy is primarily based on large-scale, export-oriented production that has historically served very narrow markets (pulp and paper, plywood, dimension timber) in the United States and Europe. The shift to flexible specialization in the forest sector, and the various challenges identified earlier, compound the difficulties facing smaller area-based tenure holders like the BLCF.

Given this large-scale orientation of the forest industry, respondents to the 2001 survey felt that the small size of the Burns Lake CFP and the related low AAC constrained them in generating sufficient revenue and competing within the forest industry. As a result, some respondents felt that accessing niche markets and providing specialty products had the potential to secure greater revenue. They saw potential opportunity in harvesting and marketing traditional botanicals, diversifying to some extent into higher value-added forest products, pursuing non-extractive activities (i.e., tourism), and working with local educational institutions for training opportunities. Some of these avenues have subsequently been developed: for instance, the BLCF has operated a mountain bike park since 2006. However, the BLCF is located relatively far from large urban markets, adding to the challenge of accessing markets for specialty products, and the dry interior forests also produce a narrower range of non-timber forest products than are available from the more varied coastal forests.

Moreover, given the severe MPB epidemic, the AAC has been dramatically increased and the BLCF has thus been forced to harvest more volume than originally intended. The provincial government was initially constrained in its efforts to ease the epidemic because of an on-again, off-again softwood trade dispute with the United States. For the interior region of British Columbia, the influx of infested wood has resulted in an oversupply of logs, which has lowered prices. Hence, the BCLF has to operate within the provincial forest economy and contend with the lower market prices for infested wood. While the BLCF has a guaranteed buyer for its wood through one of West Fraser Timber's mills, it is still obliged to harvest more than current market demand.

The BLCF operates in a complex global environment in which global and local forces influence one another (Taylor and Conti 1997; Featherstone et al. 1995; Swyngedouw 1997; Giddens 2000). The community enjoys increased local autonomy, but it does so in a highly competitive international environment in which low-cost forest products are readily available from Southeast Asia, Russia, and elsewhere. Neoliberal philosophies adopted by Canadian policy makers and others around the world have reduced barriers to trade but have also accelerated competition and reduced the role of state governments, including in resource management.

In British Columbia, the neoliberal approaches, particularly those adopted by the provincial Liberal Party (which has been in power since 2001), have meant a rejection of large-scale, Fordist-style government engagement in the forest sector and rural economic development. In the last twelve years, changes have included the scrapping of the NDP-supported Forest Renewal BC (1994–2002), which channelled tax revenue into provincially managed forest-health and wood-processing research, business-creation programs, and forest-planning strategies;

reduction of funding for the ministry responsible for forests; and greater involvement of private actors in responsibility for forest and community health. Critics of the provincial Liberals have argued that these very steps have left rural British Columbia even more vulnerable to changes (Parfitt 2012). As McCarthy (2005) and Young (2008) have argued, community forests, including the BLCF, also reflect this neoliberal shift. Indeed, as Young (2008) notes, programs designed to encourage community forestry in British Columbia represent a set of strategies designed to "encourage local self-sufficiency and entrepreneurialism at the same time that major corporate actors are 'freed' from traditional obligations to communities, labour, and the environment."

At the same time, this neoliberal shift has not been absolute, and environmental groups, Aboriginals, and resource communities situated in the periphery of both British Columbia and the Global North more generally have been able to effectively resist industrial operations and "remap" ownership and control over space (Hayter and Barnes 2012). Indeed, as part of the re-regulation of the forest sector in British Columbia, community forestry in the province represents the devolution of power and responsibility from the provincial to the local level and a more diverse management of the forest resource. While the forest tenures are embedded in global structures and are social agents who are able to answer to external restructuring from local contexts (Taylor 2000), the global forest industry, in particular, does not recognize this. Thus, two concurrent trends emerge: power has been devolved from the provincial to the local level through the CFPs, but the global economic environment may make these forest-dependent communities more vulnerable.

Significantly for community forest operations like the BLCF, market linkages for diversifying value-added and non-timber forest products suggest that existing global markets are not as responsive to local sustainability initiatives as communities might hope. Indeed, survey respondents expressed interest in accessing "green" markets by pursuing the eco-certification process; however, they also felt that the costs to implement the standards were too onerous and that there was little market value for eco-certified products. Despite this, increasing application of "sustainable production" standards and labels – such as those attained through the Forest Stewardship Council or other bodies – now accounts for the majority of forest lands in active production in British Columbia today (Hayter and Barnes 2012).

## Conclusions: Lessons from the Burns Lake Community Forest Project

British Columbia's community forest initiatives are among the most comprehensive of the many community forestry efforts that have proliferated within North America since the early 1990s. Some of the common elements of community forest efforts are flexibility, innovation, local community support, and a commitment to local benefits (McCarthy 2005). Supporters see community forestry as being "superior to centralized state control" and as something that "will lead to higher levels of economic growth, environmental protection and community stability" (McCarthy 2005, 996). One of the lessons derived from the Burns Lake CFP/BLCF example is that the community's vision of community forestry objectives (i.e., generating more profit within niche markets and testing innovative forestry management) does not necessarily match the reality of external factors (i.e., the need to supply royalties based on high-volume harvesting and the obligation to the Province to manage for the MPB) that translate into economic constraints.

The Burns Lake CFP was able to adhere to many of its initial goals as outlined in its original proposal. It secured a forest land base and timber supply because of the lack of competition in the area. Previous experience with community involvement in resource management contributed to community support for the CFP. The community's support and the level of expertise allowed them to gain the confidence and support of the government officials, which was a requirement for securing the community forest tenure. Subsequently, its capable leadership was vital to its ongoing success. Appropriate expertise enabled the Burns Lake CFP to comply with aspects of the provincial regulatory system and to generate some employment and profit. Through the CFP process, the community's awareness of the forest industry and the forest itself has increased. These factors are largely internal to the community and demonstrate the remarkable potential of the CFPs.

Initial research found that while the provincial regulatory system facilitated the Burns Lake CFP's industrial approach, resulting in employment and revenue, it also constrained it. Despite the transfer of some power and responsibility to the community through the community forest tenure, the Burns Lake CFP was obliged to manage for severe forest-health concerns, and the revenue appraisal system may not recognize the costs of forest management such as labour-intensive harvesting methods, as indicated in the Burns Lake CFP's original proposal. The provincial regulatory system is further complicated by current market prices. Despite the opportunity to generate revenue by selling logs to local mills, the Burns Lake CFP was subject to low log prices, especially for infested wood, and was indirectly affected by softwood lumber duties applied to local processing mills. Moreover, managing for products other than timber, accessing niche markets, and providing specialty products may be difficult. Given the small size and low AAC, the BLCF's efforts at being competitive may be undercut by cheaper and more abundant sources of timber from elsewhere.

The BLCF has functioned within a forest products sector that has experienced widespread restructuring, loss of employment, and dominance by multinational corporations. These changes occurred in British Columbia long before the most recent outbreak of MPB. The BLCF has met many of its goals by locally managing the resources; it has provided local employment; the majority of its harvest has been sold to local processing facilities; and it has generated revenue for the community from the forest through viable local industries such as the Lakes Outdoor Recreation Society, which maintains about thirty recreational campsites as well as trails and parks in the Lakes District (BLCF 2011). The devastating MPB infestation added considerable economic stress to the viability of this community forest, and the BLCF remains vulnerable to a host of external economic pressures. In order to address both present realities and anticipated changes, the BLCF will have to tackle difficult questions regarding how to diversify its economic base, access new and different markets, and cultivate alternative viable industries – all of which require flexibility and time to develop.

Larger firms have an advantage over community forests in grappling with the MPB: they can make strategic investments in some mills while closing others, and they may even be given some additional harvesting rights elsewhere to compensate for the change (British Columbia, Legislative Assembly 2012). Conversely, the community forestry tenure requires communities to manage their local forests, despite the fact that the province's timber supply has dwindled steadily since the MPB epidemic began, a reduction that may last up to fifty years (British Columbia, Legislative Assembly 2012, 1).

The community forest model has garnered considerable interest in British Columbia, with fifty-two active community forests and nine communities in the application process as of 2012 (British Columbia,

FLNRO 2015). There is a perception that through community forestry, communities can manage the environment more effectively and with greater legitimacy and can help rural communities more than the traditional tenures controlled by large corporations. According to the BC Community Forest Association, community forest agreements involve planning processes and requirements that are "flexible enough to accommodate broadly based community objectives and allow for innovative and unconventional forest management practices" (BCCFA 2013), a goal that is reflected in the provincial government's own objectives for community forest agreements (British Columbia, FLNRO 2011). Faced with a severe forest health crisis, however, communities may not be as responsive and flexible as the community forest tenure suggests.

Despite the advantages seen in community forestry, this model of management and control still represents a minor part of BC's land base. As of 2012, a mere 1.5 percent of the AAC was allocated to community forests (British Columbia, Legislative Assembly 2012); thus, the historic industrial forest tenures remain largely unchanged.

Clearly, Burns Lake is a town facing difficult choices, and this makes alternative models such as the BLCF all the more important. Forest-dependent communities remain stuck between two difficult choices – to remain dependent on a rapidly shifting, post-Fordist model of production, which has resulted in greater employment instability, or to move increasingly towards greater ownership over economic decisions related to their forest base while grappling with large-scale problems like the MPB that cannot easily be addressed at the community scale. Community forestry becomes more desirable as a pathway, however, *if* community forest operations are given the opportunity to expand from their current land base and *if* forest managers can successfully identify complementary forms of economic output (e.g., non-timber forest products, tourism and recreation, etc.). Forest-dependent communities have learned not to leave their social and ecological destiny in the hands of the state. Rather, they recognize the need to entrench community control through community initiatives like community forestry.

### Notes

1   The initial pilot licence was signed in 2000; Burns Lake Community Forest Ltd. was awarded Long-Term Community Forest Agreement for a term of twenty-five years in 2004 – the first of its kind in the province (http://blcomfor.com).

2  Edenhoffer and Hayter define fall-down as: "the drop in harvest levels associated with the replacement of high yield old growth forests with lower yielding second growth (141 2013).

3  During 2005 and 2009, because of the necessity of harvesting MPB-infected wood within the BLCF, both the volume harvested and the forest block size increased, thus allowing the BLCF to salvage the deteriorating pine before it lost value. These increases were followed by declines in harvesting. From 2009 to 2010, harvest volumes declined by more than 6.5 percent. From 2010 to 2011, harvests dropped by more than 40 percent (see Table 7.2). Furthermore, after 2013, the AAC is expected to be reduced further to approximately 100,000 cubic metres per year (BLCF 2011).

4  Interview with forest manager, February 2, 2012.

5  Concerning issues of representation in community forestry see also Reed and McIlveen 2006.

## References

Aspen Institute. 1996. "Measuring community capacity building: A workbook-in-progress for rural communities." *Rural economic policy program*. Queenstown, Maryland: Aspen Institute.

BC Statistics. 2014. Statistics by subject, Population Estimates. http://www.bcstats.gov.bc.ca/StatisticsBySubject/Demography/PopulationEstimates.aspx.

BCCFA (British Columbia Community Forestry Association). 2013. "Where are the community forests in BC located?" *BCCFA*. http://www.bccfa.ca/index.php/about-community-forestry/map.

Beckley, T. 1998. "Moving toward consensus-based forest management: A comparison of industrial, co-managed, community, and small private forests in Canada." *Forestry Chronicle* 74 (5): 736–44. http://dx.doi.org/10.5558/tfc74736-5.

BLCF (Burns Lake Community Forest Ltd.) 2011. "Burns Lake Community Forest Ltd annual report." In *Annual Report 2010–2011*, ComFor Management Services Ltd., 14–15. Burns Lake, BC.

Booth, A. 1998. "Putting 'forestry' and 'community' into First Nations' resource management." *Forestry Chronicle* 74 (3): 347–52. http://dx.doi.org/10.5558/tfc74347-3.

Boyd, D. 2003. *Unnatural law: Rethinking Canadian environmental law and policy*. Vancouver: UBC Press.

British Columbia. FLNRO (Forests, Lands, and Natural Resources Operations). 2011, *Government's objectives for community forest agreements.* https://www.for.gov.bc.ca/hth/timber-tenures/community/objectives.htm.

–. 2015. "History of community forests." *British Columbia.* http://www.for.gov.bc.ca/hth/community/history.htm.

British Columbia. Forests. 1998. "Legislation." https://www.for.gov.bc.ca/hth/timber-tenures/community/legislation.htm

–. 1999. "27 communities apply for new tenure under Community Forest Pilot Project." News release, 20 January. http://www2.news.gov.bc.ca/archive/pre2001/1999/1999nr/1999007.asp.

British Columbia. Forests and Range. 2007. *Timber supply and the mountain pine beetle infestation in British Columbia: 2007 update.* Victoria, BC: Ministry of Forests and Range. https://www.for.gov.bc.ca/hfp/mountain_pine_beetle/Pine_Beetle_Update20070917.pdf.

British Columbia. Legislative Assembly. 2012. *Growing fibre, growing value.* Report of the Special Committee On Timber Supply. Victoria, BC: Legislative Assembly. http://www.leg.bc.ca/cmt/39thparl/session-4/timber/reports/PDF/Rpt-TIMBER-39-4-GrowingFibreGrowingValue-2012-08-15.pdf.

British Columbia, Canada. 2015. Trade and Invest British Columbia. Key Sector Opportunities for Exporters. http://www.britishcolumbia.ca/export/key-makets/china/.

British Columbia Mid-Term Timber Supply. 2012. Lakes Timber Supply Area. https://www.for.gov.bc.ca/hfp/mountain_pine_beetle/mid-term-timber-supply project/ Lakes TSA.pdf.

Bullock, R., and K. Hanna. 2012. *Community forestry: Local values, conflict, and forest governance.* New York: Cambridge University Press.

Burda, C. 1999. "Community forestry." Discussion paper. Victoria, BC: University of Victoria.

Burns Lake Community Forest. 2011. BLCF Activity Summary, 2001–11. Report prepared by the Burns Lake Community Forest. Burns Lake, British Columbia.

Clapp, R.A. 1998. "The resource cycle in forestry and fishing." *The Canadian Geographer* 42 (2): 129–44. http://dx.doi.org/10.1111/j.1541-0064.1998.tb01560.x.

Community Forestry Forum. 2000. *US-Canadian-Mexican community forestry.* Shelton, WA: National Network for Forestry Practitioners.

District of Mission. Municipal Forest Sustainable Forest Management Plan. www.mission.ca/wp-content/uploads/mmf-2005-general-presentation.pdf

Edenhoffer, K., and R. Hayter. 2013. "Restructuring on a vertiginous plateau: The evolutionary trajectories of British Columbia's forest industries 1980–2010." *Geoforum* 44: 139–51. http://dx.doi.org/10.1016/j.geoforum.2012.10.002.

Featherstone, Mike, Scott Lash, and Roland Robertson. 1995. *Global modernities.* Thousand Oaks, California: Sage Publishers.

Giddens, Anthony. 2000. *Runaway World.* New York: Routledge.

Gunter, J. 2000. "Creating the conditions for sustainable community forestry in BC: A case study of the Kaslo and District Community Forest." Master's thesis, School of Resource and Environmental Management, Simon Fraser University, Burnaby, BC.

Hayter, R. 1976. "Corporate strategies and industrial change in Canada's forest industry." *Geographical Review* 66 (2): 209–28. http://dx.doi.org/10.2307/213581.

–. 2000. *Flexible crossroads: Restructuring of British Columbia's forest economy.* Vancouver: UBC Press.

Hayter, R., and T. Barnes. 2012. "Neoliberalism and its geographic limits: Comparative reflections from forest peripheries in the Global North." *Economic Geography* 88 (2): 197–221. http://dx.doi.org/10.1111/j.1944-8287.2011.01143.x.

Hayter, R., T. Barnes, and M. Bradshaw. 2003. "Relocating resource peripheries to the core of economic geography's theorizing: Rationale and agenda." *Area* 35: 15–23.

Hughes, J., and R. Dreyer. 2001. *Salvaging solutions: Science-based management of BC's pine beetle outbreak.* Report prepared for the David Suzuki Foundation. Vancouver: Forest Watch of BC and Canadian Parks and Wilderness Society – BC Chapter.

Klooster, Dan. 2000. "Community forestry and tree theft in Mexico: Resistance or complicity in conservation?" *Developmental Change* 31(1): 281–305.

Litke, Stephen, and J.C. Day. 1998. "Building local capacity for stewardship and sustainability: The role of community-based watershed management in Chilliwack, British Columbia." *Environments* 25(2–3): 91–109.

M'Gonigle, M. 1997. "Reinventing British Columbia: Towards a new political economy in the forest." In *Troubles in the rainforest: British Columbia's forest economy in transition*, edited by T. Barnes and R. Hayter, 37–52. Canadian Western Geographical Series No.33. Victoria, BC: Western Geographical Press.

M'Gonigle, M., and B. Parfitt. 1994. *Forestopia: A practical guide to the new forest economy.* Madeira Park, BC: Harbour.

Marchak, M.P., S.L. Aycock, and D.M. Herbert. 1999. *Falldown: Forest policy in British Columbia.* Vancouver: David Suzuki Foundation and Ecotrust Canada.

Markey, S., G. Halseth, and D. Manson. 2008. "Challenging the inevitability of rural decline: Advancing the policy of place in northern British Columbia." *Journal of Rural Studies* 24 (4): 409–21. http://dx.doi.org/10.1016/j.jrurstud.2008.03.012.

Markey, S., and M. Roseland. *Building self-reliance in BC forest-based communities.* Paper presented at the Society for Human Ecology 10th Annual Conference, "Living with the Land: Interdisciplinary Research for Adaptive Decision Making, May 1999. Vancouver: Simon Fraser Unviersity.

McCarthy, J. 2005. "Devolution in the woods: Community forestry as hybrid neoliberalism." *Environment and Planning A* 37 (6): 995–1014. http://dx.doi.org/10.1068/a36266.

–. 2006. "Neoliberalism and the politics of alternatives: Community forestry in British Columbia and the United States." *Annals of the Association of American Geographers* 96 (1): 84–104. http://dx.doi.org/10.1111/j.1467-8306.2006.00500.x.

McGillivray, B. 2010. *Geography of British Columbia: People and landscapes in transition.* Vancouver: UBC Press.

McIlveen, K. 2004. "British Columbia's community forest pilot projects: Can a localized trend survive in an increasingly globalized forest sector?" Master's thesis, Department of Geography, Simon Fraser University, Burnaby, BC.

McIlveen, K., and B. Bradshaw. 2005–6. "A preliminary review of British Columbia's community forest pilot project." *Western Geography* 15–16: 68–84.

–. 2009. "Community forestry in British Columbia, Canada: The role of local community support and participation." *Local Environment: The International Journal of Justice and Sustainability* 14 (2): 193–205. http://dx.doi.org/10.1080/13549830802522087.

McKenna, B. 2010. "BC lumber expediency rotting Canada's fair-play reputation." *Globe and Mail*, 14 November.

Natural Resource Canada. 2015. *Exports of forest products. What has changed and why?* http://www.nrcan.gc.ca/forests/industry/economic-benefits/16558.

Parfitt, B. 2005. *Battling the beetle: Taking action to restore British Columbia's interior forests.* Vancouver: Canadian Centre for Policy Alternatives.

–. 2012. *BC mills are quickly running out of wood.* Vancouver: Canadian Centre for Policy Alternatives.

Perlin, I. 1991. *A forest journey: The role of wood in the development of civilization.* Cambridge: Harvard University Press.

Reed, M., and K. McIlveen. 2006. "Toward a pluralistic civic science? Assessing community forestry." *Society and Natural Resources* 19 (7): 591–607. http://dx.doi.org/10.1080/08941920600742344.

Robson, R. 1996. "Government policy impact on the evolution of forest-dependent communities in Canada since 1880." *Unasylva* 47 (186): 53–59.

Skutsch, M. "Conflict Management and Participation in Community Forestry." 2000. *Agroforestry Systems* 48,2: 189-206.

Stiven, R. 2000. *A study of social involvement in forestry in British Columbia and Quebec.* A Millennium Award Study supported by the Millennium Forest Scotland Trust, the Millennium Commission, and the Forestry Commission.

Swyngedouw, E. 1997. "Neither global nor local: 'Glocalization' and the politics of scale." *Spaces of globalization,* edited by K.R. Cox, 137-66. New York: Guilford.

Taylor, P.L. 2000. "Producing more with less? Community forestry in Durango, Mexico in an era of trade liberalization." *Rural Sociology* 65 (2): 253–74. http://dx.doi.org/10.1111/j.1549-0831.2000.tb00028.x.

Taylor, M., and S. Conti. 1997. *Interdependent and uneven development: Global-local perspectives.* Aldershot, UK: Ashgate.

Teitelbaum, S., T. Beckley, and S. Nadeau. 2006. "A national portrait of community forestry on public land in Canada." *Forestry Chronicle* 82 (3): 416–28. http://dx.doi.org/10.5558/tfc82416-3.

Young, N. 2008. "Radical neoliberalism in British Columbia: Remaking rural geographies." *Canadian Journal of Sociology* 33 (1).

# Chapter 8    Searching for Common Ground
## An Urban Forest Initiative in Northwestern Ontario

*James Robson, Mya Wheeler, A. John Sinclair,*
*Alan Diduck, M.A. (Peggy) Smith, and*
*Teika Newton*

Kenora, in northwestern Ontario, is a former resource town where First Nation (Grand Council of Treaty #3) and municipal governments, along with the communities they represent, are striving to create new economic opportunities through sustainable development, while dealing with problems from the past and present. It is a scenario being repeated across other parts of Canada (Eeyou Istchee Framework Agreement 2011; NSRCF Action Plan 2011). Symbolic of these efforts in Kenora is Common Ground, an emerging governance initiative relating to four hundred acres of urban community forest on Tunnel Island and Old Fort Island (Waa' Say' Gaa' Bo'), a project that requires collaboration between the area's Anishinabe and settler populations.[1] The initiative is seen by some as representing a renewed local commitment to long-standing treaty rights and obligations.

The literature informs us that successful collaborative resource management, including community forestry, is associated with democratic processes that are transparent, collaborative, and accountable, and in which active communication plays a key role (Kooiman 2003; Meadowcroft 2004; Marshall 2009; Berkes 2010; Davidson 2010). This is particularly relevant in the administration of potentially contested common land (Steins and Edwards 1999), which holds a special identity for people and is also a local material resource (Rodgers et al. 2011). Previous work has shown the power of values, feelings, and sense of place connected with such land, all of which exist at both individual and community levels and can influence people's desire to participate in governance (Hay 1998; Tuan 2004; Massey 2005; Beckley et al. 2007). The strength of one's sense of place is often connected to regular physical contact with the place in question and is enhanced through personal reflection and public voicing of perceptions and knowledge (Relph 1976; Williams and Stewart 1998; Yung, Freimund, and Belsky 2003; Sampson and Goodrich 2009).

While our knowledge of urban forests and how people relate to and value such forests has increased in recent decades, most of what we know relates to rural and hinterland forests (Bengston 1994; Beckley et al. 1999; Tindall 2003; Moyer, Owen, and Duinker 2008). Urban forests in Canada generate a myriad of ecosystem, social, and economic services and values that are distinct from those associated with more rural forest environments, and considerable attention is now being paid to urban forest management (Nowak et al. 2001; Canadian Urban Forest Network 2004; Ordóñez and Duinker 2010). The research indicates that urban forest management is crucial; entails a consideration of form, function, diversity, and multiple values; and requires a governance framework that is democratic, multi-institutional, collaborative, and accountable. Such a framework empowers local users to be active agents in shaping the implementation of governance systems (Pinkerton et al. 2008; Higgins, Dibden, and Cocklin 2008; Davis 2011).

Although research focused on urban community forests is limited, a recent study of Ontario's county, municipal, and conservation authority forests (Teitelbaum and Bullock 2012) revealed that most such forests display some of the basic attributes commonly associated with community forestry, including: participatory governance, local benefits, and multiple forest use. Yet these are also forests that differ from many community forests on Crown land because of their emphasis on non-timber activities such as recreation, education, and conservation. Building on this work, in this chapter we combine insights from the literatures on environmental co-management, urban forest values, and sense of place with empirical results from interviews and surveys with people in Kenora. In doing so, we identify and discuss the opportunities and challenges facing the development of a collaborative governance regime for the Tunnel Island-Old Fort Island urban forest (see Robson et al. 2013). In dealing with non-extractive forms of forest use taking place within city limits and founded on an emergent partnership of municipal and First Nations governments, we hope to add to the understanding of community forestry models. Despite the array of experiences in Canada (Teitelbaum, Beckley, and Nadeau 2006), few community forest initiatives encompass this distinctive set of features. In addition, by giving a voice to the thoughts and concerns of local people and identifying citizen priorities for valuing, benefiting from, and administering such urban forests, these findings are of relevance to all communities of people, rural or urban, who act as stewards of local forest resources.

## Study Site and Common Ground Initiative

Kenora is located on the northern shores of Lake of the Woods, at the point where the lake flows into the Winnipeg River. The city is at a crossroads of historic trade routes: north to south, via the waterways used for trading fur and other natural resource products, and east to west over the past century or so, by means of the Trans-Canada Highway and the Canadian Pacific Railroad (Forest Capital Report 1999).

Today, the city and surrounding region are in a period of social and economic transition. The city's pulp and paper mill, one of the region's major employers, closed a decade ago. Local demographics have also shifted, with many members of surrounding First Nations taking up residence in Kenora (Wallace 2013). These changes have not only increased the ethnic diversity of Kenora but are encouraging the diversification of local and regional economic activities.

The resource governance initiative called Common Ground seeks to bring together Kenora's Aboriginal and settler populations to collaboratively manage just over four hundred acres of urban forest that lie close to the heart of downtown. Figure 8.1 shows the location of this land in relation to four of the five Common Ground partners: the City of Kenora and three adjacent First Nations – Wauzhushk Onigum, Obashkaandagaang, and Ochiichagwe'Babigo'Ining – which have longstanding links to these forest lands and surrounding waterways. The fifth partner, Grand Council of Treaty #3, is the historic and current political government of twenty-eight local Anishinabe communities within the Treaty #3 area (Grand Council Treaty #3 2011).

Both Tunnel Island and Old Fort Island exhibit the general characteristics of mixed Dry Oak Savannah and Dry White Pine Savannah vegetative communities (Forest Capital Report 1999). Like much of this region, tree species include white and red pine, white and black spruce, jack pine, poplar, and burr oak (a rare occurrence in Ontario). Old Fort Island is of particular interest because of its abundance of red pine and white pine, two species that form the focus for Ontario's old-growth policy (Ontario Ministry of Natural Resources 2003). Wildlife is abundant, with substantial bald eagle and white-tailed deer populations reported.

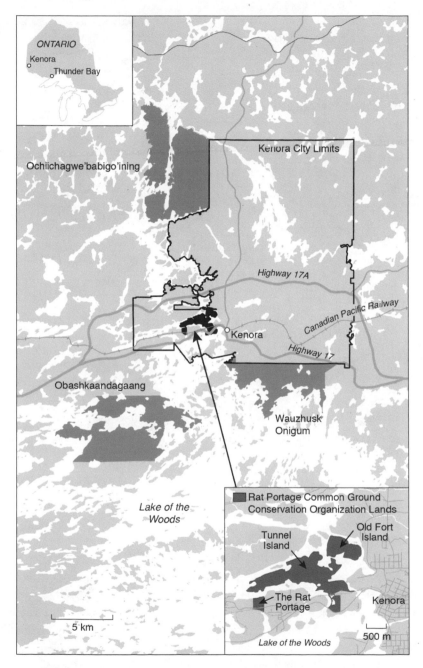

**Figure 8.1** Location of the Common Ground member communities and Tunnel Island-Old Fort Island lands

*Source:* Adapted from Sheldon Ratuski (2014), Common Ground Research Forum, with data sets retrieved from http://www.geobase.ca. Adapted by Eric Leinberger.

Of great cultural significance to local First Nations, these islands were inhabited and actively used by Aboriginal people until the late 1960s and continue to be used for occasional ceremonies and other spiritual activities (Wheeler et al. 2016). It was during the latter decades of the twentieth century that the larger and more accessible Tunnel Island began to be accessed by local city residents, who use it predominantly as a recreational space for walking, dog walking, mountain biking, and so on, with such use having increased markedly over the last ten to fifteen years (Robson, Sinclair, and Diduck 2015). Any public use of these urban forest lands, however, was de facto throughout the twentieth century, since property rights during this period were held privately by the series of owners and operators of the nearby Kenora paper mill.[2]

When the last of these owners, Abitibi-Consolidated, closed the mill down in 2005, the private sale of the company's assets on Tunnel Island and Old Fort Island was problematic because of the above-mentioned cultural status and de facto usage. Consequently, Abitibi gifted the lands to the City of Kenora and Treaty 3 governments on the proviso that a joint management corporation was established and that, once appointed, this body be given legal responsibility for administering both islands on behalf of all First Nation and non-First Nation beneficiaries. Taking the name of Rat Portage Common Ground Conservation Organization (RPCGCO), the corporation was legally constituted in 2008.[3] As of October 2015, the process of developing a vision and framework for collaboratively governing the urban forest was still in its formative stages and not yet operational.

## Study Methods

The empirical data for this chapter are drawn from two field studies undertaken by Wheeler Wiens (2011) and Robson et al. (2013). The first, conducted in fall 2010, examined how users of Tunnel Island and Old Fort Island relate to and value the area, to understand their sense of place (Wheeler Wiens 2011). The principal method involved twenty-five semi-structured interviews with local residents who had spent time on the islands. Participants included both men and women from a broad range of cultural groups (Euro-Canadian, Métis, and Anishinabe). Interview questions derived from the sense-of-place literature elicited people's connection to, perspectives about, and visions for, the urban forest on Tunnel and Old Fort Islands. Three focus groups, involving twelve of the twenty-five participants, were also conducted. The focus groups incorporated sharing circle methods, such as the use of a "talking stone" (Nabigon et al. 1999; Rothe, Ozegovic, and Carroll 2009), designed to elicit discussion in a group setting and bring collective elements to the fore.

The second study (Robson et al. 2013) focused on how local people understood *common ground*, the term chosen to signify the collaborative governance initiative affecting the urban forest on Tunnel and Old Fort Islands. Thirty-two semi-structured interviews were conducted from June to September 2011. These interviewees, who we refer to as "those in the know," can be categorized as follows:

- active members in the Common Ground governance initiative
- councilors from municipal and Treaty 3 governments
- representatives from the three partner First Nations
- representatives from community development institutions
- educators
- members of the local business community; and
- local media representatives.

A second set of interviewees consisted of members of the "general public" who participated in thirty-one sidewalk interviews conducted at the end of September 2011. Across these two groups ("those in the know" and "general public"), a total of thirty women and thirty-three men participated in the research, of whom sixteen were Anishinabeg, five identified as Métis, and forty-two were of Euro-Canadian descent.

## Study Findings

We begin our analysis by looking at connections that people have with Tunnel and Old Fort Islands. The interviews revealed varied values, perspectives, and visions regarding these places. We then consider local people's perceptions of the Common Ground initiative and of the term *common ground* more generally. This aspect of the research opens a line of inquiry into broader issues concerning cross-cultural collaboration and governance. Throughout our discussion, we use direct quotations from participants to highlight key results. The names provided are pseudonyms but do convey the interviewees' gender. The quotations are representative of what the majority of research participants told us, unless we note otherwise.

### Connection to Tunnel Island and Old Fort Island

Participants' connections with the urban forest on Tunnel and Old Fort Islands were founded on several intersecting factors, including physical experiences of the area, local cultural contexts, and personal world views. We grouped the connections into five types: 1) perceptions of the uniqueness of the islands; 2) feelings of being connected with nature and community; 3) feelings of enhanced individual and community health; 4) hopes regarding sharing of the islands; and 5) visions for the future of the islands, including concerns about becoming disconnected from these lands.

Many participants viewed the forest as "unique," and this perception enhanced their connection to the islands. Many people valued the forest being in the "heart of the city" (Laura). Others talked about the age of the trees themselves: "One thing that really sticks out in my mind was the fact that there is so much old growth forest out there and being so close, and being right in town ... old growth forest, I mean there are three hundred-year-old pines on Tunnel Island" (Harry).

Most people also spoke of the opportunities that the forest provided to reconnect with nature, care for natural environments, and be in touch with one's community. Having Tunnel Island and Old Fort Island close by brought to participants a deep sense of satisfaction and pride about having a place set aside for cultivating these benefits. One participant said, "There's an opportunity for people to get to know each other, you know, in our busy world ... a place where we can be in tune with nature, because to me, I look at it as ... it is Tunnel Island, but to me it's about Turtle Island, about the people" (Melvin). Moreover, the forest promoted positive self-identity, community engagement, and caring for where one lives. As one participant noted, "It should be an area that's used as an anchor to bring people together" (Jon).

The people we spoke with also saw the forest as a place that promotes individual and communal health or well-being. All participants noted how the forest made them feel healthier and more whole. Visiting the area, either individually or with family and friends, brought a sense of rejuvenation for many participants – a sense that was almost spiritual, in some cases.

While most participants viewed Tunnel and Old Fort Islands as being common to all and of benefit to the wider community, there was a lack of consensus as to how the land should be shared. For some, the islands provide a clear example of shared space, both in the past and the present, whereby the "power of the place" was enough to bring people together. Furthermore, the land, as a place "to be respected" (Elaine), would ensure that people cared for it. However, others were less able to envisage the land ever being shared in a way that would respect all users' preferences and purposes. These participants expressed a concern that adopting one group's interpretation of place or preferred usage would exclude others: "I think the danger of interpreting [place] is whose story do you tell? Because everyone that comes there [to Tunnel Island] probably experiences it in a different way and to impose then an interpretation of that place diminishes, I think, the stories of other people" (Ophelia).

Evidence of this concern was especially apparent in people's visions of the future, which included consideration of how to "take care" of Tunnel Island and Old Fort Island in a way that enhances experiences on the islands while keeping the land "as it is" (Gerald). We categorized visions for the islands into four broad themes: economic value, respect for the community, community inclusion, and fear of disconnection. The first three are related to participants seeing the future of the land as a place where the community is included, valued, respected, and possibly able to benefit from increased tourism. However, the fourth theme, fear of disconnection, highlights a concern held by numerous participants about being excluded if and when formal management decisions are made. In particular, participants expressed anxiety about being refused access to these lands or having their access restricted. They shared the belief that if access was denied or limited, their sense of place would change to reflect this disconnection, something that would lead to damaging consequences for both themselves as individuals and the community at large.

The urban forest on Tunnel and Old Fort Islands also featured in stories about how change affected the way people constructed their sense of place. For the research participants of Aboriginal descent, memories of both islands functioning as seasonal and annual gathering places contrasted with the significant changes in culture and lifestyle that they have experienced in recent times. While inclusivity concerning the land was expressed – "that's how our people have always been, they've always wanted to share ... we've always wanted to be inclusive to everybody and we are still like that" (Pauline) – many Aboriginal participants felt resentment towards city authorities for no longer allowing them to use the islands as they once had. The long-term and intergenerational attachment held by many Aboriginal participants was a key reason why many wished to be involved in decisions about how Tunnel and Old Fort Islands are accessed and used. Consequently, there was dismay over a lack of involvement to date, a feeling that was particularly acute among Métis participants:

> There's lots of talk about First Nations and the City of Kenora and not mentioning the Métis, and it seems like we're being left out, but I know that the Métis are part of it ... we all own it. It's not just First Nations people and Kenora, you know the City of Kenora, they have to look at it in a broader perspective and say yes, we are all owners of this land. Period. And that's where it needs to come from. (Melvin)

Such concerns have become apparent since the Common Ground initiative began in 2004, largely because participants felt that they have been "left in the dark" with regard to current plans for the two islands. Furthermore, they talked about being uncertain, even borderline suspicious, as to what might have been decided without their knowledge or input: "Like, is somebody at some point going to say, 'Okay, well nobody is getting their act together to do anything about it [Tunnel Island] so, you know, the City's broke so let's just sell it,' because it's been sitting in limbo ever since it was gifted from Abitibi" (Shane).

**Perceptions of "Common Ground" and Cross-Cultural Collaboration**
We asked people in Kenora what *common ground* signified to them and whether they thought that the wider community shared in the visions and perspectives that they expressed. In doing so, we were in a position to speculate, in a more informed manner, about the degree to which the term actually facilitates collaboration locally.

Interview data indicated that *common ground* was used in Kenora in reference to governance initiatives as early as 2000 and 2001. This usage pertained to talks between the city's mayor and the Grand Chief of the Grand Council of Treaty #3. Since then, it has been used to front several other (and seemingly separate) initiatives, including the emerging collaborative resource governance initiative concerning Tunnel and Old Fort Islands (see Table 8.1).[4]

The interviews made clear that despite its varied local adoption, there had been no widespread and/or formal discussions as to what the term was meant to signify. Our interviewees noted three broad meanings or understandings that were in circulation in the community. One popular understanding focused on the physical, of a place that is used and shared by many. This sense of the term is one that is interchangeable with what common (or shared) land would mean to many

TABLE 8.1  Usage of the term *common ground* among local
initiatives in Kenora

| Year(s) of use | Initiative | Context |
| --- | --- | --- |
| 2000–1 | Common Land, Common Ground | A series of meetings between Treaty 3 governments and the City of Kenora to identify areas of mutual interest |
| 2004–present | Tunnel Island Common Ground Project* | Efforts to govern Bigsby's Rat Portage, Tunnel Island, and Old Fort Island collaboratively as a shared resource |
| 2005–present | Common Ground Storytelling Series* | An annual storytelling event hosted by the Lake of the Woods Museum for citizens of Kenora to tell stories about place and people |
| 2006–present | Common Ground Feasts | Held each spring and fall on Tunnel Island and open to the public. |
| 2010 | Finding Common Ground through Creativity | A collaborative project aiming to bridge the gap between Aboriginal and non-Aboriginal people through participatory art |
| 2010–15 | Common Ground Research Forum | A university-community research alliance developing research projects to generate information to aid the Rat Portage-Tunnel Island project |

* See Wallace (2013) for a detailed description of this "common ground" initiative.

people. Approximately one-third of interview participants understood the term in this way. In contrast, just over a quarter of each group of interviewees understood the term as a metaphor for building and developing relationships. This is what we have called the "big picture" connotation of the term, and is one shared by several different initiatives using "Common Ground" in their name (Table 8.1). The third meaning in circulation is best described as an integrated integrated understanding of the term, where *common ground* is both about the land that people share or have common ties to and about the relationships between people, particularly across cultures, but not exclusively so.

This broader understanding of *common ground* was shared by two-fifths of our "those in the know" participants. Some among this group had clearly taken time to think about what *common ground* meant to them and provided well-developed responses. This was especially the case among those heavily involved in one or more of the Common Ground initiatives. As one participant explained, "[It's a space] where people can come together and have a feeling of comfort to engage in dialogue, resolve issues, make plans to move forward ... a neutral comforting area that can facilitate conversation" (Roy).

In terms of commonalities and differences along ethnic lines, a consistent theme to emerge among Anishinabe respondents was the linking of *common ground* to treaty and what they considered to be the original spirit and intent of that agreement: "[It was] originally based around the land, and it still is, it is about sharing land and resources, [which are] then the basis for relationship-building" (Craig). This coincided with the views of key players involved in the Common Ground initiative regarding collaborative governance of Tunnel and Old Fort Islands.[5]

Other Anishinabe respondents, however, remained suspicious and cynical of this same initiative. Some viewed the project as a politically savvy land grab: "[It means] encroachment into our treaty areas, it's another way of sneaking in without going through the whole Treaty 3 assembly" (Kevin). Still others voiced concerns about how earlier decisions regarding Tunnel and Old Fort Islands had been made: "It's just a name, you can call it what you want. I'm just interested in what's there and the question of what we are going to do with it. It's the principle of what took place and how it came to be what it is now" (Albert).

Given the connotations of *common ground* held by "those in the know," which broadly signified sharing and inclusivity, we were interested to see how this compared to the views of the general public. Our sidewalk interviews showed that fewer than half of the respondents knew anything about the specific governance initiative concerning Tunnel and Old Fort Islands. In general, the responses from the public illustrated a less nuanced understanding, either focusing on denotations of the two words (*common* and *ground*) that make up the term or expressing a diversity of connotations, such as "making something more of a well-known public area" (SR [Sidewalk Respondent] 007) or "dealing with social division, setting the boundaries of community in terms of property" (SR 004). Not one member of the general public explicitly understood the term as both a reference to shared land and a metaphor for building relations. Regarding differences in people's broad understanding along ethnic lines, while a majority of Euro-Canadians couched their responses in terms of access to and development of a physical piece of land, the views of First Nation participants often carried a clearer social dimension, with a focus on relations between people.

When asked about cross-cultural collaboration, roughly three-quarters of the sidewalk interview respondents were positive about the idea of the different cultures in Kenora working together to improve relations among cultural groups. A number of Euro-Canadian and Métis respondents were very enthusiastic, often making reference to "not wanting to go back to how things were" (SR 015) and saying that "there has not been enough of that in the past" (SR 004). Others were less convinced – "It will be good if it works out, but there is always a danger it could cause more problems" (SR 010) – and some expressed rather more discriminatory tones – "That's great if they want to share things but it involves give and take and I'm skeptical about how it would work out here" (SR 002).

## Discussion

Not one of the communities in Kenora is currently dependent on the urban forest at Tunnel Island and Old Fort Island or on its biodiversity for livelihood support, and none of these lands have, to date, been a legal commons. Nevertheless, historical access rights for First Nations, enhanced accessibility for city residents, and both differentiated and shared processes of creating place have endowed this area with great significance for the communities it serves, in part because the land and its resources hold great promise (economic and otherwise) for the Kenora region.

Our findings make clear the strong connections that local people have to Tunnel and Old Fort Islands, which clearly form a dynamic place within the community. People are connected not only via a physical geographic location but also through social and cultural discourse and personal identification with the islands. The results also show the diversity of benefits that accrue from cultivating connection to place, consistent with other studies in the field (Agnew and Duncan 1989; Hay 1998; Casey 2001; Tuan 2004; Smaldone, Harris, and Sanyal 2005). Furthermore, the results highlight the overall significance of urban forests to local populations and confirm that users attach multiple and often divergent values to such forests and have an interest in being active agents in their management (Canadian Urban Forest Network 2004; Pinkerton et al. 2008; Ordóñez and Duinker 2010; Davis 2011). Many participants noted their concerns about being excluded from management decisions and disconnected from place. In the minds of many, the urban forest on Tunnel and Old Fort Islands is and should remain a community forest in the most basic sense of that term: a public forest area managed by the community for the benefit of the community (Baker and Kusel 2003; Teitelbaum, Beckley, and Nadeau 2006).

For some users, their connection to the islands is a spiritual one, while for others, the forest is a place to respect and learn about nature and community. The responses of interviewees show that these different views impact how people think the forest and land should be governed. People's connections and perspectives are framed by their activities on the islands and tied to the emotions they feel when spending time there. These results reflect the power of place (Massey 1993, 2005; Latour 1999; Cruikshank 2005; De la Cadena 2010), whether through relationship-building experiences with family and new friends or acquaintances or through individual experiences that help create new understandings of one's relationship to place. The results also confirm the importance of community forestry's emphasis on managing for multiple uses, whereby place shapes formative and diverse experiences, values, aspirations, and desires and expectations related to forest use.

These strong connections to place stood in contrast to the participants' understandings of *common ground*. Interviews with "those in the know" and others in the community revealed multiple uses of, and meanings for, the term. "Common ground" has become an umbrella term manifest in an assortment of practical initiatives, including the one pertaining to collaborative governance of the urban forest on Tunnel and Old Fort Islands. However, the various meanings underlying the term have rarely been publicly articulated and discussed, at least not locally. While there may be a presumption that people will have similar understandings of the term *common ground*, our findings show that its multiple uses in the region have resulted in different meanings. Consequently, the term is taking centre stage in promoting an ideology of collaboration that, as our street survey shows, is not always recognizable among the wider Kenora public.

From a governance perspective, if connection to and engagement with place is to be maintained, it is important to examine shared perspectives of that place, as well as visions for its future. In addition, the causes and effects of "loss of place" (Relph 1976; Billig 2005; Carter, Dyer, and Sharma 2007) are important considerations. In one study on displacement, people's connections were severed either by being physically denied access or by not having their voices heard regarding development of the area (Carter, Dyer, and Sharma 2007). This type of loss has occurred previously on Tunnel and Old Fort Islands. Aboriginal people, for example, were told by local authorities to permanently leave their camps and seasonal meeting spots back in the 1960s. More recently, we have heard both Aboriginal and non-Aboriginal people express feelings of exclusion and anger from having been shut out (in their view) from decision-making regarding the islands. Such findings show the importance of both physical access and inclusion in governance for maintaining sense of place and suggest that such factors are of greater importance to local people than having actual property rights (as Lawrence, Molteno, and Butterworth [2010] found regarding community wildlife sites in the United Kingdom). The fear expressed by many regular users was that the urban forest, which currently functions as a de facto recreational and spiritual commons, would become overregulated as governance becomes formalized (Robson, Sinclair, and Diduck 2015).

Studies conducted elsewhere suggest that a lack of local understanding of the vision that underpins the Common Ground initiative on Tunnel and Old Fort Islands may affect the chances for successful collaborative governance. Such research highlights the need for deliberative processes and transparency in communication (Meadowcroft 2004; Wilding 2011), with governance actors being aware of the perspectives and conceptualizations of other affected individuals (Steins and Edwards 1999; Kooiman 2003). These findings suggest that when a term like *common ground* begins to circulate and is wielded to accomplish goals, it is important to be clear about how it may be understood by different people. For instance, Wight (2005) has shown how planning that fails to pay heed to cultural context or the social construction of meaning is restricted in its ability to shape the building of cross-cultural relations and discourse. Furthermore, such an approach limits the building up of local institutional capacity via discussion of alternative interpretations of reality (Healey 2006). In addition, without broad public engagement involving diverse perspectives, the possibilities for social learning (Keen, Brown, and Dyball 2005; Blackmore 2007; Sinclair, Diduck, and Fitzpatrick 2008) about place and community through, for example, participatory (or civic) science endeavours in community forests (Bagby and Kusel 2003; Reed and McIlveen 2006), are limited to a small group of people rather than a wider-ranging public.

If governance processes are developed for Tunnel and Old Fort Islands, it appears crucial that the community at large have the opportunity to participate. If planning followed a communicative model (e.g., Friedmann 1987; Healey 2006), for example, then Aboriginal and non-Aboriginal partners could be engaged in jointly establishing and renewing sense-of-place connections. They could also be engaged in collaboratively developing the primary connotation of *common ground* associated with the islands. Such outcomes could help construct shared experiences and meanings that transcend social and cultural differences. It is through these kinds of efforts that the crucial relationship between bonding social capital (shared interests and networks that hold cultural groups together) and bridging social capital (shared interests that exist between cultural groups) can be strengthened (Eames 2005). If this does not take place, then the danger exists that *common ground*, as applied in Kenora, would simply come to reify an ideal set of social relations employed by those in a position of power, with local people unable to find a meaningful voice to express their everyday experience. Moreover, our results show that people want their voices heard. Numerous participants were deeply concerned about losing their positive connections to Tunnel and Old Fort Islands. It is clear that if governance decisions either restrict access or put too many rules in place about what people can or cannot do on the islands, there will probably be protest and conflict – the very antithesis of what *common ground* denotes for many residents of the Kenora region.

Ultimately, our findings suggest that cross-cultural engagement is entirely practicable in the Kenora area, given the public's deeply felt connection with Tunnel and Old Fort Islands and a resonance with the philosophy of *common ground*. The use of the term to signify multiple initiatives may confound its meaning, yet all such usages share a desire to improve relations among the cultures that inhabit the Kenora area and hold connections to the urban forest on Tunnel and Old Fort Islands. Indeed, the dominant strand that connects people's understanding of *common ground* concerns this powerful idea of sharing and inclusivity, about which there was a great deal of positivity among the research respondents. This suggests a strong platform on which to build.

## Conclusion

The urban forest on Tunnel and Old Fort Islands has become the focal point for an experiment in collaborative governance between municipal and First Nation governments. In many respects, the experiment approximates what Nelles and Alcantara (2011) called a decolonization agreement between municipal and First Nations authorities. Such agreements recognize that First Nation signatories historically occupied lands that are now under municipal jurisdiction, establish long-term cooperative relationships, and represent commitments to build equal and respectful relationships between the parties.

Planning for Tunnel and Old Fort Islands brings together a diverse citizenry in the governance of a place of both contest and possibility – one that constitutes a recreational space, shared resource, and political landscape where ideals of cross-cultural relations, social cohesion, and nature are being tried and tested. Whatever governance regime emerges, it will have potential to advance notions of community forestry in the Kenora region more broadly. As Baker and Kusel (2003) argue, the very essence of community forestry is the restructuring of relations between forest-dependent people and communities and the forests they depend on, in a manner that advances equity and promotes investment in natural and social capital. In several ways, the Tunnel Island and Old Fort Island governance initiative brings together the four principles – participatory governance, rights, local benefits, and ecological stewardship – around which community forestry practices in Canada are situated (Teitelbaum, introduction to this volume). As an urban forest, it also provides a new avenue for the participation of Aboriginal people in forest management (Beaudoin 2012), one that is likely to become more common as demographics shift and more Aboriginal people move off reserves to reside in small cities such as Kenora.

However, numerous practical questions remain regarding the collaborative governance model being contemplated for Tunnel Island and Old Fort Island. For instance, will mechanisms be developed that are able to resolve tensions among cultural values and between recreational and spiritual uses of the land, and among the various governance priorities that are sure to emerge? It remains to be seen how the joint management body, the Rat Portage Common Ground Conservation Organization (RPCGCO), crafts and implements its mandate. Despite its formation seven years ago, the RPCGCO has yet to begin meeting and functioning as an active entity. Having a slow start could become problematic, since the community forestry literature suggests the importance of timely definitions of goals and objectives and early formalization of governance arrangements (Bullock, Hanna, and Slocombe 2009). The relationships that connect law, governance institutions, and customary usage are particularly critical given the multicultural setting; the governance rules for the urban forest will need to reflect the nature of both Aboriginal and non-Aboriginal resource use, place connection, and valuation. In terms of rule-making, it is those who use the resource who should ideally take a leading role in the crafting and enforcement of new institutional arrangements (Ostrom 1990), reflecting Krogman and Beckley's (2002) community forestry continuum based on ideals of local control and benefit.

The challenges are therefore considerable, and especially so for a Euro-Canadian society where the primary connotation of *common ground* – with its focus on collaboration, community benefit, and sustainability – involves a very different narrative than the conventional one. Yet despite such obstacles, the people of the Kenora area, as inheritors of a complex cultural legacy, appear willing to negotiate the diverse and sometimes conflicting objectives in pursuit of a potentially unifying goal. To make this happen, a paradigm of resource governance is needed that views planning and management prescriptions in a different, more adaptive way, one that acknowledges uncertainty and allows community members and groups to drive the processes forward.

Overcoming such obstacles is a challenge shared by other models of community forestry in Canada. As one of our participants noted, "I see Tunnel Island as the perfect microcosm of our entire history that has played out. The marginalization, the conflicts, the fighting over limited natural resource spaces, figuring out how to share those natural resources, how to manage things, what role sustainability has" (Elaine). Time will tell whether Kenora's residents, in their collective desire to build community through the sharing of place, are able to develop a form of resource governance that can help to achieve this. Whatever the result, their experience will no doubt generate some important lessons for other community-based efforts in Canada that look to promote and strengthen forest resource stewardship through participatory and democratic processes.

### Notes

1  The name Waa' Say' Gaa' Bo' was revealed through ceremony to a Naotkamegwaning (Whitefish Bay) elder, Ken Kakeeway, and is understood to refer to the spirit that guides the relationship central to the Common Ground landmanagement initiative on Tunnel Island-Old Fort Island. This spirit is embodied in the gift of the thunderbird feather, which appears at all feasts and ceremonies associated with the initiative. The land falls under this relationship and is therefore an important aspect of Waa' Say' Gaa' Bo', but the relationship exists irrespective of any physical place or locality. Another Ojibwe name, Kagapekeche,

or "a place to stay over," has been used to name Tunnel Island as a distinct locale.

2  Newton's (forthcoming) research provides a detailed picture of tenure rights for Tunnel Island from the late 1800s to the early 2000s. It appears that timber rights for the islands (including Tunnel Island and Old Fort Island) north of Sabaskong Bay in Lake of the Woods, were awarded in 1875 to John Mather's Keewatin Lumbering and Manufacturing Company. In 1891, the so-called Mather Settlement revoked this earlier lease but granted John Mather the option to purchase, among other things, all of Tunnel Island and to develop the west channel for waterpower. In the mid-1890s, Mather decided to establish a paper mill. The Keewatin Power Company was formed and took over ownership of Tunnel Island. The company remained the landlord until Ernest Wellington Backus, through his Backus-Brookes Co., bought out all remaining assets of the Keewatin Power Company in the early twentieth century. Again, rather than Tunnel Island itself, the prized asset was the water power of the Norman Dam, which was to provide energy to the new Kenora paper mill being constructed on the mainland. For the remainder of the twentieth century, the different companies that owned and operated the mill held property rights over Tunnel Island and Old Fort Island. For these owners, the value of the islands was always in their water rather than their forest resources. Dominated by white and red pine, the forest on Tunnel Island was of some value to the early sawmilling companies, but this was not the case for the subsequent paper mills that needed not pine but spruce, of which there was far less on the island.

3  A working group was created in early 2006; a memorandum of understanding was signed by Abitibi, the City of Kenora, and Grand Council of Treaty #3 in November 2006; Tunnel Island was transferred in March 2007, Old Fort in 2008; and the RPCGCO was formally created in 2008.

4  The six initiatives identified in Table 8.1 are those that were name-checked by multiple interviewees. One participant, however, believed that as many as eleven separate Kenora-based events have used "Common Ground" for their name and/ or inspiration. Because the additional initiatives or activities were not mentioned by any of the other interviewees, they are not included here.

5  Indeed, the connections to treaty had been made clear in interviews given at the time of the land transfer between Abitibi and the five partner organizations (Aiken 2010), and it was further cemented in June 2007 and June 2008, when the city's mayor and the Grand Chief of Grand Council of Treaty #3 marched together from Tunnel Island to the Kenora waterfront and signed a Joint Proclamation on Renewed Treaty-Based Relations.

## References

Agnew, J.A., and J.S. Duncan. 1989. *The power of place: Bringing together geographical and sociological imaginations*. Boston: Unwin Hyman.

Aiken, M. 2010. *After the mill: From confrontation to common ground in Kenora*. Winnipeg, MB: Aboriginal Issues Press.

Bagby, K., and J. Kusel. 2003. *Civic science partnerships in community forestry: Building capacity for participation among under-served communities*. Taylorsville, CA: Pacific West Community Forestry Center.

Baker, M., and J. Kusel. 2003. *Community forestry in the United States: Learning from the past, crafting the future.* Washington, DC: Island Press.

Beaudoin, J.M. 2012. "Aboriginal economic development of forest resources: How can we think outside the wood box?" *Forestry Chronicle* 88 (5): 571–77. http://dx.doi.org/10.5558/tfc2012-108.

Beckley, T.M., P.C. Boxall, L.K. Just, and A.M. Wellstead. 1999. *Forest stakeholder attitudes and values: Selected social-science contributions.* Information report NOR-X-362. Edmonton, AB: Canadian Forest Service, Northern Forestry Centre.

Beckley, T.M., R.C. Stedman, S. Wallace, and M. Ambard. 2007. "Snapshots of what matters most: Using resident-employed photography to articulate attachment to place." *Society and Natural Resources* 20 (10): 913–29. http://dx.doi.org/10.1080/08941920701537007.

Bengston, D.N. 1994. "Changing forest values and ecosystem management." *Society and Natural Resources* 7 (6): 515–33. http://dx.doi.org/10.1080/08941929409380885.

Berkes, F. 2010. "Devolution of environment and resources governance: Trends and future." *Environmental Conservation* 37 (4): 489–500. http://dx.doi.org/10.1017/S037689291000072X.

Billig, M. 2005. "Sense of place in the neighborhood in locations of urban revitalization." *GeoJournal* 64 (2): 117–30. http://dx.doi.org/10.1007/s10708-005-4094-z.

Blackmore, C. 2007. "What kinds of knowledge, knowing, and learning are required for addressing resource dilemmas? A theoretical overview." *Environmental Science and Policy* 10 (6): 512–25. http://dx.doi.org/10.1016/j.envsci.2007.02.007.

Bullock, R., K. Hanna, and D.S. Slocombe. 2009. "Learning from community forestry experience: Challenges and lessons from British Columbia." *Forestry Chronicle* 85 (2): 293–304. http://dx.doi.org/10.5558/tfc85293-2.

Canadian Urban Forest Network. 2004. *Canadian Urban Forest Strategy 2004–2006.* Ottawa: Canadian Urban Forest Network, Tree Canada Foundation.

Carter, J., P. Dyer, and B. Sharma. 2007. "Displaced voices: Sense of place and place identity on the Sunshine Coast." *Social and Cultural Geography* 8 (5): 755–73. http://dx.doi.org/10.1080/14649360701633345.

Casey, E.S. 2001. "Between geography and philosophy: What does it mean to be in the place-world?" *Annals of the Association of American Geographers* 91 (4): 683–93. http://dx.doi.org/10.1111/0004-5608.00266.

Cruikshank, J. 2005. *Do glaciers listen? Local knowledge, colonial encounters, and social imagination.* Vancouver: UBC Press.

Davidson, D. 2010. "The applicability of the concept of resilience to social systems: Some sources of optimism and nagging doubts." *Society and Natural Resources* 23 (12): 1135–49. http://dx.doi.org/10.1080/08941921003652940.

Davis, E. 2011. "Resilient forests, resilient communities: Facing change, challenge, and disturbance in British Columbia and Oregon." PhD diss., Department of Geography, University of British Columbia, Vancouver.

De la Cadena, M. 2010. "Indigenous cosmopolitics in the Andes: Conceptual re-flections beyond 'politics.'" *Cultural Anthropology* 25 (2): 334–70. http://dx. doi.org/10.1111/j.1548-1360.2010.01061.x.

Eames, R. 2005. "Partnerships in civil society: Linking bridging and bonding social capital." In *Social learning in environmental management: Building a sustainable future*, edited by M. Keen, V. Brown, and R. Dyball, 78–90. London, UK: Earthscan.

Eeyou Istchee Framework Agreement. 2011. *Framework agreement between the Crees of Eeyou Istchee and the Government of Quebec on governance in the Eeyou Istchee James Bay Territory*. Quebec, Canada.

Forest Capital Report. 1999. "Forest capital 1999: Tunnel island legacy project." Final Report, May 1999. Winnipeg, Canada: Hildeman Thomas Frank Cram.

Friedmann, J. 1987. *Planning in the public domain: From knowledge to action*. Princeton, NJ: Princeton University Press.

Grand Council Treaty #3. 2011. *"We have kept our part of the treaty": The Anishinaabe understanding of Treaty 3*. Kenora, ON: Offices of Grand Council Treaty #3.

Hay, R. 1998. "A rooted sense of place in cross-cultural perspective." *The Canadian Geographer* 42 (3): 245–66. http://dx.doi.org/10.1111/j.1541-0064. 1998.tb01894.x.

Healey, P. 2006. *Collaborative planning: Shaping places in fragmented societies*. Vancouver: UBC Press.

Higgins, V., J. Dibden, and C. Cocklin. 2008. "Neoliberalism and natural resource management: Agri-environmental standards and the governing of farm practices." *Geoforum* 39 (5): 1776–85. http://dx.doi.org/10.1016/j.geoforum.2008. 05.004.

Keen, M., V. Brown, and R. Dyball. 2005. "Social learning: A new approach to environmental management." In *Social learning in environmental management: Building a sustainable future*, edited by M. Keen, V. Brown, and R. Dyball, 3–21. London, UK: Earthscan.

Kooiman, J. 2003. *Governing as governance*. London, UK: Sage.

Krogman, N., T. Beckley. 2002. "Corporate 'bail-outs' and local 'buyouts': Pathways to community forestry?" *Society and Natural Resources* 15 (2): 109–27.

Latour, B. 1999. *Pandora's hope: Essays on the reality of science studies*. Cambridge, MA: Harvard University Press.

Lawrence, A., S. Molteno, and T. Butterworth. 2010. "Community wildlife sites in Oxfordshire: An exploration of ecological and social meanings for green spaces." *International Journal of the Commons* 4 (1): 122–41.

Marshall, G. 2009. "Polycentricity, reciprocity, and farmer adoption of conservation practices under community-based governance." *Ecological Economics* 68 (5): 1507–20. http://dx.doi.org/10.1016/j.ecolecon.2008.10.008.

Massey, D. 1993. "Power-geography and a progressive sense of place." In *Mapping the futures: Local cultures, global change*, edited by J. Bird, B. Curtis, T. Putnam, G. Robertson, and L. Tickner, 60–70. London, UK: Routledge.

–. 2005. *For space*. London, UK: Sage.

Meadowcroft, J. 2004. "Deliberative democracy." In *Environmental governance reconsidered: Challenges, choices and opportunities*, edited by R.F. Durant, D.J. Fiorino, and R. O'Leary, 183–217. Cambridge, MA: MIT Press.

Moyer, J.M., R.J. Owen, and P.N. Duinker. 2008. "Forest values: A framework for old-growth forest with implications for other forest conditions." *Open Forest Science Journal* 1 (1): 27–36. http://dx.doi.org/10.2174/18743986008010 10027.

Nabigon, H., R. Hagey, S. Webster, and R. MacKay. 1999. "The learning circle as a research method: The Trickster and Windigo in research." *Native Social Work Journal* 2 (1): 113–37.

Nelles, J., and C. Alcantara. 2011. "Strengthening the ties that bind? An analysis of Aboriginal-municipal inter-governmental agreements in British Columbia." *Canadian Public Administration* 54 (3): 315–34. http://dx.doi.org/10.1111/j.1754-7121.2011.00178.x.

Newton, T. Forthcoming. "A Capital Wilderness: The transformation of heritage space in the rise of modern industry at Kenora, Ontario." Friesens: Altona, MB.

Nowak, D., M. Noble, S. Sisinni, and J. Dwyer. 2001. "People and trees: Assessing the US urban forest resource." *Journal of Forestry* 99 (3): 37–42.

NSRCF Action Plan. 2011. "Bringing balance to resource decision making through recognition and reconciliation." *Journal of Public Deliberation* 9(2) (2013): Art. 7, Draft NSRCF vision statement and master action plan. Chapleau, ON: Northeast Superior Regional Chiefs' Forum (NSRCF). http://www.publicdeliberation.net/jpd/vol9/iss2/art7.

Ontario Ministry of Natural Resources (OMNR). 2003. Old Growth Policy for Ontario's Crown Forests. Forest Policy Series, Version 1. Sault Ste. Marie, Ontario, Canada.

Ordóñez, C., and P.N. Duinker. 2010. "Interpreting sustainability for urban forests." *Sustainability* 2 (6): 1510–22. http://dx.doi.org/10.3390/su2061510.

Ostrom, E. 1990. *Governing the commons: The evolution of institutions for collective action*. Cambridge: Cambridge University Press. http://dx.doi.org/10.1017/CBO9780511807763.

Pinkerton, E., R. Heaslip, J.J. Silver, and K. Furman. 2008. "Finding 'space' for co-management of forests within the neoliberal paradigm: Rights, strategies, and tools for asserting a local agenda." *Human Ecology* 36 (3): 343–55. http://dx.doi.org/10.1007/s10745-008-9167-4.

Ratuski, S. 2014. *Cultural landscapes of the common ground: Mapping traditional Anishinaabe relationships to the land.* Master's Thesis, University of Manitoba, Winnipeg, MB.

Reed, M.G., and K. McIlveen. 2006. "Toward a pluralistic civic science? Assessing community forestry." *Society and Natural Resources* 19 (7): 591–607. http://dx.doi.org/10.1080/08941920600742344.

Relph, E. 1976. *Place and placelessness*. London, UK: Pion Books.

Robson, J.P., A.J. Sinclair, I.J. Davidson-Hunt, and A.P. Diduck. 2013. "What's in a name? The search for 'common ground' in Kenora, northwestern Ontario." *Journal of Public Deliberation* 9 (2): art. 7.

Robson, J.P., A.J. Sinclair, and A. Diduck. 2015. "A study of institutional origins and change in a Canadian urban commons" *International Journal of the Commons* 9 (2): 698–719.

Rodgers, C.P., E.A. Straughton, A.J.L. Winchester, and M. Pieraccini. 2011. *Contested common land: Environmental governance past and present*. London: Earthscan.

Rothe, J.P., D. Ozegovic, and L.J. Carroll. 2009. "Innovation in qualitative interviews: 'Sharing circles' in a First Nations community." *Injury Prevention* 15 (5): 334–40. http://dx.doi.org/10.1136/ip.2008.021261.

Sampson, K., and C. Goodrich. 2009. "Making place: Identity construction and community formation through 'sense of place' in Westland, New Zealand." *Society and Natural Resources* 22 (10): 901–15. http://dx.doi.org/10.1080/08941920802178172.

Sinclair, A.J., A.P. Diduck, and P.J. Fitzpatrick. 2008. Conceptualizing learning for sustainability through environmental assessment: Critical reflections on 15 years of research. *Environmental Impact Assessment Review* 28 (7): 415–522.

Smaldone, D., C. Harris, and N. Sanyal. 2005. "An exploration of place as a process: The case of Jackson Hole, WY." *Journal of Environmental Psychology* 25 (4): 397–414. http://dx.doi.org/10.1016/j.jenvp.2005.12.003.

Steins, N.A., and V.M. Edwards. 1999. "Platforms for collective action in multiple-use common-pool resources." *Agriculture and Human Values* 16 (3): 241–55. http://dx.doi.org/10.1023/A:1007591401621.

Teitelbaum, S., T. Beckley, and S. Nadeau. 2006. "A national portrait of community forestry on public land in Canada." *Forestry Chronicle* 82 (3): 416–28. http://dx.doi.org/10.5558/tfc82416-3.

Teitelbaum, S., and R. Bullock. 2012. "Are community forestry principles at work in Ontario's county, municipal, and conservation authority forests?" *Forestry Chronicle* 88 (6): 697–707. http://dx.doi.org/10.5558/tfc2012-136.

Tindall, D.B. 2003. "Social values and the contingent nature of public opinion and attitudes about forests." *Forestry Chronicle* 79 (3): 692–705. http://dx.doi.org/10.5558/tfc79692-3.

Tuan, Y.-F. 2004. "Sense of place: Its relationship to self and time." In *Reanimating places: A geography of rhythms*, edited by T. Mels, 45–56. Burlington, VT: Ashgate.

Wallace, R. 2013. *Merging fires: Grassroots peacebuilding between Indigenous and non-Indigenous peoples*. Winnipeg, MB: Fernwood Publishing.

Wheeler, M.J., A.J. Sinclair, P. Fitzpatrick, A.P. Diduck, and I.J. Davidson-Hunt. 2016. "Place-Based Inquiry's Potential for Encouraging Public Participation: Stories from the Common Ground Land in Kenora, Ontario." *Society and Natural Resources*. Published Online First, 10.2.2016. DOI: 10.1080/08941920. 2015.1122130.

Wheeler Wiens, M.J. 2011. *Imagining Possibilities for Shared Place: Sense of Place Investigations into Local Connections and Visions for the Common Ground Land on Tunnel Island, Kenora, Ontario*. Master's Thesis, University of Manitoba, Winnipeg, MN.

Wight, I. 2005. "Placemaking as applied integral ecology: Evolving an ecologically wise planning ethic." *World Futures* 61 (1–2): 127–37. http://dx.doi.org/10.1080/02604020590902407.

Wilding, N. 2011. *Exploring community resilience in times of rapid change.* Dunfermline, Fife: Fiery Spirits Community of Practice and Carnegie UK Trust.

Williams, D.R., and S.I. Stewart. 1998. "Sense of place: An elusive concept that is finding a home in ecosystem management." *Journal of Forestry* 96 (5): 18–23.

Yung, L., W.A. Freimund, and J.M. Belsky. 2003. "The politics of place: Understanding meaning, common ground, and political difference on the Rocky Mountain Front." *Forest Science* 49 (6): 855–66.

Chapter 9    **Community Forestry and Local Development at the Periphery**
Four Cases from Western Quebec

*Édith Leclerc and Guy Chiasson*

In Quebec, community-based forestry has long been seen as a lever for regional development. In the 1960s, rural communities in the Bas-Saint-Laurent and Gaspésie regions proposed community-based models of forest management as an alternative to the modernist ideas predominant in regional planning at the time. Since the 1990s, a number of provincial government programs and policy statements have emerged that are oriented towards community forestry. These policies have raised hopes in rural places that community-based forest management will fulfill its potential for contributing to local development (Chiasson and Leclerc 2013).[1] Thus, in the discourse of both regional actors and government in Quebec, there is a link between local governance of forests and positive social and economic outcomes for forest-dependent communities. For example, Quebec's Sustainable Forest Development Act (2013) foresees socio-economic development through decentralization to regional institutions and community-based management via the proximity forests program.[2] According to the government, the new forest policy regime in Quebec provides "broad latitude to take local aspirations and needs into account" (Québec, MRNF 2012). However, little research has yet been devoted to examining whether these aspirations to derive additional benefits from the forest through local governance models are being realized.

The relationship between governance and local development has been a major theme in the contemporary social sciences. In metropolitan areas, the literature examining the linkage between local governance schemes and local development is mixed: some describe positive relationships (Scott 2001; Cooke and Lazaretti 2008), while the findings of others are less rosy (Scott 2001; Cox 2004; Brenner and Keil 2006). However, the bulk of research to date has focused on major cities (Sassen 2001; Storper 2013), leaving less substantive evidence from smaller and more peripheral regions (Hayter and Barnes 2001; Monsson 2014). This, despite the fact that peripheral regions in Quebec, as elsewhere in Canada and internationally, are having increased governance responsibilities devolved to them by central governments (Belley 2005), including in the area of forest management (Blais and Chiasson 2005; Andersson, Benavides, and Léon 2014).

There is a growing literature on community forestry and its impact on community well-being. These studies largely assess economic outcomes (Vermeulen, Nawir, and Mayers 2008; Lawrence and Ambrose-Oji 2014), community well-being (Beckley 1995; German et al. 2010), or social relevance of community-led forest initiatives (Bouda et al. 2011; Bullock, Hanna, and Slocombe 2009). However, there are very few comparable studies on community forestry in Quebec that allow for an assessment of its impact on local development.

The main objective of this paper is to reflect on the contribution that local forest-governance approaches are making to local development in rural areas in Quebec. We analyze four distinct cases, all local forestry initiatives located in peripheral regions of Quebec: Abitibi-Témiscamingue, the Outaouais region, and the Northern Quebec administrative region. These are all regions where community economic development is a major concern, and a challenge, and where forestry has been an important economic engine for a long time (Chiasson and Leclerc 2013). Three of these cases are forestry cooperatives in small rural municipalities: the Coopérative de solidarité de Duhamel, the Coopérative forestière de Beaucanton, and Roulec 95. Roulec 95 and our fourth case, the Corporation de gestion de la Forêt de l'Aigle (CGFA), were given pilot project status by the Quebec government in the mid-1990s (see Figure 9.1).[3]

## Method

These four cases are part of a broader research project that examined six cases of local forest governance in Quebec conducted between 2004 and 2006.[4] In this chapter, we revisit the four cases that fit most closely with the overarching theme of this book. Analyzing these four cases through the lens of local development can inform the present debates on community participation in forestry as Quebec is actually revising its overall forest policy as well as its proximity forest program.

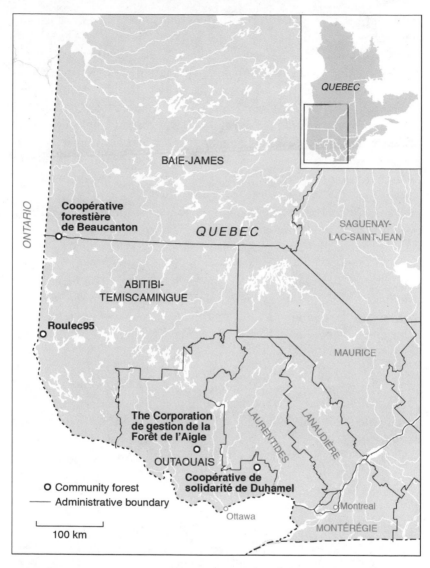

**Figure 9.1**   Location in western Quebec of the four community forestry cases.
*Source:* Google Maps. Adapted by Eric Leinberger.

Our research methodology is an adaptation of a methodological framework elaborated by Yvan Comeau (2003). This framework is intended to provide a road map for conducting case studies of local governance initiatives, with a view to evaluating social relations as well as institutional and organizational elements. It provides guidance on thematic content and data sources. Comeau's grid identifies three distinct data-gathering techniques: document analysis (local and regional press coverage, official and internal documents, etc.), interviews, and participant observation. We only used the first two, since participant observation was impossible in some of the cases. Interviews were conducted with a variety of participants in these initiatives, including workers, board members, representatives from partner organizations, and municipal councillors. Five to eight interviews were conducted at each case study site, depending on the number of partners involved with the initiative. This was complemented by document analysis, including policies, management plans, press clippings, and so on. All field research was conducted in 2005.

Comeau's methodology also set outs five potential research themes, including the prevailing context at the launch of the initiative, the actors involved in the case, institutional aspects such as legal rules and internal/external power relationships, and organizational elements (e.g., means of production, industrial relations, etc.). The fifth aspect is integrative in the sense that it includes elements from the previous four, while addressing more directly the particular research question being explored by the project – in this case, the impact of local governance on different facets of local development.

We have chosen a qualitative lens for this study. While a quantitative approach would have produced more generalizable insights (Yin 2009), we believe that a qualitative approach better captures local specificities that are important for understanding the complex links between local governance and development. Among other things, a qualitative approach allows us to pay attention to how local actors themselves perceive their actions, as well as their contribution to local development and its different dimensions.

In the first section of this chapter, we clarify the concept of local development as it is conceived in this study. In the second section, we analyze the four cases in terms of their contribution to different facets of local development; in so doing, we underline some of the difficulties faced by local governance initiatives operating in peripheral regions. In the third and final section, we explore a series of factors that may help explain the vulnerable nature of these local governance arrangements.

## Local Development and Its Components

The concept of local development emerged in the 1980s as a new approach to regional development (Jean 1989). From the end of the

Second World War until the 1980s, regional development in most Western countries had been heavily associated with centralized state action (Higgins and Savoie 1997). Its focus was regional policies of the state, the main concern being industrial growth, as many have already pointed out (Murdoch 1997; Polèse 1999). In other words, as Juan-Luis Klein (1995) has aptly explained, regional development was the spatial component of the welfare state. However, in the 1980s, state-led regional redistribution policies came under increasing criticism, perceived by some as too centralized and inflexible ("one size fits all"), leaving insufficient space for the aspirations of regional social movements (Dionne 1989; Joannis and Martin 2005). Other critics saw regional policies as inefficient and even counterproductive. For instance, Savoie (1986) argued that states could no longer steer market forces and should instead adopt a different role – namely, that of supporting the dynamism of local and regional entrepreneurs. These criticisms, along with the financial problems of the 1980s, drove many states, including the federal and Quebec governments, into major policy revisions regarding the development of local and regional territories (Morin 1991; Chiasson 1999). Another outcome was a renewal of theories of regional development, in both Europe and North America (Jean 1989; Blakely 2009). These new theories, often grouped under the umbrella term "local development," stress a set of endogenous factors – social capital, networks, governance, and others – as contributing to successful revitalization in some regions (Joyal 2001; Becattini et al. 2003; Blair and Carroll 2009).

The concept of local development implies an important shift in both the nature of development and the means necessary to achieve it. As suggested above, regional development in the decades following the Second World War was centred on the redistribution of industrial growth. It was assumed that regional well-being derived quite naturally from jobs in heavy industry. In the post-1980s era, however, many felt that local development would not be guaranteed by industrial jobs alone, especially in a context where many jobs in heavy industries were relocating elsewhere (Mercure 1996; Hayter 2000). Over time, more complex definitions of development emerged, which are multifaceted and recognize a diversity of values (Hayter 2003; Stedman, Parkins, and Beckley 2005; Kelly and Bliss 2012). Theories of regional development shifted from a focus on macroeconomic policies to a focus on local actors as the central agents of regional and local development (Fauré, Labazée, and Kennedy 2005; Leloup, Moyart, and Pecqueur 2005).

This effort to rethink development has also taken place at the international level. Sen's (1999) capability approach and the associated human development index are good examples of efforts to measure development with a wider range of indicators. A number of theorists have also attempted to propose a more holistic notion of development at the regional and local level that integrates multiple quality-of-life aspects (Storper 1995, 1997a, 1997b).

Our own notion of local development is grounded in these ongoing efforts to adopt a richer concept of development that captures a broader range of community values. In our attempt to identify local forest governance's contribution to local development, we have chosen to focus on three components:

- Work: We opted to look at not only the creation and maintenance of jobs but also more qualitative aspects of work, such as the types of jobs being created (professional, manual, seasonal, etc.), the quality of jobs, and how jobs are valued by the community (Chiasson and Leclerc 2013; Ternaux 2006).
- Environment: In this study, environment has a broad significance. We wanted to know whether forest activities conformed to local expectations regarding the maintenance of environmental quality and whether forestry supported a diversity of ecosystem services. We also investigated the extent to which forest management remained tied to an industrial pattern of forest use (pulp, paper, softwood lumber, etc.) or integrated a wider set of values and activities (Costanza et al. 1997).
- Participation of marginalized actors: This component involved looking at whether local governance processes allowed for meaningful participation of social groups often excluded from decision-making processes. In our study, the focus was on women and First Nation peoples (Chiasson, Boucher, and Martin 2005; Kelkar, Nathan, and Walter 2003).

We recognize that this list is not exhaustive and that other valuable aspects could be included in the analysis of local development. However, we chose attributes that are often cited as yardsticks for local development in research evaluating local governance generally (Jouve and Booth 2004; Papadopoulos and Warin 2007) and local forest governance in particular (Bray et al. 2003; Kelkar, Nathan, and Walter 2003; Parkins 2006). In the next section, we will present the four local governance initiatives and evaluate their performance according to these three aspects of local development.

## Assessment of Local Forest Governance in Peripheral Regions

Our sample included initiatives with different organizational structures and legal statuses. Three out of four are cooperatives: the Coopérative forestière de Beaucanton (in the northern Quebec region), the Coopérative de solidarité de Duhamel (in the Outaouais region), and Roulec 95 (in the Abitibi-Témiscamingue region). In the cases of Duhamel and Beaucanton, both were developed as worker cooperatives, meaning that local forestry workers created a new organization in order to gain more control over their working conditions and to be able to bid on larger contracts. The members of the cooperative shared contracts, machinery, and maintenance costs in order to improve their bargaining power, enhance their market share, and improve incomes. The third case, Roulec 95, is a cooperative founded by residents in the municipalities of Roulier and Nedelec and is best known for its status as one of fourteen Inhabited Forest pilot projects. The inhabited forests were part of a provincial government program in the 1990s and early 2000s that sought to experiment with new forms of community participation in forest management. The fourth case, the CGFA, is an organization that was also granted inhabited forest status. It is governed by a board of directors composed of representatives from seven different stakeholder groups (municipality, First Nation band council, research institute, ATV association, hunting and fishing clubs, and private woodlot owner's group).

### The Two Cooperatives: Making It Work?
As mentioned, our analysis of these initiatives covers three facets of local development: work, environment, and participation of marginalized groups.

Both Duhamel and Beaucanton were created as worker cooperatives in order to obtain contracts from large companies operating mills, thereby maximizing job opportunities for local workers. Both cooperatives were hiring workers from other communities – and, in some cases, other regions of Quebec – to do harvesting work. Thus, these two organizations had a very clear job-creation orientation. However, both cooperatives ran into problems fairly quickly.

In the case of Duhamel, after a first phase as a worker cooperative, founding members opted to change the governance structure to a *coopérative de solidarité.*[5] This meant that membership was no longer restricted exclusively to workers but was expanded to include representatives from the wider community. This change was part of a broader effort to mobilize local resources in order to diversify activities beyond the harvesting of timber. The cooperative chose to develop a new niche market of dried firewood production, destined for campgrounds. This new form of production required the use of new technologies, which in turn required the training of local workers for a number of more technical jobs. However, the cooperative faced difficulties – first, in finding financial partners to support the project, and second, in finding and training labourers to undertake the work. After just a few rocky months, the project promoters decided to shut down operations. Thus, despite the adoption of a modified governance structure, followed by a concerted effort to move away from a simple primary-production role, the lack of human and technical resources became an important stumbling block for this cooperative.

The Beaucanton cooperative also faced challenges in matching its activities with an appropriate workforce. This local community, part of the James Bay Municipality (a very large administrative entity in northern Quebec), first faced challenges in securing contracts but was eventually successful in acquiring a logging contract from the municipality on intramunicipal public land (IPL).[6] However, according to one of the project proponents, the cooperative ran into trouble because the work was not executed properly and thus did not entirely comply with logging regulations set out by the Ministère des Ressources naturelles et de la Faune. This situation quickly led to the termination of the contract by the municipality. As with Duhamel, the problem proved to be one of recruiting qualified workers locally, and like Duhamel, this eventually led to the cessation of the cooperative's activities. Thus, these two cooperatives, both of which had the goal of securing and diversifying the local workforce, proved short-lived.

The other two aspects of local development examined in this study – environment and participation of marginalized groups – were not well developed within these two case studies. Neither of the cooperatives had explicitly adopted goals or objectives in these areas, which did not appear to be driving concerns. For the environmental dimension, the aim of the cooperatives was to meet provincial requirements, but no additional efforts were made to enhance environmental quality or protect specific forest values. The same was true for the participation of marginalized groups, such as women and First Nations. Interviewees did not describe the cooperatives as places to facilitate women's involvement in decision-making, and First Nations were not directly involved in either project. Overall, it appears that the short lifespan of these projects limited opportunities to go beyond their initial work-related objectives.

### Roulec 95 and the Corporation de gestion de la Forêt de l'Aigle: Examples of Diversification?

Our assessment of the impacts of the two other community forest projects is markedly different. The first, Roulec 95, was set up in the mid-1990s by the municipalities of Nedelec and Roulier (Témiscamingue Regional County Municipality) to manage intramunicipal public land (IPL). In addition, the managers of Roulec 95 negotiated a deal with Tembec, a large forest company with a large mill and licence, giving Tembec exclusive purchasing rights to the wood harvested by Roulec 95 on the IPL in exchange for an exclusive contract allowing Roulec 95 to carry out forestry operations (harvesting and silviculture) on Tembec's Crown forest licence located nearby. The early results of this deal were quite impressive for Roulec 95. In 2000, the cooperative hired more than thirty full-time workers for the totality of its forest operations (Chiasson and Gaboury 2000). During the same period, it also launched a blueberry farm on the IPL, which increased employment and contributed to a diversification in the types of jobs offered locally. More recently, Roulec 95 has diversified its activities still further, introducing consulting services, which provide integrated resource management services for municipal and private woodlots (Roulec 95 2008).

The CGFA is one of the best-known community forestry initiatives in Quebec, in part because of the broad suite of activities developed in the forest. It is a well-documented case, recognized for its innovative approach to planning and forest management (Teitelbaum 2014). While the CGFA is facing organizational challenges at the time of writing, several authors have described it as a model in terms of community-based governance, contributions to rural and local development (De Blois Martin 2002), and collaboration with and integration of First Nation communities in forest management (Chiasson, Boucher, and Martin 2005). The CGFA benefitted from the allocation of a large contiguous territory in the northern part of the Outaouais region.[7] It also inherited a quality wood supply, since the territory had not been allocated to an industrial licensee for over a hundred years and included some of the last stands of old-growth pine in Quebec. The absence of an industrial licence on the land gave the CGFA enhanced flexibility, allowing it to develop a complex management plan that divided the territory into distinct geographic zones based on different uses (intensive logging, forest-based tourism, conservation). Profits from timber harvesting were channelled into a series of other forest-based activities, including recreation (rental cottages, ziplining, interpretive trails, etc.) and research.

The multi-use approach put forward by the CGFA had many positive outcomes in terms of local development. From an employment perspective, Forêt de l'Aigle managers were able to hire a diversified team of professionals (biologist, recreation/tourism specialist, forest ecologist) in addition to a number of forest engineers and technicians. This specialized team was in charge of planning, operations, conservation, and recreation. Moreover, most of the logging work was subcontracted to local businesses. In diversifying activities towards a suite of recreation activities, the organization was also able to enhance local employment through seasonal hiring. Indeed, integration of recreation activities allowed the CGFA to hire individuals with very different backgrounds (not necessarily trained in forestry), which proved to be an important benefit.

Forêt de l'Aigle's integrated approach should be seen as a *step* towards sustainable forest management. Research has shown that zoning at a landscape level can be an important contributor to sustainability (Torquebiau and Taylor 2009). For example, the zoning approach at Forêt de l'Aigle has allowed some areas to be excluded completely from logging and reserved for uses such as conservation or recreation. There has been debate about the sustainability of harvesting rates at Forêt de l'Aigle, with one research study indicating that logging was too predominant (Gélinas 2000). Local interviewees, however, defend the CGFA's practices, pointing out that revenues from logging have been needed to cover costs associated with activities (and the necessary staff) in zones reserved for conservation and tourism.

Finally, the CGFA also made significant inroads in terms of integrating First Nation interests into local forest governance. The Kitigan Zibi Anishinabeg First Nation, whose ancestral lands include the Forêt de l'Aigle, was one of the seven partners represented on the CGFA's board of directors. While this did not result in traditional Anishinabeg knowledge being incorporated into planning in any obvious way (Chiasson, Boucher, and Martin 2005), it did represent an opportunity for the Kitigan Zibi community to take part in negotiations leading to the development and implementation of the forest-management plan.

Overall, then, it can be said that our two community forest projects had significant impacts with regard to the aspects of environment and the integration of marginalized groups in decision-making. However, more recently, the Forêt de l'Aigle has experienced considerable up-heaval, which has jeopardized many of the benefits created during its first decade of operations. In the midst of the forestry and financial crisis in 2008, the CGFA began to face serious financial difficulties. This eventually led the Kitigan Zibi band council to withdraw its support of the organization and finally resulted in bankruptcy of the CGFA. At the time of writing, although some of the original partners were working with new ones to reinvent a local governance structure for Forêt de l'Aigle, the governance of this emblematic territory remains uncertain.

## Local Governance's Contribution to a Multi-faceted Development

As explained earlier, our study relied on a definition of local develop-ment with multiple facets. The CGFA was the only project that had explicit objectives in terms of environmental sustainability and inte-gration of marginalized groups, at least in its initial phase. Roulec 95, after an initial phase dedicated mostly to creating local jobs in estab-lished activities (logging, thinning, etc.), was able to add consulting services in integrated management for private woodlot owners and municipalities. For Roulec 95, integrating First Nations and women into the decision-making process was not an explicit objective, al-though First Nation members were hired from time to time. The same was true of Duhamel and Beaucanton. Actually, in some cases, First Nation communities were more likely to be seen as competitors than as members of the community to be integrated into the initiative.

Three of the initiatives studied did not stand the test of time. Two (the Coopérative forestière de Beaucanton and the Coopérative de solidarité de Duhamel) had a fairly short existence while the third, the CGFA, managed to established itself as a leader in forest local govern-ance before falling apart because of financial difficulties. The uncer-tainties surrounding Forêt de l'Aigle at the time of writing serve as a powerful reminder that local governance of forests is fraught with challenges for peripheral communities. This is especially the case in the context of a forestry crisis in which even long-established and resourceful multinational corporations face difficulties (Barré and Rioux 2012).

Table 9.1 summarizes the four local governance initiatives in terms of their projected contribution to local development. As is shown, all the initiatives had objectives related to work. This is unsurprising, since these communities are in peripheral regions where local jobs are scarce. In one case (Beaucanton), the main objective was simply to

TABLE 9.1   Case study contributions towards different aspects of local development

|  | Work | Environment | Integration |
|---|---|---|---|
| Coopérative forestière de Beaucanton | Providing forestry work for local people | – | – |
| Coopérative de solidarité de Duhamel | *Phase 1:* Providing forestry work for local people<br><br>*Phase 2:* Creating more specialized jobs | – | – |
| Roulec 95 | *Phase 1:* Providing forestry work for local people<br><br>*Phase 2:* Creating more specialized jobs | Phase 2: Providing consulting services for integrated resource management | – |
| Corporation de gestion de la Forêt de l'Aigle | Creating more specialized jobs | Integrated resource management (logging, conservation, and recreation) | First Nation participation in governance |

keep jobs in the community – in other words, to counter a trend of companies hiring people from outside the community and region. In another case (Duhamel), the objective was to change the nature of work by creating more skilled jobs in forestry. Finally, in the case of Roulec 95 and the CGFA, local actors began with an objective of keeping jobs locally but later moved on to creating more specialized jobs.

It should be remembered that two of these initiatives had a fairly short existence, which hampered their opportunities to make a long-term contribution to local development. This explains some of the empty cells in Table 9.1. Roulec 95 and the CGFA had a more lasting impact on local development, not only because they were able to persist over time but also because they were able to make contributions that went beyond the single issue of local job creation. The results presented in Table 9.1 suggest the need to be cautious when exploring the linkages between local forest governance and local development. While studies elsewhere have suggested that decentralization of forest or resource management can contribute to more sustainable and inclusive management (Ostrom 1990; Agrawal and Chhatre 2007; Kelkar, Nathan, and Walter 2003; Ribot 2008), our case studies suggest that work remains a key driver for community forestry in peripheral regions of Quebec. To help explain the difficulties faced by the two cooperatives compared with some of the achievements of the two community forests, we go on to explore a number of factors.

## Framing Local Governance

Studies of local forest governance (Chiasson, Andrew, and Perron 2006), like broader studies of governance (Moulaert and Mehmood 2008), often explain change by focusing on endogenous territorial dynamics. Factors like local social capital (Flora et al. 1997), local networks (Angeon and Caron 2009; Orozco-Quintero and Berkes 2010), and public participation (Sheppard 2005; Parkins 2006) are seen as drivers for successful local governance and, ultimately, for local development. These are, of course, important elements that contribute to the successes and challenges of initiatives such as the ones studied here, but many case studies tend to overlook another set of factors that are related to the broader framework within which local governance is embedded (McGuirk 2003, 2007; Faure et al. 2007). Factors such as tenure (ownership of the land and the forest), support received from public policies, and the geographical location of the communities are all contextual elements of this broader framework. Our aim is to explore two of these factors — geographical location and land tenure — in order to analyze how they played out in our four cases.

### Governance at the Periphery

Unlike studies of regional development in the decades after the Second World War, studies of local development in the same period have paid little attention to the issue of remoteness. For local development theorists, remoteness (the fact of being at the periphery) is considered less important than having the appropriate set of territorial dynamics in place (social capital, entrepreneurial culture, dense networks for innovation, etc.). Elsewhere, we have argued that remoteness should be reinstated as a major issue for local development in order to better understand the difficulties and barriers of governance at the periphery (Chiasson, Leclerc, and Andrew 2010). Many of the difficulties faced by the cooperatives in Duhamel and Beaucanton can clearly be linked to their geographic position as peripheral communities. Indeed, not only are these two communities located in peripheral resource-dependent regions, but they are also far from major cities within these regions. In other words, they are at the *periphery of the periphery* (Gumuchian

1990). For these cooperatives, being far from regional cities meant that many of the resources necessary to promote innovative uses of local forests were not readily available in or near the communities. Factors such as the availability of a qualified workforce, technical expertise, and financial capital are all crucial, according to the literature on innovation (Veltz 1996; Doloreux and Dionne 2007), but these were difficult to find for the organizers of these two cooperatives. In Beaucanton, the lack of qualified labour turned out to be a major stumbling block, while in Duhamel, technical expertise had to be imported from outside the region, and funding proved difficult to find.

These problems related to remoteness are not insurmountable. As Doloreux and Dionne (2007) have shown in their study of La Pocatière's local innovation system, disadvantages linked to peripheral location can be at least partially mitigated by dynamic institutional action and support. However, issues related to remoteness were not the only problems the two cooperatives faced.

The CGFA and Roulec 95, although also located in peripheral areas, could rely on institutional support from different levels. Both had access to some support coming from provincial funds, in the form of a specific program supporting integrated management of local forests (Québec, MRNF 2012). Also, as a part of the Inhabited Forest pilot project, the two cases were part of a select group of initiatives across Quebec that received government support, and indeed, they were the only ones within their respective regions. This meant that they were, in a sense, regional champions and could count on support from regional institutions (the Regional Development Council and, later, the Regional Elected Officials Councils). The former director of the CGFA pointed out publicly that support from regional institutions was very important for the development of the organization (Beaudoin 2000).

### Rights to the Land

A second factor related to the framework within which local governance is embedded is that of entitlement to the land in the form of formal rights. Indeed, this was the factor that distinguished the two stronger cases of Roulec 95 and the CGFA from the weaker cases in Beaucanton and Duhamel. Roulec 95 and the CGFA, as part of the Inhabited Forest pilot project, had the benefit of exclusive timber-harvesting rights on the forest territory that they managed, which was not the case for all fourteen pilot projects. In the case of Roulec 95, rights to manage intra-municipal public land (IPL) were granted to the municipalities of

Nedelec and Roulier by the Ministère des Ressources naturelles et de la Faune (Chiasson and Gaboury 2000). For Roulec 95, the timber rights allocated on the IPL proved to be the major leveraging point allowing them to negotiate with Tembec for exclusivity of contracts for the company's wood-extraction activities. Roulec 95 was also able to use parts of the IPL to develop other activities. Eventually, Roulec 95 was able to rely on the expertise and the resources developed in the management of the IPL to venture into the service side and offer integrated forest-management consulting services to private woodlot owners and other municipalities.

The CGFA was granted a direct tenure by the Ministère des Ressources naturelles et de la Faune, giving them exclusive rights to manage and harvest timber resources. While cottage leases were attributed separately on the territory, there were no other competing leases. This gave the CGFA considerable leeway to develop and implement a fairly elaborate integrated management plan without having to negotiate with other rights holders, such as forest companies with overlapping licences. It allowed the organization to experiment with an innovative planning practice in which intensive logging alternated with less intensive logging and full conservation areas (Teitelbaum and Saumure 2010). Thus, in both community forests, land tenure and exclusivity of land use were key factors that allowed for the realization of certain organizational goals related to local development.

Our two other cases, the cooperatives in Duhamel and Beaucanton, had no legal rights to the forest on the territory in which they operated. We can clearly see, at least in Beaucanton, that the lack of rights to the forest constituted a vulnerability for the organization. The cooperative was created in order to undertake contractual work on the IPL of James Bay, but the municipality – a large entity not unlike a regional government – did not demonstrate clear commitment to the proponents. Although the cooperative was able to secure a contract with the municipality, when the work was deemed insufficient, the municipality quickly ended the contract, leaving the cooperative with little choice but to stop its operations. Furthermore, the cooperative had very little influence over forest management, being hired only for forest-harvesting purposes. Thus, there were few opportunities to develop other forest uses or to become actively involved in other management responsibilities.

In brief, it seems that the rights to the land that Roulec 95 and the CGFA could count on were major boosters to their operations. In both initiatives, rights to the land and clear access to forest resources allowed local governance actors not only to experiment with alternative uses and more integrated management of the land (especially the CGFA) but also to develop and consolidate the integration of new professional expertise into forest management. Actors such as the Beaucanton cooperative, which had contractual access to the land, were more constrained. In that context, it proved much more difficult for these actors to go beyond more classical forms of forestry. This does not mean that a local governance form of community forestry that has full rights to a local forest will always promote more sustainable and integrated management – as studies on IPL (Chiasson and Gaboury 2000) have shown. However, given appropriate local social conditions, full rights to the forest territory can greatly help a local governance initiative to introduce innovative practices.

## Conclusion

Our cases are far from the success stories that are often presented in local development studies (Benko and Lipietz 1992). However, analysis of these four local governance initiatives does point out some of the limitations and opportunities associated with community forestry with regard to enhancing local development. As we have argued, the contribution of local governance can be innovative but, under certain conditions, not easily achieved in the context of peripheral regions, especially those that are far from regional central cities.

Our three components of local development (work, environment, and integration of marginalized groups) allowed us to see how local governance initiatives evolve over time in terms of their projected contribution to local development. The primary concern of the four initiatives examined in this chapter was work – specifically, to ensure that wood harvested locally would support local jobs. The CGFA, with its early environmental and social integration objectives, was something of an exception. In the case of Beaucanton and Duhamel, moving beyond traditional forms of forestry work to more value-added activities or more integrated forms of management proved very difficult. Many of the difficulties stemmed from the fact that these organizations did not have legal management rights to the forest territories in which they operated, but their peripheral geographic location was also an important factor. In the end, the short life of these projects did not give them time to experiment with other facets of local development.

The two community forests were able to make a more sustainable contribution to local development. While the failure of the CGFA in 2008 can serve as a reminder that, even under good conditions, local governance of forest lands remains fragile, it should be remembered that the CGFA's problems came after more than a decade of successful operation. Better conditions in terms of guaranteed rights to the land allowed these two projects to experiment with more complex forms of forest management and to engage in more diversified local development.

Overall, our four case studies confirm that local governance of forests – and, to some extent, community forestry – not only needs proper social conditions but is also greatly influenced by a broader, exogenous framework within which these conditions operate. This means that we cannot rely simply on local resources to promote innovative forestry-based local development. Proper public policies creating the right conditions and giving adequate support for innovation are also essential aspects.

## Notes

1 Our view of local development is similar to that described by Storper (2013). In his view, economic geography, institutions, innovation, and justice are "the keys to understanding city-region development. For each of them, we have abundant existing tools of analysis" (228). Storper's work follows others who also see local development as relying on local social dynamics and conducive to analysis using spatial theories (Brenner and Keil 2006; Shearmur 2011).

2 According to the new Sustainable Forest Development Act, forest communities were able to propose small woodlot development projects to the government. Respecting a number of conditions, communities could be granted management rights for local woodlots and financial support to reinforce the link between the community and its nearby forest. The communities were supposed to submit projects in April 2013, but in March 2013, the government decided to defer the program for two years.

3 The Quebec government's Inhabited Forest program will be discussed later in this chapter.

4 This research program was funded by the Social Sciences and Humanities Research Council of Canada, and its results are presented in Chiasson and Leclerc (2013).

5 Coopérative de solidarité is one of the statuses recognized by the cooperative law in Quebec, which allows members of the broader community to be represented on the cooperative's board of directors (St-Martin and Côté 1999).

6 IPLs are forest territories where regional county municipalities have the responsibility of managing the forest.

7   Of the fourteen pilot projects, the CGFA is the only one to have received legal tenure of a forest lot. Thus, it also had the distinct advantage of exclusive logging rights in this forest area.

## References

Agrawal, A., and A. Chhatre. 2007. "State involvement and forest co-governance: Evidence from the Indian Himalayas." *Studies in Comparative International Development* 42 (1–2): 67–86. http://dx.doi.org/10.1007/s12116-007-9004-6.

Andersson, K., J.P. Benavides, and R. Léon. 2014. "Institutional diversity and local forest governance." *Environmental Science and Policy* 36: 61–72. http://dx.doi.org/10.1016/j.envsci.2013.07.009.

Angeon, V., and A. Caron. 2009. "How does proximity impact both the emergence and permanence of sustainable natural resources management systems?" *Nature Sciences Sociétés* 17 (4): 361–72. http://dx.doi.org/10.1051/nss/2009065.

Barré, P., and C. Rioux. 2012. "L'industrie des produits forestiers au Québec: La crise d'un modèle socio-productif." *Recherches Sociographiques* 53 (3): 645–69. http://dx.doi.org/10.7202/1013460ar.

Beaudoin, M. 2000. "La Forêt de l'Aigle: Les leçons d'une expérience réussie nouveaux modes d'exploitation et d'aménagement des forêts au Québec?" In *Actes du colloque sur la forêt habitée, qui s'est tenu du 18 au 20 octobre 2000*, Corporation de gestion de la Forêt de l'Aigle, 12–16.

Becattini, G., M. Bellandi, G. Del Ottati, and F. Sforzi. 2003. *From industrial districts to local development: An itinerary of research*. Cheltenham, UK: Edward Elgar.

Beckley, T.M. 1995. "Community stability and the relationship between economic and social well-being in forest-dependent communities." *Society and Natural Resources* 8 (3): 261–66. http://dx.doi.org/10.1080/08941929509380919.

Belley, S. 2005. "La recomposition des territoires locaux au Québec: Regards sur les acteurs, les relations intergouvernementales et les politiques depuis 1990." In *Jeux d'échelle et transformation de l'état: Le gouverment des territoires au Québec et en France*, edited by L. Bherer, J.-P. Collin, É. Kerrouche, and J. Palard, 203–30. Saint-Nicolas, QC: Les Presses de l'Université Laval.

Benko, G., and A. Lipietz. 1992. "Le nouveau débat régional: Positions." In *Les régions qui gagnent: Districts et réseaux – Les nouveaux paradigmes de la géographie économique*, edited by G. Benko and A. Lipietz, 13–32. Paris: Presses Universitaires de France.

Blair, J.P., and M. Carroll. 2009. "Social capital and local economic development." In *Theories of local economic development: Linking theory to practice*, edited by J. Rowe, 265–84. Burlington, VT: Ashgate.

Blais, R., and G. Chiasson 2005. "L'écoumène forestier canadien: État, techniques, et communautés – L'appropriation difficile du territoire. " *Revue canadienne des sciences régionales* 28 (3): 487–512.

Blakely, E.J. 2009. "The evolution of American (spatial) local and regional economic development policy and planning." In *Theories of local economic development: Linking theory to practice*, edited by J.E. Rowe, 39–62. Burlington, VT: Ashgate.

Bouda, H.-N., P. Savadogo, D. Tiveau, and B. Ouedraogo. 2011. "State, forest and community: Challenges of democratically decentralizing forest management in the Centre-West Region of Burkina Faso." *Sustainable Development* 19 (4): 275–88. http://dx.doi.org/10.1002/sd.444.

Bray, D., B.L. Merino-Pérez, P. Negreros-Castillo, G. Segura-Warnholtz, J.M. Torres-Rojo, and H. Vester. 2003. "Mexico's community-managed forest as a global model for sustainable landscapes." *Conservation Biology* 17 (3): 672–77. http://dx.doi.org/10.1046/j.1523-1739.2003.01639.x.

Brenner, N., and R. Keil. 2006. *The global cities reader*. London, UK: Routledge.

Bullock, R., K. Hanna, and D.S. Slocombe. 2009. "Learning from community forestry experience: Challenges and lessons from British Columbia." *Forestry Chronicle* 85 (2): 293–304. http://dx.doi.org/10.5558/tfc85293-2.

Chiasson, G. 1999. "La gouvernance locale à la fois risquée et favorable pour l'intérêt public." *Economie et Solidarites* 30 (2): 7–20.

Chiasson, G., C. Andrew, and J. Perron. 2006. "Développement territorial et forêts: La création de nouveaux territoires forestiers en Abitibi et en Outaouais." *Recherches Sociographiques* 47 (3): 555–72. http://dx.doi.org/10.7202/014658ar.

Chiasson, G., J.L. Boucher, and T. Martin. 2005. "La forêt plurielle: Nouveau mode de gestion et d'utilisation de la forêt, le cas de la Forêt de l'Aigle." *Vertigo* 6 (2): 115–25.

Chiasson, G., and G. Gaboury. 2000. "Les lots intramunicipaux en Abitibi: Un bilan mitigé." In *Québec 2001: Annuaire politique, social, économique, et culturel*, edited by R. Côté, 262–67. Montréal: Fides.

Chiasson, G., and É. Leclerc. 2013. *La gouvernance locale des forêts publiques québécoises: Une avenue de développement des régions périphériques?* Québec: Les Presses de l'Université du Québec.

Chiasson, G., É. Leclerc, and C. Andrew. 2010. "La multifonctionnalité forestière à l'épreuve de la distance: Réflexions à partir de deux localités de la périphérie québécoise." In *La multifonctionnalité de l'agriculture et des territoires ruraux: Enjeux théoriques et d'action publique*, edited by B. Jean and D. Lafontaine, 161–76. Rimouski, QC: Édition du GRIDEQ et du CRDT.

Comeau, Y. 2003. *Guide de collecte et de catégorisation des données pour l'étude d'activités de l'économie sociale et solidaire*. 2nd ed. Gatineau: Université du Québec en Outaouais et Chaire de recherche du Canada en développement des collectivités.

Cooke, P., and L. Lazaretti. 2008. *Creative cities, cultural clusters, and local development*. Northampton, MA: Edward Elgar.

Costanza, R., R. D'Arge, R.S. de Groot, S. Farber, M. Grasso, B. Hannon, Karin Limberg et al. 1997. "The value of the world's ecosystem services and natural capital." *Nature* 387 (6630): 253–60. http://dx.doi.org/10.1038/387253a0.

Cox, K.R. 2004. "Globalization and the politics of local and regional development: The question of convergence." *Transactions of the Institute of British*

*Geographers* 29 (2): 179–94. http://dx.doi.org/10.1111/j.0020-2754.2004. 00124.x.

De Blois Martin, C. 2002. "Émergence d'une nouvelle économie rurale." In *L'annuaire du Québec 2003*, edited by R. Côté and M. Venne, 241–59. Montréal: Fides.

Dionne, H. 1989. "Développement autonome du territoire local et planification décentralisée." *Revue Canadienne des Sciences Régionales* 12 (1): 61–73.

Doloreux, D., and S. Dionne. 2007. *Évolution d'un système local d'innovation en région rurale: Le cas de La Pocatière dans une perspective historique (1827–2005)*. Rimouski, QC: Édition du GRIDEQ et du CRDT.

Faure, A., J.-P. Leresche, P. Muller, and S. Nahrath. 2007. *Action publique et changements d'échelles: Les nouvelles focales du politique*. Paris: L'Harmattan.

Fauré, Y.A., P. Labazée, and L. Kennedy. 2005. *Productions locales et marché mondial dans les pays émergents: Brésil, Inde, Mexique*. Paris: Editions IRD-Karthala.

Flora, J.L., J. Sharp, C. Flora, and B. Newlon. 1997. "Entrepreneurial social infrastructure and locally initiated economic development in the nonmetropolitan United States." *Sociological Quarterly* 38 (4): 623–45. http://dx.doi.org/10.1111/j.1533-8525.1997.tb00757.x.

Gélinas, N. 2000. "Les expériences de forêts habitées: L'évaluation des participants." Paper presented at the Colloque sur la forêt habitée: Nouveaux modes d'exploitation et d'aménagement des forêts au Québec?, 18–20 October, Maniwaki, QC.

German, L., W. Mazengia, H. Taye, M. Tsegaye, S. Ayele, S. Charamila, and J. Wickama. 2010. "Minimizing the livelihood trade-offs of natural resource management in the Eastern African Highlands: Policy implications of a project in 'creative governance.'" *Human Ecology* 38 (1): 31–47. http://dx.doi.org/10.1007/s10745-009-9291-9.

Gumuchian, H. 1990. *À la périphérie de la périphérie: L'espace rural et le concept de fragilité en Abitibi*. Montréal: Université de Montréal.

Hayter, R. 2000. *Flexible crossroads: The restructuring of British Columbia's forest economy*. Vancouver: UBC Press.

–. 2003. "'The War in the Woods': Post-Fordist restructuring, globalization, and the contested remapping of British Columbia's forest economy." *Annals of the Association of American Geographers* 93 (3): 706–29. http://dx.doiorg/10.1111/1467-8306.9303010.

Hayter, R., and T.J. Barnes. 2001. "Canada's resource economy." *The Canadian Geographer* 45 (1): 36–41. http://dx.doi.org/10.1111/j.1541-0064.2001.tb01165.x.

Higgins, B., and D. Savoie. 1997. *Regional development theories and their applications*. New Brunswick, NJ: Transaction.

Jean, B. 1989. "Le développement régional à l'heure du développement local: 'Le temps des incertitudes.'" *Revue Canadienne des Sciences Régionales* 12 (1): 9–24.

Joannis, M., and F. Martin. 2005. *La dimension territoriale des politiques de développement économique au Québec: Enjeux contemporains*. Montréal: Centre interuniversitaire de recherche en analyze des organisations.

Jouve, B., and P. Booth. 2004. *Démocraties métropolitaines: Transformations de l'état et politiques urbaines au Canada, en France, et en Grande-Bretagne.* Sainte-Foy, QC: Presses de l'Université du Québec.

Joyal, A. 2001. "Économie alternative et filière bois en milieu rural: Au Québec et en Limousin (France)." Unpublished paper.

Kelkar, G., D. Nathan, and P. Walter. 2003. *Gender relations and forest societies in Asia: Patriarchy at odds.* New Delhi: Sage.

Kelly, E.C., and J.C. Bliss. 2012. "From industrial ownership to multifunctional landscapes: Tenure change and rural restructuring in central Oregon." *Society and Natural Resources* 25 (11): 1085–101. http://dx.doi.org/10.1080/08941920.2012.656183.

Klein, J.-L. 1995. "De l'État providence à l'État accompagnateur dans la gestion du social: Le cas du développement régional au Québec." *Lien Social et Politiques* (33): 133–41. http://dx.doi.org/10.7202/005133ar.

Lawrence, A., and B. Ambrose-Oji. 2014. "Beauty, friends, power, money: Navigating the impacts of community woodlands." *Geographical Journal*, 22 July. http://dx.doi.org/10.1111/geoj.12094.

Leloup, F., L. Moyart, and B. Pecqueur. 2005. "La gouvernance territoriale comme nouveau mode de coordination territoriale?" *Géographie, Économie, Société* 4 (7): 321–32.

McGuirk, P.M. 2003. "Producing the capacity to govern in global Sydney: A multi-scaled account." *Journal of Urban Affairs* 25 (2): 201–23. http://dx.doi.org/10.1111/1467-9906.t01-3-00006.

–. 2007. "The political construction of the city-region: Notes from Sydney." *International Journal of Urban and Regional Research* 31 (1): 179–87. http://dx.doi.org/10.1111/j.1468-2427.2007.00712.x.

Mercure, D. 1996. *Le travail déraciné: L'impartition flexible dans la dynamique sociale des entreprises forestières au Québec.* Montréal: Boréal.

Monsson, C.K. 2014. "Development without a metropolis: Inspiration for non-metropolitan support practices from Denmark." *Local Economy* 29 (4–5): 295–308. http://dx.doi.org/10.1177/0269094214532903.

Morin, C. 1991. "La forêt de l'est du Québec: Un apport économique à valoriser." In *Enjeux forestiers*, edited by P. Larocque and J. Larrivée, 37–63. Rimouski: Université du Québec à Rimouski et GRIDEQ.

Moulaert, F., and A. Mehmood. 2008. "Analysing regional development: From territorial innovation to path dependent geography." In *The Elgar companion to social economics*, edited by J. Davis and W. Dolfsma, 607–31. Cheltenham, UK: Edward Elgar. http://dx.doi.org/10.4337/9781848442771.00051.

Murdoch, J. 1997. "The shifting territory of government: Some insights from the rural white paper." *Area* 29 (2): 109–18. http://dx.doi.org/10.1111/j.1475-4762.1997.tb00013.x.

Orozco-Quintero, A., and F. Berkes. 2010. "Role of linkages and diversity of partnerships in a Mexican community-based forest enterprise." *Journal of Enterprising Communities* 4 (2): 148–61. http://dx.doi.org/10.1108/17506201011048059.

Ostrom, E. 1990. *Governing the commons: The evolution of institutions for collective action*. Cambridge, UK: Cambridge University Press. http://dx.doi.org/10.1017/CBO9780511807763.

Papadopoulos, Y., and P. Warin. 2007. "Are innovative, participatory, and deliberative procedures in policy making democratic and effective?" *European Journal of Political Research* 46 (4): 445–72. http://dx.doi.org/10.1111/j.1475-6765.2007.00696.x.

Parkins, J.R. 2006. "De-centering environmental governance: A short history and analysis of democratic processes in the forest sector of Alberta, Canada." *Policy Sciences* 39 (2): 183–202. http://dx.doi.org/10.1007/s11077-006-9015-6.

Polèse, M. 1999. "From regional development to local development: On the life, death, and rebirth (?) of regional science as a policy relevant science." *Canadian Journal of Regional Science* 22 (3): 299–314.

Québec. MRNF (Ministère des Ressources naturelles et de la Faune). 2012. *The forests of Québec vast and fascinating*. http://www.mern.gouv.qc.ca/english/publications/international/forests.pdf.

Ribot, J.C. 2008. *Building local democracy through natural resources interventions: An environmentalist's responsibility*. Policy brief. Washington, DC: World Resources Institute.

Roulec 95. 2008. "Accueil." *Roulec*. http://www.roulec.com.

Sassen, S. 2001. *The global city: New York, London, Tokyo*. Princeton, NJ: Princeton University Press. http://dx.doi.org/10.1515/9781400847488.

Savoie, D. 1986. *Regional economic development: Canada's search for solutions*. Toronto: University of Toronto Press.

Scott, A.J. 2001. *Les régions et l'économie mondiale*. Paris: L'Harmattan.

Sen, A. 1999. *Development as freedom*. Oxford, UK: Oxford University Press.

Shearmur, R. 2011. "Innovation, regions, and proximity: From neo-regionalism to spatial analysis." *Regional Studies* 45 (9): 1225–43. http://dx.doi.org/10.1080/00343404.2010.484416.

Sheppard, S.R.J. 2005. "Participatory decision support for sustainable forest management: A framework for planning with local communities at the landscape level in Canada." *Canadian Journal of Forest Research* 35 (7): 1515–26. http://dx.doi.org/10.1139/x05-084.

Stedman, R.C., J.R. Parkins, and T.M. Beckley. 2005. "Forest dependence and community well-being in rural Canada: Variation by forest sector and region." *Canadian Journal of Forest Research* 35 (1): 215–20. http://dx.doi.org/10.1139/x04-140.

St-Martin, N., and M. Côté. 1999. *La coopérative de solidarité: Guide de formation*. Sherbrooke, QC: Université de Sherbrooke.

Storper, M. 1995. "The resurgence of regional economies, ten years later: The region as a nexus of untraded interdependencies." *European Urban and Regional Studies* 2 (3): 191–221. http://dx.doi.org/10.1177/096977649500200301.

–. 1997a. *The regional world: Territorial development in a global economy*. New York: Guilford Press.

–. 1997b. "Territories, flows, and hierarchies in the global economy." In *Spaces of globalisation: Reasseting the power of the local*, edited by K. Cox, 19–44. New York: Guilford Press.

–. 2013. *Keys to the city: How economics, institutions, social interactions, and politics shape development*. Princeton, NJ: Princeton University Press.

Teitelbaum, S. 2014. "Criteria and indicators for the assessment of community forestry outcomes: A comparative analysis from Canada." *Journal of Environmental Management* 132: 257–67. http://www.sciencedirect.com/science/article/pii/S030147971300697X.

Teitelbaum, S. and E. Saumure. 2010. *Local People Managing Local Forests: An Information Guide to Community Forestry in Quebec*, Solidarité rurale du Québec. http://www.ruralite.qc.ca/fichiers/guides/2309_-_serie_action_-_foresterie_english.pdf.

Ternaux, P. 2006. "Mutation des marchés du travail et régulation des territoires." *Espaces et Sociétés* 124–25: 169–83. http://dx.doi.org/10.3917/esp.124.0169.

Torquebiau, E., and R.D. Taylor. 2009. "Natural resource management by rural citizens in developing contries: Innovations still required." *Biodiversity and Conservation* 18 (10): 2537–50. http://dx.doi.org/10.1007/s10531-009-9706-3.

Veltz, P. 1996. *Mondialisation, villes et territoires: L'économie d'archipel*. Paris: Les presses Universitaires de France.

Vermeulen, S., A.A. Nawir, and J. Mayers. 2008. "Rural poverty reduction through business partnerships? Examples of experience from the forestry sector." *Environment, Development, and Sustainability* 10 (1): 1–18. http://dx.doi.org/10.1007/s10668-006-9035-6.

Yin, Robert K. 2009. *Case study research: Design and methods*. 4th ed. Thousand Oaks, CA: Sage.

Chapter 10 **Striking the Balance**
Source Water Protection and
Organizational Resilience in
BC's Community Forests

*Lauren Rethoret, Murray Rutherford,*
*and Evelyn Pinkerton*

One of the core benefits of community forestry is the opportunity for
an area's residents to determine for themselves what resources in their
surrounding forests are assigned the highest values in forest management (Anderson and Horter 2002; Benner, Lertzman, and Pinkerton
2014). This opportunity has been received with enthusiasm in British
Columbia, where the specific mandates of local organizations responsible for the province's community forests vary as widely as the characteristics and personalities of the communities themselves. For example,
some BC community forests have established a primary mandate of
revenue generation through harvest of the timber resource (McIlveen
and Bradshaw 2005). Others see their land base as an area where traditional activities and cultural connection can take place (Larsen
2003). For some community forests, the land base is managed, first and
foremost, as a source of drinking water. This chapter focuses on two BC
community forests for which protection of source watersheds – those
from which households or communities draw their drinking water – is
a key element of the community forest's raison d'être.

The BC Community Forest Agreement Program emerged as a provincial policy initiative in the late 1990s amidst significant conflict in
and around BC's forest sector. This War in the Woods, as it has been
named, centred on three main issues: the growing perception among
the BC public that forest-management practices in the province were
unsustainable (Lertzman, Rayner, and Wilson 1996); the recurrence of
boom-and-bust economic cycles that had come to characterize the industry (Clapp 1998; McIlveen and Bradshaw 2005); and a concentration

of power over forests in the hands of a few major tenure holders, leaving most citizens of British Columbia with little direct say in how "their" forests were managed (Ambus, Davis-Case, and Tyler 2007). Many lobbyists argued that devolution of power to smaller, community-based tenure holders would address some, if not all, of these issues (Ambus, Davis-Case, and Tyler 2007). In 1998, the provincial government granted community forest pilot agreements to seven communities. As the program grew, many short-term pilot or probationary agreements were converted to long-term community forest agreements, signalling the government's intention to have community forests remain a feature of BC's forest tenure system. As of 2015, fifty community forest agreements have been issued in British Columbia, collectively managing almost 1.4 million hectares of forest (British Columbia, FLNRO 2015).

During the War in the Woods and the early stages of the community forest movement in British Columbia, the demand for local control over source watersheds emerged as a common rallying point in the resistance to conventional logging. Lakes, creeks, and rivers that supply water to drinking water systems are often situated on land that is part of the provincial timber-harvesting land base. The quality, quantity, and timing of water flow from both surface and ground sources are sensitive to disturbance by timber harvesting, road building, and fires (Gluns and Toews 1989; Herbert 2007; Boon 2008). Thus, logging in source watersheds has provoked much tension between BC's forest sector and the communities in and around its operational area. History demonstrates that the BC government appears to have approached community forests as a route through which to simultaneously mitigate water-related conflict on Crown land and access timber from highly constrained land bases – those on which timber harvests are impeded by the presence of sensitive natural or socially valued features (McCarthy 2006; Pinkerton et al. 2008). Indeed, land bases granted under the Community Forest Agreement tenure have often encompassed environmentally sensitive terrain, including source watersheds.

Not surprisingly, this practice of selecting politically and environmentally sensitive watersheds for community forests has caused controversy in some areas. In Sechelt, for example, inclusion of the Chapman and Grey Creek watersheds, the town's primary water sources, within the original boundaries of the Sunshine Coast Community Forest sparked a multi-year debate among community groups, various levels of government, and local First Nations. The conflict culminated in the provincial government removing major parts of the watersheds from the community forest's tenure in 2008 (Bouman and Scott 2009). In other communities, however, residents actively sought inclusion of their source watersheds within the community forest land base, in the hope that a community forest agreement, as an alternative to a tenure held by a conventional logging operation, would provide residents with at least a measure of control over how the watershed would be managed.

In this chapter, we assess the success of two cases, the Harrop-Procter and Creston community forests, in using the Community Forest Agreement as a tool to protect their source watersheds. As will be justified below, we consider "success" to include not only objectives related to forest practices and associated water quality and quantity issues but also the sustainability of the community forest organization as a business, community-based organization, and licensee under BC's forest tenure system.

The Harrop-Procter and Creston community forests have both prioritized the "ecological sustainability" principle of the conceptual framework around which this book is oriented. However, for residents of these communities, an important motivation for prioritizing forest sustainability is to ensure that the drinking water supply is protected; thus, there is also a link to the principle of "local benefits and use." These community forests also acknowledge and profit from the potential local benefits of job creation and revenue opportunities that accompany community forestry. As discussed in the introductory chapter to this volume, in order to consistently deliver these ecological and local benefits over time, both community forests, in addition to being economically viable, also had to achieve a certain minimum threshold level of performance on the other three principles underlying the conceptual framework: participatory governance, local management, and rights over the forest.

The Harrop-Procter Community Cooperative and the Creston Valley Forest Corporation (CVFC) are among the earliest examples of community forestry in British Columbia. They have both been operational since the late 1990s and have therefore had time to mature in their approaches to governance, financial management, and forest stewardship. Both community forest organizations have also evolved significantly over the course of their existence. Their experiences illustrate potential obstacles facing other communities hoping to achieve similar objectives. Their stories also highlight certain effective strategies for overcoming challenges and developing organizational resilience.

## Methodology

The findings in this chapter are based on a study conducted in 2009 and 2010 by the lead author, Lauren Rethoret, which was part of a larger interdisciplinary research project conducted by a team of researchers from Simon Fraser University and the University of British Columbia (see Pinkerton et al. 2008; Rethoret 2010; Mealiea 2011; Benner 2012). Members of the research team travelled to six community forests in the central and southern interior of British Columbia to gather data through semi-structured interviews and site visits to community forest tenures (see Figure 10.1). Interviewees included a diverse range of stakeholders: community forest staff, members of the community forests' boards of directors, local politicians, loggers, sawyers, customers, representatives of nearby conventional logging operations, government regulators, and local opinion leaders. The research team conducted approximately twenty to twenty-five interviews in each community. The interviews covered topics such as forest practices, forest-management objectives, governance arrangements, regulatory systems, perspectives on forestry and forest values, and barriers and bridges to the community forest's success. To corroborate and supplement findings from the interviews, the team examined forest planning documents (e.g., forest stewardship plans, management plans), hydrological studies, and public information from regulatory bodies (e.g., Forest Practices Board evaluations, water quality notices, legislation). The Harrop-Procter and Creston community forests were also the subject of several previous studies (Teitelbaum, Beckley, and Nadeau 2006; Ambus 2008; Bullock, Hanna, and Slocombe 2009). These studies provided critical historical context and were useful in identifying key stakeholders.

Rethoret's (2010) research evaluated community forests' approaches to, and success in, source water protection by assessing the experiences of each case study against an identified set of goals, objectives, and criteria relating to watershed protection and long-term viability. These benchmarks are discussed and justified below. The evaluation protocol assumed that in addition to meeting objectives of protecting water quality, quantity, and timing of flow through sound forest-management practices, successful community forests also had to maintain financial viability, a social licence to operate, and a legal licence to operate under the provincial timber tenure system. Failure in any one of these realms could compromise the ongoing legal and social authority of the community forest to act as a steward of the watershed. This evaluation

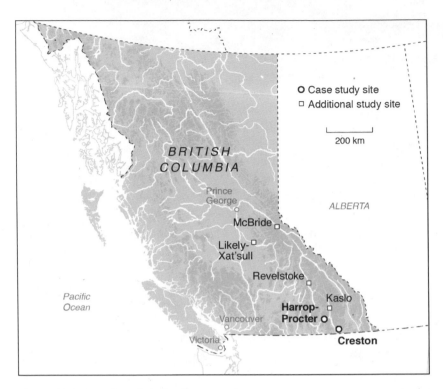

**Figure 10.1**   Location of study sites.
*Source:* Canada, NRCan 2003. Adapted by Eric Leinberger.

approach is supported by Floress et al. (2009), who state that evaluators of watershed management initiatives must measure organizational success by assessing both the longevity of the entity and the management outcomes for which it is responsible.

Our evaluation protocol included four main objectives, with supporting criteria and measures, as detailed in Table 10.1. Each objective and criterion was drawn from relevant provincial legislation or policy documents or from the literature on collaborative and community-based resource management, especially that which specifically pertains to water-management initiatives. The first objective, to engage in forest planning and practices that promote source water protection, acknowledges that ecological outcomes are perhaps the most important indicators of success for community-based resource-management initiatives (Kenney 2001; Leach, Pelkey, and Sabatier 2002). The second objective, to adopt appropriate and effective governance arrangements, is based on several authors' findings that strong decision-making structures and stakeholder-engagement strategies are key determinants in the success of community-based resource-management initiatives (Kenny et al. 2000; Menzies 2004; Ivey et al. 2006). The third objective, to achieve financial stability and maintain funding for water-management initiatives, recognizes that access to stable funding is a primary factor affecting the ability of collaborative watershed-management organizations to achieve their goals (Sommarstrom 2000; Leach and Pelkey 2001). Finally, the fourth objective, to fulfill tenure requirements in order to maintain legal authority over the watershed, is drawn from BC's Forest Act (R.S.B.C. 1996, c. 157), which states that a community forest agreement can be suspended or cancelled if the agreement holder does not adhere to the provisions of their agreement and to those of the Forest Act and the Forest and Range Practices Act (S.B.C. 2002, c. 69).

Data were analyzed through a qualitative coding process that used both inductively and deductively generated codes, merging approaches from grounded theory methodologies (Charmaz 2006; Corbin and Strauss 2008) with those more traditionally applied to evaluative studies (Berg 2004). Using this approach, we were able to examine how the data answered pre-established questions without imposing pre-established notions regarding what the responses might be.

We graded the community forest's achievement on each objective using the scores "met," "partially met," or "not met." Scoring decisions were made qualitatively but systematically. If all criteria were satisfied, we awarded a score of "met." If most of the criteria were satisfied, we awarded a score of "partially met." If most of the criteria were not satisfied, we awarded a score of "not met."

In the remainder of this chapter, we provide an overview of the Harrop-Procter and Creston community forests, summarize our findings from the evaluation described above, and discuss in detail the objectives that proved to be the greatest strength and weakness of each community forest, as per the evaluation. We also explore how performance on these objectives influenced the ability of these community forests to implement their source water protection mandate. Although our focus is on the Harrop-Procter and Creston community forests, we also draw on the experiences of the other community forests examined by the initial research team. Where appropriate and feasible, we have updated findings to reflect changes that have taken place since the conclusion of the 2010 research.

**TABLE 10.1 Protocol for evaluating community forests' success in achieving four primary objectives**

| Objective | Criteria | Measures |
|---|---|---|
| Engage in forest planning and practices, for the following activities, that promote source water protection:<br><br>• timber harvest<br>• reforestation<br>• road building<br>• pest/disease management<br>• interface fire management | Water quality, quantity, and timing of flow conditions within the community forest land base have been considered satisfactory by water users since the community forest's inception. | • Level of stakeholder satisfaction with watershed conditions (interviews)<br>• Watershed conditions described in reports by regulators and water-monitoring initiatives |
| | Efforts to monitor the effects of forest activities on watershed conditions are undertaken by the community forest. | • Number and type of monitoring initiatives (forest planning documents, interviews) |
| | Responses to threats (or perceived threats) to watershed conditions have been addressed by the community forest to a degree that satisfies all stakeholders. | • Level of stakeholder satisfaction with responses to watershed threats (interviews) |
| | Forest planning and practices in source watersheds meet accepted standards for logging activities that protect source water quality. | • Forest practices described in forest planning documents<br>• Forest practices described in interviews<br>• Forest practices revealed by site visits<br>• Performance in Forest Practices Board audits and Compliance and Enforcement evaluations |
| Adopt effective governance arrangements, including sound decision-making structures and stakeholder-engagement strategies | The community forest demonstrates governance arrangements that serve the common interest. Additional protocols for effective community-based governance are also met. | • Governance arrangements established by community forest documents (in comparison with literature standards)<br>• Governance arrangements described in interviews (in comparison with literature standards) |
| | Confidence in governance arrangements is expressed by community forest staff/board members/community members. | • Level of satisfaction with governance arrangements (interviews, minutes of AGM, and board meetings) |

| Objective | Criterion | Measures |
|---|---|---|
| | Level of conflict between community forest and other community groups is manageable and does not affect the community forest's potential for success. | • Level and type of conflict (interviews)<br>• Number and type of documented complaints about the community forest |
| | Level of public engagement with community forest is high. | • Level of public engagement described in interviews<br>• Level of attendance and participation revealed by minutes of meetings |
| Achieve financial stability and funding for water-management initiatives | Financial stability is demonstrated by community forest. | • Financial status described in annual reports and audits (if available)<br>• Financial status described in interviews<br>• Level of community satisfaction with the financial stability of the community forest (interviews) |
| | Commitment to implementing promising funding strategies is shown among community forest staff/board members. | • Number and type of funding strategies proposed, adopted, and implemented (interviews, planning documents, annual reports, minutes of meetings)<br>• Effectiveness of funding strategies (interviews, comparison with literature on specific funding strategies)<br>• Progress in implementing funding strategies (interviews, annual reports, minutes of meetings) |
| Fulfill legal requirements in order to maintain authority over watershed | Environmental management requirements, including harvest commitments, are met by community forest. | • Level of compliance with environmental management requirements (regulatory records, interviews) |
| | Legislated planning and payment requirements are met by community forest. | • Level of compliance with legislated requirements (regulatory records, interviews) |
| | Performance on official audits and evaluations is satisfactory. | • Number and severity of Forest Practices Board complaints<br>• Number and severity of Compliance and Enforcement contraventions |

Source: Rethoret (2010).

## Community Portraits

### Harrop-Procter

The villages of Harrop and Procter are two small, adjacent, unincorporated communities on Kootenay Lake, thirty minutes from the nearby city of Nelson and roughly equidistant between Vancouver, to the west, and Calgary, to the east. The two villages, with a combined population of less than seven hundred, are accessible only by ferry. As a result, they are somewhat physically, socially, and economically segregated from other nearby population centres.

The villages have been involved for decades in a battle to protect their surrounding environment from the potentially damaging effects of logging. Beginning in 1985, when the provincial government announced plans to log the Lasca Creek drainage, a watershed just west of the villages, concerned residents came together to form the Harrop-Procter Watershed and Community Protection Committee. The committee lobbied the local office of the Ministry of Forests, Lands, and Natural Resource Operations (MFLNRO) to develop policies to minimize the implementation of destructive logging practices in the area.[1] The strategy proved ineffective, since the plan to log Lasca moved forward, largely unchanged (HPFP 2015). In response, residents of several Kootenay Lake communities, including Harrop and Procter, organized a series of blockades and protests. In the early 1990s, activists promoted Lasca Creek as a candidate for protection under the BC government's new Protected Areas Strategy (WCWC 1992). A proposal to create the West Arm Wilderness Park achieved success in 1995. The park protected the area surrounding Lasca Creek, but it did not include the mountain slopes and source watersheds directly above the villages of Harrop and Procter (HPFP 2015).

Shortly thereafter, the Harrop-Procter Watershed Protection Society (HPWPS) formed as a collective of citizens with the specific objective of protecting local source watersheds. The BC government had recently announced its intention to commence the Community Forest Pilot Project, and the HPWPS saw the potential for community forestry to bring nearby forested land under local control. The society submitted an application for a community forest pilot agreement, which the MFLNRO initially refused. The HPWPS spent the next two years gathering local support, embarking on public education campaigns, and working with the Silva Forest Foundation, a local organization committed to ecosystem-based forest management, to develop a plan to manage the neighbouring watersheds (Silva Forest Foundation 1999).

Subsequently, the MFLNRO invited the HPWPS to submit a new application for a pilot agreement, and in 1999, the five-year tenure was finally granted (HPFP 2015). In 2007, the MFLNRO converted the Harrop-Procter Community Forest (HPCF) pilot agreement to a twenty-five-year long-term community forest agreement.

At the time of our research, the communities of Harrop and Procter did not have a centralized water filtration or treatment system. Most residents drew drinking and irrigation water directly from surface sources, primarily from three creeks within the community forest land base. Source water protection was thus of critical importance to the villages. Such a high level of direct reliance on watershed conditions was an important driver of management strategies undertaken by the community forest.

Our interviews revealed that, perhaps in part because of the isolated nature of the communities, the local population was relatively cohesive in its specific vision for the community forest. The source water protection objective of the HPCF was well-supported by local residents, who demonstrated consistently high levels of engagement with the community forest. Indeed, to local residents, the initiative was more than simply a forestry operation – it was an important civic organization and a source of community pride.

The Harrop-Procter Community Forest is relatively small, with an area of 10,860 hectares. At the time of our research, the forest was operating with a negotiated allowable annual cut (AAC) of just over 2,600 cubic metres. However, since that time, the HPCF's AAC has been raised to 10,000 cubic metres. This revision was based on an evolving understanding of the forest and its relation to the Harrop-Procter community and is explored in detail in Chapter 12 of this volume. In general terms, a community forest's AAC is the amount of timber that it is expected to cut in the forest each year. When community forests were first established in British Columbia, provincial policies formally required that licensees harvest a volume within 10 percent of their AAC over a five-year period; if this was not achieved, their AAC would be reduced accordingly (Anderson and Horter 2002). This provision was recently eliminated, but the MFLNRO still puts substantial pressure on community forests to cut their full AAC. The original AAC of the HPCF reflected early assessments of the highly constrained nature of the land base, which consists of mostly steep terrain and several source watersheds; however, the AAC also reflected the HPCF's unique stewardship objectives and ecosystem-based management approach, themes that drove lengthy negotiations between the community and regulators regarding limiting annual harvest requirements within the boundaries of the community forest.

The HPCF, the first community forest in British Columbia to achieve certification from the Forest Stewardship Council, is managed by the Harrop-Procter Community Cooperative. The cooperative maintains a board of directors, with membership open to all community members. In an attempt to ensure the ongoing influence of the community's original forest vision, the cooperative requires that half of its directors be members of the board of the HPWPS.

**Creston**

The Creston Community Forest surrounds Arrow Creek, which supplies drinking water to the Town of Creston and the neighbouring community of Erickson. Prior to the inception of the community forest, the Arrow Creek watershed had not been logged since the early 1970s because of strong local opposition to industrial forestry. In fact, the pervasive nature of the conflict in Creston's forests motivated the provincial government to establish, in 1977, a local Public Advisory Committee to the Forest Service (now the MFLNRO) – the first of its kind. The committee, consisting of Creston-area residents, functioned for twenty-four years to provide advice to the provincial government regarding controversial resource-management issues; however, it did not eliminate concern among a vocal portion of the community that remained opposed to conventional logging. A community forest emerged as a potential solution that might allow some logging to occur in the watershed with the approval of the community.

In 1997, the district manager for the Kootenay Lake Forest District announced the availability of a forest licence for a community-based organization in the Creston region. In response, a group of stakeholders consisting of representatives from the Town of Creston, the regional district, a local development authority, a neighbouring First Nation, and a prominent Kootenay-based environmental non-profit organization assembled and submitted an application. The group was initially awarded a non-replaceable forest licence (a conventional form of tenure) with an AAC of fifteen thousand cubic metres (CVFC 2014). In 2008, the MFLNRO converted the licence to a five-year probationary community forest agreement, which, following amendments to the Forest Act, was converted to a twenty-five-year community forest agreement in 2009. The tenure covers a land base encompassing eighteen thousand hectares.

The Creston Community Forest (CCF) adjoins a relatively populated area that includes the town of Creston and several nearby communities. The community forest therefore manages numerous community watersheds. In addition to Arrow Creek, five other creeks within the CCF supply water to communities surrounding Creston. Several domestic watersheds, serving water licences for individual households, exist in the CCF as well (CVFC 2008). A multi-million-dollar water treatment plant, including filtration, was constructed on Arrow Creek in 2005, but the community forest still plays an important role in managing water quantity in the watershed. Most households that are served by the smaller community and domestic watersheds within the CCF land base do not have centralized water treatment or filtration and therefore remain highly dependent on favourable watershed conditions. Consequently, as confirmed by our interview results, many local water users remain apprehensive regarding logging operations in the CCF.

The Creston area economy is highly dependent on clean, plentiful water. The most significant economic activities in the area centre on agriculture and the Columbia Brewery, which is owned by a major international brewing company (CVFC 2008). Perhaps because of the resource-oriented nature of the Creston economy in general, many in the local population do not support outright preservation of Creston's source watersheds. In fact, interviewees told us that during a land-management planning process in the 1990s, agriculturalists and loggers in Creston actively opposed the idea of establishing a park in the Arrow Creek watershed because of the perceived loss of economic opportunities that could result from the removal of this area from the timber-harvesting land base. Unlike in Harrop-Procter, where many residents said they would rather have their watersheds preserved outright, many respondents in Creston indicated that they would prefer that the community maintain opportunities for more diverse local use of the watershed.

Our interviews revealed that the Creston-area population, perhaps because of its size (over five thousand residents), is less socially cohesive than that of Harrop-Procter. A smaller percentage of local residents engaged with the CCF on a regular basis. That being said, many residents recognized the important role that the community forest has played in watershed protection, both in regard to forest conservation and active fuel management for wildfire protection.

At the time of our research, the CCF's tenure was held by a corporation with five shareholders representing various community interests.

The original group of shareholders comprised the Town of Creston, the Regional District of Central Kootenay, Wildsight (a local environ-. mental non-profit organization), the Creston Valley Development Corporation, and the Lower Kootenay Indian Band. Recently, the latter two organizations relinquished their shares and two alternative shareholders joined the board: the Erickson Community Association and the Kitchener Valley Recreation and Fire Protection Society. Both new shareholders represent residents in the rural areas surrounding Creston.

## Evaluating Source Water Protection and Organizational Resilience

### Harrop-Procter Community Forest

The Harrop-Procter Community Forest was founded with a mandate to protect the quality and quantity of water from local watersheds. Our evaluation found that the HPCF had generally fulfilled this mandate. In the following paragraphs, we discuss the community forest's performance on each of the objectives in our evaluation framework; then, we elaborate on the HPCF's primary strengths and weaknesses.

The HPCF partially met the objective of engaging in forest planning and practices that promote source water protection. Interview results indicate that the quality, quantity, and timing of water flow in the watersheds managed by the HPCF have been satisfactory to stakeholders since the inception of the community forest. We confirmed the presence of operational efforts to protect source watersheds through a threefold review: we examined water-quality monitoring documents produced, in part, under a monitoring program undertaken by the community forest (Quamme 2009); a third-party audit undertaken to obtain Forest Stewardship Council certification (Jones 2002); and a Forest Practices Board audit, which found no significant instances of non-compliance (FPB 2008). The HPCF took a careful and restrained approach to logging, evidenced by its frequent use of the intermediate cutting and shelterwood silvicultural systems (see Mealiea 2011), and by its use of unconventional forest practices, including building relatively narrow roads whenever possible (to minimize sources of sediment), using long skid trails (to eliminate roads altogether), leaving high amounts of forest structure in logged areas (including large, mature trees and snags), and transporting logs by cable yarding (to minimize soil disturbance).[2] All of these practices can reduce the detrimental impacts of logging on water sources. Despite this commitment to careful logging, at the time of our research some interviewees thought that the community forest had failed to adequately consider wildfire protection in its harvest plans. Interface fire management is an important component of source water protection in forested areas (Herbert 2007), yet our research indicated that the HPCF had only recently begun to plan for and implement firebreaks on its land base.

Harrop-Procter met the evaluation objective of adopting effective governance arrangements. As detailed below, the community forest demonstrated a strong commitment to adaptive governance and sought to serve the common interest, especially as that interest related to source water protection. As a result of the community's confidence in the organization and the organization's approach to governance, levels of engagement and volunteerism were high.

On the objective of achieving financial stability and maintaining funding for source water protection, the HPCF's performance was mixed. The community forest struggled after its inception to achieve financial stability, but a changeover in management staff in 2005 sparked progress in moving out of its historic position of dependence on grant funding and in-kind contributions from the community. The HPCF's financial difficulties are discussed in more detail below.

The HCPF fully met the objective of satisfying legal requirements as a timber licensee. Interviews with regulators and a review of official compliance documents indicate that Harrop-Procter's fulfillment of its tenure requirements has been exemplary. One HPCF representative explained this as follows: "We're so far beyond what regulations there are in BC that we never ever have a problem ... The [BC] Forest Practices Board came here; we didn't have one infraction. And they said that rarely happens and because we're not doing that much and everything we do is so precautionary that there's just not going to be the same issues."

*Key Strength: Effective Governance Arrangements*
The Harrop-Procter Community Forest is operated by the Harrop-Procter Community Cooperative (HPCC), which receives policy guidance from the Harrop-Procter Watershed Protection Society (HPWPS). The intersection of the two boards creates a unique institutional environment and dynamic for community-based resource governance. Membership in both organizations is open to all local residents, and both organizations have an elected board of directors who make decisions regarding activities within the scope of the organization. The community forest also has a small number of staff members, who vary depending on the funding available and the type of activities in which

the HPCF is engaged at any given time. The HPCC board maintains accountability to the community through annual general meetings, open houses, semi-annual newsletters, and a website. The staff maintains accountability to the board through informal monthly meetings.

A strong leadership presence has greatly enhanced the HPCF's capacity to achieve its mandate of source water protection. Respondents from inside and outside the organization acknowledged the importance of the community forest's leadership in helping the community to achieve its environmental objectives despite operating within an industry and tenure system that are not strongly oriented towards such objectives. The forest manager, especially, was appreciated by many community members and regulators as an individual who has been able to effectively bridge the gap between the interests of the community and those of the provincial government. Some interviewees spoke of a time earlier in the community forest's evolution when such balanced and effective leadership was missing. Those respondents said that the new leadership has helped develop the community forest into an entity that is able to find a way to work within the provincial tenure system to achieve its unconventional objectives.

Strong leadership and engagement are also present among Harrop-Procter's volunteer community, which greatly assists the community forest in maintaining effective governance. Interview results indicate that several board members have been involved with the community forest for many years because of their commitment to its mission. It was apparent that such high levels of continuity in leadership helped facilitate institutional learning. These long-term volunteers have also adopted unofficial roles within the organization that were recognized by some respondents as important factors in contributing to well-ordered, equitable, and effective decision-making. These roles were described by one person as belonging to five categories: the "eagles," who steadfastly advocate the original vision of the HPWPS; the "monitors," who ensure that the organization adheres to its process rules; the "bridge-builders," who understand competing perspectives and help groups with different world views to understand each other; the "communicators," who help build the profile of the HPCF in the community; and the "bulldogs," who work relentlessly with regulators, lenders, and other groups to help the HPCF access the support it needs to achieve its vision.

Some interviewees viewed the HPCF's unique governance structure, involving two organizations with intertwined boards of directors (the HPCC as tenure holder and "manager" and the HPWPS as "steward"), as an advantage in achieving the community forest's source water protection mandate, but others saw it as a hindrance. The HPWPS was created specifically to ensure that decisions and outcomes for the community's watersheds were compatible with the goal of protecting source water quality. The rule that at least 50 percent of HPCC directors must also be members of the board of the HPWPS was designed to guarantee the latter's influence in this regard. However, some respondents saw the relationship between the two boards as limiting the number of local residents willing to get involved with community forest governance. The rule requiring some members to sit on both boards places a heavy burden on volunteers. Thus, board representation has remained more static than might be desirable for an organization that requires such a strong volunteer commitment and fresh energy among its directors in order to carry out innovative and well-planned forestry.

*Key Weakness: Financial Stability*
The HPCF's approach to business management has evolved significantly over time, from ad hoc strategies focused on exploiting nontimber resources in the community forest to well-planned initiatives that centre on the forest's most valuable resource (timber) while still adhering to the ethic of the organization. Original visioning documents produced by the community forest discussed plans to implement innovative commercial strategies that, in theory, would have provided an initial block of funding to then begin performing careful, ecosystem-based forestry. These schemes included an ecotourism operation and a business unit that produced and sold botanicals. While the HPCF made genuine attempts to implement both of these strategies, they were ultimately abandoned because, according to long-term board members, they required too much volunteer effort and resulted in only minimal economic gain. Instead, community members focused their energy on obtaining a number of grants and loans that allowed the HPCF to fulfill the planning and business start-up requirements that were necessary to begin logging operations. These financial inputs were sizable, leading some to question the long-term sustainability of the HPCF's business model.

Though value-added wood processing was a part of Harrop-Procter's business plan from the inception, since the changeover in management mentioned above, the community forest has put substantial effort into developing a value-added strategy that, if effectively implemented, would help the HPCF to attain a higher return for its timber. Recent developments in the community forest's own mill show promise in this direction. *Value-added* is a term used within the forest industry to describe manufacturing processes or other practices that generate forest products with a higher market value and a higher potential financial return than could be obtained by simply converting raw logs into dimensional lumber. Almost all respondents agreed that for a community forest that has such a low AAC yet is committed to a type of forestry that is so expensive, ensuring that a considerable portion of harvested wood is sold at a price premium is a necessity. Value-added strategies have been pursued by many community forests (Charnley and Poe 2007; Anderson and Horter 2002), and implementing such strategies is widely recognized as an approach that could help stabilize resource-dependent communities and improve the state of BC's forest economy as a whole (Hoberg 2001a; DeLong, Kozak, and Cohen 2007). Value-added implementation in the forest industry has proven challenging, however, as many businesses have struggled to find the resources and capital necessary to finance expansion, do market research, and adequately train workers (DeLong, Kozak, and Cohen 2007). In addition, recent research suggests that BC sawmills that specialize in value-added products lack adequate access to timber (Pinkerton and Benner 2013).

Perhaps the most important factor inhibiting the HPCF's financial success has been a lack of economies of scale. The Harrop-Procter land base consists of steep slopes and several sensitive watersheds, which preclude harvests of large volumes of timber, even without the additional constraint of managing for source water protection. In addition, the HPCF land base is very small. Gunter (2000) confirms that such factors are a problem for many community forests in British Columbia. A larger land base would allow for a larger AAC and would also allow the community forest to minimize harvests in drinking watersheds while still producing timber from elsewhere on its tenure. Anderson and Horter (2002) similarly assert that community forests are often "ghettoized," being forced to operate only in socially contentious areas without less constrained or more productive forests to augment available economic opportunities.

The high level of community engagement with the HPCF has played an important role in helping the community forest to survive its financial struggles. Interview results indicate that community and board members have been extremely generous with their time and with other support for the HPCF's money-making strategies. Some residents even invested their own funds in the community forest during its early stages in order to demonstrate the access to financial capital that was required to secure other sources of funding. HPCF staff have also helped the community forest through financially difficult periods, sometimes working without pay or under the assumption that they would be paid at a later date.

**Creston Community Forest**

The Creston Community Forest (CCF) did not perform as well as the IIPCF on our evaluation criteria. However, the CCF did overcome some crucial challenges in its early developmental stages, and at the time of our research, it was maturing into an organization with increased resilience and capacity to successfully carry out its mandate of source water protection.

The CCF met the objective of engaging in forest planning and practices that promote source water protection. In our assessment, these fundamental aspects of community forestry were the CCF's greatest strength, as will be discussed in more detail below. Throughout its history, the CCF has demonstrated a strong commitment to the type of forest practices that are generally known to maintain, and sometimes improve, the integrity of source watersheds.

In contrast, the CCF only partially met the objective of adopting effective governance arrangements. This critical component of any community-based resource-management initiative was a key weakness for the CCF, as will be discussed below. The organization has struggled since its inception with the need to effectively engage the community and maintain its support.

Similar to Harrop-Procter, the Creston Community Forest encountered financial difficulties for many years. However, it has steadily moved over time towards a more stable and viable economic model, thereby partially meeting the objective of achieving financial stability and funding for water-management initiatives. At the time of our research, a debt of more than $500,000 remained from the community forest's early years, when a forest planning error resulted in unanticipated stumpage charges and silvicultural costs of more than $700,000. At the time of our research, the CCF had begun to export some of its lowest-quality logs to the United States in order to increase revenue. This strategy was highly beneficial economically to the community forest, given its close proximity to the border and the higher log prices available in the United States at that time. Export of raw logs, however, was also a highly controversial issue in the Creston area because of perceptions that the practice reduces opportunities for local jobs and economic gain through value-added processing. As a result, several community members discussed the CCF's involvement in the export market as an important factor affecting the level of local support for the community forest. After our field research was completed, further discussions with CCF staff indicated that the financial position of the community forest had greatly improved because of successive profitable harvests, demonstrating that moderate debt loads are not insurmountable for community forests when market conditions are favourable and sufficient timber is available to harvest.

The CCF also only partially met the objective of fulfilling tenure requirements in order to maintain legal rights to manage the watershed. Although our interviews with regulators and a review of annual provincial reports on compliance and enforcement did not reveal any ongoing failures of the CCF to meet its legislated requirements, a 2008 Forest Practices Board audit found that the community forest had failed to meet provincial stocking standards on more than 170 hectares of harvested land (FPB 2009).[3] CCF representatives stated to us that they had intended for most of that land to remain sparsely vegetated for the purposes of wildfire management in and around the Creston watersheds, but there were also areas where that was not the case. Following the audit, the community forest was required to immediately regenerate some sites and to submit revised stocking standards for the sites that would be managed as firebreaks (FPB 2009).

### Key Strength: Forest Practices

Research results confirm that the CCF has focused on the adoption of forest practices that promote source water protection, sometimes to the detriment of its achievement of other objectives more related to organizational resilience. The CCF's logging policies demonstrate adherence to standard protocols for careful forestry in source watersheds. Interviews and site visits suggest that, similar to the Harrop-Procter Community Forest, Creston has generally taken an ecosystem-based approach to logging, guided by a landscape-level plan prepared by the Silva Forest Foundation (Leslie, Bradley, Hammond 2003). According to interviews with CCF staff, the organization had a hydrological assessment done before harvesting any block located in an identified source watershed. While harvest strategies shifted somewhat in the year prior to our research, the CCF used predominantly shelterwood or selection silvicultural systems in its initial cut blocks.[4] Hand falling or cable harvesting were often employed, which resulted in a low level of site disturbance. Harvests were also preferentially completed during the winter months, also promoting minimal site disturbance. In the year leading up to research, the CCF had prescribed some clear-cuts with reserves in order to remove large stands of beetle-killed timber.

The CCF's focus on managing threats from pests and corresponding wildfire risk emerged repeatedly in our research, through multiple data sources. Interviewees confirmed that interface fire management was a primary concern of the community forest, and many of the more environmentally knowledgeable respondents discussed it as an important theme in source watershed protection. CCF representatives spoke of several harvests that were completed with the specific objective of reducing wildfire risk. In some cases, in order to install firebreaks around the town and its water source, the community forest had chosen not to replant logged areas in a manner that met conventional stocking standards described in provincial regulations. As discussed in the preceding section, a 2008 Forest Practices Board audit highlighted this management strategy as a significant act of noncompliance (FPB 2009). During our research, CCF representatives discussed the links between this issue and a more general lack of flexibility within BC's regulatory system to allow for forest management for alternative, non-timber values.

Many of our research respondents acknowledged the CCF's adoption of sound forest practices through their general perception that watershed conditions and associated quality, quantity, and timing of water flow were satisfactorily managed in the community forest. In fact, CCF representatives said that logging activities in the Arrow Creek watershed, which had historically been a contentious area in the debate surrounding the environmental impacts of conventional logging, were undertaken by the community forest with the general support of the citizens of Creston.

The CCF's commitment to managing the forest, first and foremost, for source water protection has challenged its achievement of all three remaining evaluation objectives assessed by our study. As an example of how the CCF's performance on the objective of maintaining community support has been compromised, in the months leading up to our research, a decision to log a pest-affected area within a small, local domestic watershed was a concern for affected water users and therefore affected the public image of the community forest initiative. Despite the CCF's intentions to engage in logging to *improve* the health of the watershed by addressing a significant wildfire risk, certain water users protested the decision because of their belief that logging activities represent a *threat* to source water, no matter the circumstances under which that logging occurred.

### Key Weakness: Governance Arrangements

While the CCF has built trust among portions of the Creston community for its forest operations, other portions remain unsupportive of active management (i.e., timber harvesting) of their source watersheds. This tension embodies some of the CCF's persistent challenges with engagement of stakeholders and participatory governance. Despite the fact that the formal structure of the board of directors is designed to encourage community participation – with five shareholder representatives on the board balanced by five additional board members selected from among community applicants – the CCF has not engaged sufficiently with the broader community in Creston and the surrounding areas.

Brunner (2002) discusses the need to continuously involve community members in decision-making in order to clarify and serve the common interest. Although CCF representatives told us of their desire to see stakeholder involvement in community forest decisions, in practice this involvement has been lacking to a large extent. Awareness of the role of the CCF in managing local forests is low among the Creston population, limiting interest and involvement in public meetings and other engagement opportunities. As previously discussed, the larger size of the Creston population has probably contributed to the community forest's difficulties in raising public awareness. These challenges seem to be exacerbated by the board of directors' "hands-off" approach to forest management. Interviewees said that the board tends to leave most decisions to the discretion of the staff forest manager. Many of our respondents, including forest-management personnel, felt that this policy is detrimental, insofar as it limits the ability of the organization to function as a true community-based entity. Other BC community forests have found value in specifically recruiting leaders who are deeply connected to their region's civic society and who are willing to engage those connections to further the mandate of the community forest (Gunter 2000). CCF staff also acknowledged that greater engagement of multiple community interests could help educate the public regarding important source water protection issues (e.g., interface fire management) and could perhaps help manage public expectations about how the community forest could realistically contribute to achievement of the Creston region's social, economic, and environmental priorities.

The opportunity to educate through engagement is especially relevant to the Creston case, where the most "engaged" citizens have been those who are most vocally opposed to the logging activities of the CCF. In interviews, water users served by creeks within the CCF's land base clearly articulated the risks they perceived to be associated with logging in source watersheds, without fully acknowledging the arguably more significant risks posed by increasing wildfire hazards in these same areas.

Since the completion of our field research, the CCF has made multiple adjustments to its governance arrangements and its approach to community engagement that may help address some of the issues discussed above. First, staff and board members have made initial steps towards increasing the profile of the community forest by investigating opportunities to adopt a more significant role in a prominent local issue – interface fire management on both public and private land. Second, by recruiting representatives of the Erickson Community Association and the Kitchener Valley Recreation and Fire Protection Society (both representing rural areas where some residents remain resistant to logging in source watersheds) to the CCF's board of directors, the community forest has created opportunities for ongoing dialogue that could help resolve past tensions with water users. While the withdrawal of the Lower Kootenay Indian Band as a shareholder and participant on the board is a concern for an organization that seeks a broad community mandate (Bullock and Hanna 2012), the CCF has left the door open for future involvement of the band in community forest governance.

## Conclusion

This volume is based on the presumption that in order to be viable, community forest initiatives must achieve a minimum threshold level of performance on five core principles of community forestry: participatory governance, local management, rights, local benefits, and ecological sustainability. Our research supports this claim. In order for community forests to be successful in managing watersheds for drinking water quality and quantity they must, of course, engage in forest planning and practices that promote source water protection. But they must also ensure that they survive as organizations with sufficient authority over the watersheds to achieve their goals. The HPCF and CCF cases illustrate that striking a sustainable balance between source water protection and organizational resilience requires attention to all five of the core principles. Long-term source water protection entails ecological sustainability. Organizational resilience requires financial viability and ongoing community support, achieved through participatory governance, stable and effective local management, and the provision of local benefits. Legal authority and social licence to operate in the watersheds (rights) are attained and sustained by fulfilling tenure obligations and performing well in all the other areas covered by the core principles.

The concept of rights deserves further discussion here, as the rights granted under the BC community forest tenure have played a critical role in these cases, enabling but also limiting Harrop-Procter's and Creston's attempts to take an alternative approach to managing the timber-harvesting land base. The Community Forest Agreement, like all forms of timber tenure in British Columbia, transfers a significant array of rights to agreement holders, including opportunities to exclusively manage timber resources on a defined landscape. These rights are what attracted concerned water users to consider such an unconventional approach – a logging licence – to protecting their source watersheds.

Yet the regulatory system out of which the community forest agreement emerged imposes certain limitations on community forests' land-management rights that have had important implications for both Harrop-Procter and Creston. Harrop-Procter was refused the right to manage its watersheds without cutting timber. Creston's right to manage wildfire risk in one of its watersheds was limited by legislation that prescribes standards for restocking designed primarily to sustain timber harvests, leaving little room to implement alternative ecological objectives. Both of these limitations on local discretion arise from the provincial government's focus on maintaining the overall level of timber harvest and economic return from BC's forestry land base, a long-time driver of forest policy in the province (Hoberg 2001b). The conflict between rigid top-down standards designed to address provincial economic concerns and the site-specific flexibility needed to meet local concerns about ecological sustainability and other non-timber values has been a major issue for community forestry in British Columbia (Ambus 2008; Ambus and Hoberg 2011).

For Harrop-Procter, like some other community forests in the province, the limited size of its tenure has also challenged its resilience as an organization (Ambus 2008). Undertaking forest planning and practices that promote source water protection is more costly than conventional forestry, and, given the size and nature of the land base under Harrop-Procter's tenure, the community forest has had limited opportunity to recover those additional costs. Interviewees in Harrop-Procter and elsewhere expressed a need to increase the size of the community forest land base, acknowledging that attempts to implement source water protection or other alternative forest-management objectives are not meaningful without tenure rights that permit community forests to remain viable over the long term.

Over the course of their twelve- to fifteen-year existence, the Harrop-Procter and Creston community forests have sought to achieve a delicate balance between their goal of source water protection and the need to maintain resilience as community-based organizations. Our research indicates that both community forests have made progress in meeting this challenge, but their positions are by no means secure. Their routes towards organizational resilience have been longer than expected and fraught with unforeseen obstacles. However, throughout their evolutionary process, the community forests have been able to benefit from lessons learned from each other and from similar initiatives across British Columbia. Their experiences have also informed decisions made by communities that have more recently applied for tenure under the Community Forest Agreement Program. This knowledge-sharing process, supported by organizations such as the BC Community Forest Association and by studies like this one, will be crucial to the future success of BC's community forests.

### Notes

1 Since the inception of BC's community forests program, the provincial government ministry with regulatory responsibility for the program has changed names from the Ministry of Forests, to the Ministry of Forests and Range, to the Ministry of Forests, Lands, and Natural Resource Operations. Throughout this chapter, the current name is used, regardless of the time period in discussion.

2 Intermediate cuttings are small harvests that occur before a more significant harvest that leads to regeneration of the stand. Intermediate cuts are usually designed to enhance the quality of growing conditions for timber (Mealiea 2011). A shelterwood silvicultural system is a "system in which trees are removed in a series of cuts designed to achieve a new even-aged stand under the shelter of remaining trees" (British Columbia, Forests and Range 2008, 93).

3 Provincial stocking standards define the legally required density and spacing of healthy trees that are of a desirable species after harvest. Stocking standards can be met by leaving a certain amount of standing timber after harvest or by replanting the area following more extensive timber removal (British Columbia, Forests and Range 2008).

4  A selection system is "a silvicultural system that removes mature timber either as single scattered individuals or in small groups at relatively short intervals, repeated indefinitely, where the continual establishment of regeneration is encouraged and an uneven-aged stand is maintained" (British Columbia, Forests and Range 2008, 92).

### References

Ambus, L. 2008. "The evolution of devolution: Evaluation of the community forest agreement in British Columbia." Master's thesis, Faculty of Forestry, University of British Columbia, Vancouver.

Ambus, L., D. Davis-Case, D. and S. Tyler. 2007. "Big expectations for small forest tenures in British Columbia." *BC Journal of Ecosystems and Management* 8 (2): 46–57.

Ambus, L., and G. Hoberg. 2011. "The evolution of devolution: A critical analysis of the community forest agreement in British Columbia." *Society and Natural Resources* 24 (9): 933–50. http://dx.doi.org/10.1080/08941920.2010.520078.

Anderson, N., and W. Horter. 2002. *Connecting lands and people: Community forests in British Columbia*. Victoria, BC: Dogwood Initiative.

Benner, J. 2012. "Social contracts and community forestry: How can we design policies and tenure arrangements to generate local benefits in the forestry sector?" Master's thesis, School of Resource and Environmental Management, Simon Fraser University, Burnaby, BC.

Benner, J., K. Lertzman, and E. Pinkerton. 2014. "Social contracts and community forestry: How can we design forest policies and tenure arrangements to generate local benefits?" *Canadian Journal of Forest Research* 44 (8): 903–13. http://dx.doi.org/10.1139/cjfr-2013-0405.

Berg, B. 2004. *Qualitative research methods for the social sciences*. 5th ed. Boston: Pearson Education.

Boon, S. 2008. "Impact of mountain pine beetle infestation and salvage harvesting on seasonal snow melt and runoff." Mountain pine beetle working paper 2008–24. Victoria, BC: Natural Resources Canada.

Bouman, D., and A. Scott. 2009. *The people's water: The fight for the Sunshine Coast's drinking watersheds*. Sechelt, BC: Sunshine Coast Conservation Association.

British Columbia. FLNRO (Forests, Lands, and Natural Resource Operations). 2015. "History of community forests." *FLNRO*. https://www.for.gov.bc.ca/ftp/HTH/external/!publish/web/timber-tenures/community/CURRENT-cfa-status-report.xlsx.

British Columbia. Forests and Range. 2008. *Glossary of forestry terms in British Columbia*. Victoria: Ministry of Forests and Range, Government of British Columbia.

Brunner, R. 2002. "Problems of governance." In *Finding common ground: Governance and natural resources in the American west*, edited by R. Brunner, C.H. Colburn, C.M. Cromley, and R.A. Klein, 1–87 New Haven, CT: Yale University Press.

Bullock, R., and K. Hanna. 2012. "A 'watershed' case for community forestry in British Columbia's interior: The Creston Valley Forest Corporation." In *Community forestry: Local values, conflict, and forest governance*, 82–99. Cambridge, UK: Cambridge University Press. http://dx.doi.org/10.1017/CBO978 0511978678.005.

Bullock, R., K. Hanna, and S. Slocombe. 2009. "Learning from community forestry experience: Challenges and lessons from British Columbia." *Forestry Chronicle* 85 (2): 293–304. http://dx.doi.org/10.5558/tfc85293-2.

Charmaz, K. 2006. *Constructing grounded theory: A practical guide through qualitative analysis.* Trowbridge, UK: Cromwell Press.

Charnley, S., and M. Poe. 2007. "Community forestry in theory and practice: Where are we now?" *Annual Review of Anthropology* 36 (1): 301–36. http://dx.doi.org/10.1146/annurev.anthro.35.081705.123143.

Clapp, A. 1998. "The resource cycle in fishing and forestry." *The Canadian Geographer* 42 (2): 129–44. http://dx.doi.org/10.1111/j.1541-0064.1998tb 01560.x.

Corbin, J., and A. Strauss. 2008. *Basics of qualitative research: Techniques and procedures for developing grounded theory.* 3rd ed. Thousand Oaks, CA: Sage.

CVFC (Creston Valley Forestry Corporation). 2014. "About us: History of the Creston Community Forest." *Creston Community Forest.* http://www.creston communityforest.com/who-we-are/about-us.

–. 2008. *Management plan.* Creston, BC: Creston Valley Forest Corporation.

DeLong, D., R. Kozak, and D. Cohen. 2007. "Overview of the Canadian value-added wood products sector and the competitive factors that contribute to its success." *Canadian Journal of Forest Research* 37 (11): 2211–26. http://dx.doi.org/10.1139/X07-027.

Floress, K., J. Mangun, M. Davenport, and K. Williard. 2009. "Constraints to watershed planning: Group structure and process." *Journal of the American Water Resources Association* 45 (6): 1352–60. http://dx.doi.org/10.1111/j.1752 -1688.2009.00368.x.

FPB (Forest Practices Board). 2008. *Audit of forest planning and practices in the Kootenay Lake Forest District: Harrop-Procter Community Forest and Kaslo and District Community Forest Society.* Victoria, BC: Forest Practices Board.

–. 2009. *Audit of forest planning and practices in the Kootenay Lake Forest District: Creston Valley Forest Corporation, Forest Licence A54214.* Victoria, BC: Forest Practices Board.

Gluns, D., and D. Toews. 1989. "Effect of a major wildfire on water quality in southeastern British Columbia." In *Proceedings of the Symposium on Headwaters Hydrology*, edited by D.F. Potts and W.W. Woessner, 487–99. Bethesda, MD: American Water Resources Association.

Gunter, J. 2000. "Creating the conditions for sustainable community forestry in BC: A case study of the Kaslo and District Community Forest." Master's thesis, School of Resource and Environmental Management, Simon Fraser University, Burnaby, BC.

Herbert, E. 2007. "Forest management by West Coast water utilities: Protecting the source?" *Journal: American Water Works Association* 99 (2): 91–106.

Hoberg, G. 2001a. "'Don't forget government can do anything': Policies towards jobs in the BC forest sector." In *In search of sustainability: British Columbia forest policy in the 1990s*, edited by B. Cashore, G. Hoberg, M. Howlett, J. Rayner, and J. Wilson, 207–31. Vancouver: UBC Press.

–. 2001b. "The 6 percent solution: The Forest Practices Code." In *In search of sustainability: British Columbia forest policy in the 1990s*, edited by B. Cashore, G. Hoberg, M. Howlett, J. Rayner, and J. Wilson, 69–75. Vancouver: UBC Press.

HPFP (Harrop-Procter Forest Products). 2015. "The Harrop-Procter Watershed Protection Society." *HPFP*. http://www.hpcommunityforest.org/about-hpwps/.

Ivey, J., R. de Loe, R. Kreutzwiser, and C. Ferreyra. 2006. "An institutional perspective on local capacity for source water protection." *Geoforum* 37 (6): 944–57. http://dx.doi.org/10.1016/j.geoforum.2006.05.001.

Jones, K. 2002. *Woodmark International Forest certification public report: Harrop-Procter Community Cooperative*. Bristol, UK: Soil Association Woodmark.

Kenney, D. 2001. "Are community-based watershed groups really effective? Confronting the thorny issue of measuring success." In *Across the great divide: Explorations in collaborative conservation and the American West*, edited by P. Brick, D. Snow, and S. Bates, 188–94. Washington, DC: Island Press.

Kenny, D., S. McAllister, W. Caile, and J. Peckham. 2000. *The new watershed source book*. Boulder, CO: Natural Resources Law Centre, University of Colorado School of Law.

Larsen, S. 2003. "Promoting Aboriginal territoriality through interethnic alliances: The case of the Cheslatta T'en in northern British Columbia." *Human Organization* 62 (1): 74–84. http://dx.doi.org/10.17730/humo.62.1.63afcv8 lk0dh97cy.

Leach, W., and N. Pelkey. 2001. "Making watershed partnerships work: A review of the empirical literature." *Journal of Water Resources Planning and Management* 127 (6): 378–85. http://dx.doi.org/10.1061/(ASCE)0733-9496 (2001)127:6(378).

Leach, W.D., N.W. Pelkey, and P.A. Sabatier. 2002. "Stakeholder partnerships as collaborative policymaking: Evaluative criteria applied to watershed management in California and Washington." *Journal of Policy Analysis and Management* 21 (4): 645–70. http://dx.doi.org/10.1002/pam.10079.

Lertzman, K., J. Rayner, and J. Wilson. 1996. "Learning and change in the British Columbia forest policy sector: A consideration of Sabatier's advocacy coalition framework." *Canadian Journal of Political Science* 29 (1): 111–33. http:// dx.doi.org/10.1017/S0008423900007265.

Leslie, E., T. Bradley, and H. Hammond. 2003. *Creston Valley Forest Corporation initial ecosystem-based plan*. Slocan Park, BC: Silva Forest Foundation.

McCarthy, J. 2006. "Neoliberalism and the politics of alternatives: Community forestry in British Columbia and the United States." *Annals of the Association of American Geographers* 96 (1): 84–104. http://dx.doi.org/10.1111/j.1467 -8306.2006.00500.x.

McIlveen, K., and B. Bradshaw. 2005. "A preliminary review of British Columbia's Community Forest Pilot Project." *Western Geography* 15–16: 68–84.

Mealiea, D. 2011. "Comparing forest management practices under community-based and conventional tenures in British Columbia: An ecological perspective." Master's thesis, School of Resource and Environmental Management, Simon Fraser University, Burnaby, BC.

Menzies, N. 2004. "Communities and their partners: Governance and community-based forest management." *Conservation and Society* 2 (4): 449–56.

Natural Resources Canada. 2003. Atlas of Canada data.

Pinkerton, E.W., and J. Benner. 2013. "Small sawmills persevere while the majors close: Evaluating resilience and desirable timber allocation in British Columbia, Canada." *Ecology and Society* 18 (2): art. 34. http://dx.doi.org/10.5751/ES-05515-180234.

Pinkerton, E., R. Heaslip, J. Silver, and K. Furman. 2008. "Finding 'space' for comanagement of forests within the neoliberal paradigm: Rights, strategies, and tools for asserting a local agenda." *Human Ecology* 36 (3): 343–55. http://dx.doi.org/10.1007/s10745-008-9167-4.

Quamme, D. 2009. *Water quality update for Harrop (Mill) and Narrows Creeks: 2005–08*. Nelson, BC: Integrated Ecological Research.

Rethoret, L. 2010. "Evaluating BC's Community Forest Agreement Program as a tool for source water protection." Master's thesis, School of Resource and Environmental Management, Simon Fraser University, Burnaby, BC.

Silva Forest Foundation. 1999. *Ecosystem-based forest use plan for the Harrop-Procter watersheds*. Slocan Park, BC: Silva Forest Foundation.

Sommarstrom, S. 2000. "Evaluating the effectiveness of watershed councils in four western states." In *Proceedings of the Eighth Watershed Management Council Conference*, edited by R. Coates. Riverside, CA: Center for Water Resources, University of California.

Teitelbaum, S., T. Beckley, and S. Nadeau. 2006. "A national portrait of community forestry on public land in Canada." *Forestry Chronicle* 82 (3): 416–28. http://dx.doi.org/10.5558/tfc82416-3.

WCWC (Western Canada Wilderness Committee). 1992. *Save Lasca*. Educational Report 11(5). WCWC – West Kootenay Branch. https://www.wildernesscommittee.org/sites/all/files/publications/1992%2006%20Lasca%20Creek.pdf.

Chapter 11    **Practicing Participatory Governance through Community Forestry**
A Qualitative Analysis of
Four Canadian Case Studies

*Sara Teitelbaum*

In Canada, community forestry has figured periodically in debates around forest policy reform, primarily because of the interventions of civil society organizations, including environmental groups and some rural constituents and organizations (Coon 2004; Solidarité rurale 2008). The appeal of community forestry, at least in principle, stems from its association with wider social, economic, and ecological ideals and objectives. The literature frequently makes reference to community forestry's potential to foster more participatory approaches in decision-making, produce better ecological outcomes, and create more locally centred economic development approaches (Duinker et al. 1994; McCarthy 2006; Albert 2007).

However, the Canadian discourse on community forestry has rarely gone beyond the level of generalities. On the one hand, there are the bold statements of civil society organizations, which have tended to adopt an uncritical stance towards the concept of community forestry and its potential to remedy many of the ills of current forest-management practice (Coon 2004; Sierra Club of Canada, 2004). On the other hand, we have an academic literature, which, while generally supportive of community forestry, has pointed to a series of institutional barriers that have limited the success and proliferation of community forestry thus far. Chief among these is lack of support by governments, which are not facilitating the development of community forestry through the allocation of tenures or appropriate tenure conditions, the provision of financial and technical support, or economic incentives (Anderson and Horter 2002; Bradshaw 2003; McIlveen and Bradshaw

2005–6). While the literature gives us a sense that there is a considerable gap between "rhetoric" and "reality," few studies go as far as to detail the experience of particular community forests in their pursuit of the goals described above. We therefore know little about the extent to which these ideas are driving community forestry practice, how they are interpreted or put into practice by community forest organizations, and what the concrete results of these efforts are.

This chapter investigates the coherence between participatory governance, one of the theoretical concepts driving the community forestry movement, and the practical realities of four community forest organizations. In this effort "to make explicit the gap between theory and practice" (Beckley 1998, 737), I examine a number of dimensions of participatory governance, including representation, participation, and empowerment. I seek to describe governance experiences, not only in order to point to results achieved but also to highlight the specific approaches and strategies these community forests have employed and the factors that have either facilitated or hindered their success. The intention is to help create a more nuanced and contextually bound understanding of what the ideas behind community forestry mean when they are applied in diverse forestry contexts in Canada. It is my hope that this may help to inform the public about the state of practice in community forestry, possibly realign expectations, and provide useful lessons in terms of implementation.

## Methods

This chapter stems from doctoral research that sought to evaluate performance, through a comparative case study analysis, in terms of three key concepts: participatory governance, local economic benefits, and multiple forest use (Teitelbaum 2014). I selected these four case studies to represent the diversity of initiatives taking place across Canada. Previous to this study, a national survey of community forests identified 116 initiatives corresponding to the following definition of *community forest*: "a public forest area managed by the community, as a working forest, for the benefit of the community." (Teitelbaum, Beckley, and Nadeau 2006). The definition aimed to capture all community-based organizations (more than 50 percent local decision-making) managing either Crown or local government-owned land with an explicit mandate to benefit the community. These 116 cases became the pool from which I selected these four case studies. While they were not chosen in an explicitly random or statistically representative fashion, they were selected to represent diversity in terms of four characteristics: organizational structure, decision-making structure, land and tenure type, and political context.

In the doctoral study, a social indicator analysis was performed and fifty-two semi-structured interviews were conducted. The interviews took place over a two-year period, between June 2004 and December 2006. The thematic content for both the social indicator analysis and the interviews was derived from a broad literature review, which included not only sources from the community forestry literature but also related topics such as collaboration, social forestry, co-management, public participation, and community economic development (see Teitelbaum [2014] for an extensive list of sources). Different types of documents were reviewed, including academic literature, certification standards, governmental and industry reports. This literature review allowed for the establishment of a series of widely recognized evaluation criteria for each of the three themes of participatory governance, local economic benefits, and multiple forest use. These criteria were used as the basis for the development of specific indicators and helped to inform the interview protocol. For the interview segment of this project, a protocol was established that sought to capture two types of information: important background information about each case – origins, participants, orientations, operations, challenges, and so on – and specific information related to each criterion established in the indicator framework. In the case of participatory governance, this included issues of representation, inclusivity, transparency, authority of decision-making, and public participation.

This chapter is based on the data generated from the interview portion of the doctoral project. In total, fifty-two semi-structured interviews were conducted. Interviews were conducted with a wide variety of stakeholders, including community forest employees and board members, government representatives, recreational users, tourism operators, local politicians, economic development officers, members of First Nations, environmentalists, and forestry contractors. Findings were triangulated with data from written documents such as policies, annual reports, consultant reports, newspapers, and newsletters.

Interviews were all fully transcribed and coded, on an ongoing basis, using a qualitative software. Categorical aggregation was used to organize the interview data. This involved identifying main themes from the text with a word or short phrase. The thematic categories were oriented around the three main research concepts and also covered important contextual issues. The categories were not static but changed over the course of the coding process. In this chapter, all quotations by interviewees were translated from French to English by the author.

## Introducing the Case Studies

The case studies span three Canadian provinces: British Columbia, Ontario, and Quebec (see Figure 11.1). All four cases have an organizational structure that is not driven by profit, have a decision-making structure made up of 100 percent local representatives, and manage an area of public land of similar size. A summary of characteristics is presented in Table 11.1. What follows is a brief description of the four case studies.

The first case study is a community forest organization called the Creston Valley Forest Corporation (CVFC) situated in the town of Creston, in the Kootenay region of British Columbia. The community of

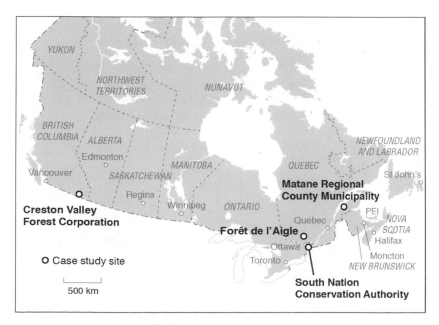

**Figure 11.1**   Map of case study sites.
*Source:* Google Maps. Adapted by Eric Leinberger.

**TABLE 11.1  Background information on community forest case studies**

| | Creston Valley Forest Corporation | Corporation de gestion de la Forêt de l'Aigle | Matane RCM | South Nation Conservation Authority |
|---|---|---|---|---|
| Location of community forest | Kootenay region of British Columbia | Southwest of the town of Maniwaki, in the Gatineau region of Quebec | Near Matane, in the Gaspésie region of Quebec | South Nation watershed in eastern Ontario |
| Organizational type | Private corporation with not-for-profit status | Private corporation with not-for-profit status | Regional county municipality (RCM) | Conservation authority |
| Year community forest started | 1997 | 1996 | 1999 | 1997 |
| Size (ha) | 15,000 | 14,000 | 13,000 (approx. 25 parcels) | 3,500 owned and 14,000 managed for local governments |
| Type of land | Crown | Crown | Crown | Land owned fee simple by conservation authority |
| Tenure | Forest licence in transition to community forest agreement | Forest management contract | Territorial management agreement | None |
| Average harvest, 2005–6 (m3) | 15,000 | 39,000 | 13,000 | 2,185 |

close to five thousand is located near the US border, between the Purcell and Selkirk mountain ranges. The Lower Kootenay Indian Band (pop. 206) has a reserve just south of the town of Creston. The CVFC was formed in 1996 and has operated under a Crown land licence since 1997. In 2008, the organization's forest licence was transferred to a community forest agreement. The CVFC's operating area is largely within the community's watershed (Arrow Creek) and includes some lands adjacent to the community. The primary objective of the CVFC is to implement ecosystem-based management in the watershed.

The second case study, the Corporation de gestion de la Forêt de l'Aigle (also referred to as the Forêt de l'Aigle) is situated in the Outaouais region of Quebec between the municipality of Cayamant and the town of Maniwaki (pop. 4,102). This region has a long history of forest-industry dependence, in part because of the high diversity of trees and the proximity to markets in Gatineau and Montreal. However, the region has seen the closures of several large sawmills and paper mills in recent years. The Kitigan Zibi Anishinabeg First Nation has a large reserve (18,437 ha) adjacent to Maniwaki. The Corporation de gestion de la Forêt de l'Aigle was formed in 1996, specifically to take over management of a fifteen-thousand-hectare tract of Crown land situated just south of Maniwaki, land that had not been under industrial tenure. The mission of the Forêt de l'Aigle, a not-for-profit corporation, is to manage the forest for multiple values and to contribute to the socio-economic development of the Outaouais region. Since the time of research, the organization has gone through major restructuring and is now called the Coopérative de solidarité de la Forêt de l'Aigle.

The third case study is the Matane Regional County Municipality (also referred to as the Matane RCM) in the Gaspésie region of Quebec. This region borders the St. Lawrence River, just north of the Bas-Saint-Laurent region. The Matane RCM is an administrative region that comprises eleven municipalities and has a population of 22,646. This region is characterized by both its connection to the St. Lawrence – Matane serves as a port and has a large shrimp-processing factory – and its connection to the forest. The Matane RCM is a regional government organization with a decision-making structure composed of the mayors of the eleven municipalities in the region. The community forest is made up of "intramunicipal lots," which are scattered parcels of Crown land within municipal boundaries. The forest area totals thirteen thousand hectares in twenty-five parcels. These were under the authority of the Ministère des Ressources naturelles et de la Faune and managed by joint management organizations until 1999, when they were transferred to the RCM through a territorial management agreement. Publicized by the Government of Quebec as a step towards "decentralization of powers" (Quebec, MRNF 1999), the agreement gives the RCM jurisdiction over many activities, including forest management, infrastructure, leasing and selling of land, permitting, and some monitoring. Currently, the Matane RCM has allocated operational responsibilities to two municipal organizations but does the planning in-house.

The fourth case study, the South Nation Conservation Authority (also referred to as South Nation), was formed in 1947 and is a type of para-municipal environmental organization that manages its own private forests for the public interest. Conservation authorities are community-based environmental agencies organized on a watershed basis with boards of directors made up of municipal representatives. Their mandate includes environmental protection, water resource management, forest management, and environmental education. The South Nation River watershed, located in eastern Ontario, comprises almost four thousand square kilometres and flows into the Ottawa River; its southern border is just north of the St. Lawrence, between Brockville and Morrisburg. The South Nation Conservation Authority, a conservation organization with more than forty permanent employees, has divisions in water, forestry, planning and engineering, sewage, and works. The community forest comprises forest lands that were once held under the Agreement Forest Program and managed by the provincial government; however, South Nation took over management of its approximately 3,500 hectares of forests in 1997. South Nation has since been hired by two united counties in the watershed to manage their agreement forests.

### Participatory Governance: Theory

Community-based management approaches such as community forestry have been put forward as a way to reinvigorate local democracy and make decision-making more responsive to local concerns (Beckley 1998; Robinson, Robson, and Rollins 2001). According to Leach (2004, 1), "proponents see devolution as a tremendous opportunity to reinvigorate American democracy by engaging local officials and ordinary citizens in the stewardship of natural resources, and by spurring a more thoughtful, less partisan, policy dialogue." These ideas resonate with the notion of participatory governance, which is an approach to governance that seeks to involve citizens more closely with the political institutions that affect their lives (Gaventa 2002). Kearney et al.

(2007, 2) define participatory governance as "the effort to achieve change through actions that are more effective and equitable than normally possible through representative government and bureaucratic administration by inviting citizens to a deep and sustained participation in decision-making." Participatory governance is therefore conceived as a way of overcoming political apathy and reinvigorating citizen engagement in issues of societal importance (Fung and Wright 2001). Proponents contend that this will widen the perspective of participants, encouraging them to consider a wider range of issues, reduce conflict, build social networks and foster social learning, and improve overall natural resource management outcomes (Schusler et al. 2003; Reed and McIlveen 2006; Newig and Fritsch 2009).

The study of participatory governance is part of a larger scholarship on governance more generally, a scholarship that is also concerned with fostering collaboration and civil society participation in order to increase democratization (Carrier and Jean 2000; Ayeva 2003; Turnhout et al. 2010). Improving people's capacity to participate in governance is described as a way "to rebuild a sense of community and to restore a capacity for community self-efficacy among publics in those communities" (Overdevest 2000, 686). Although participatory governance is often contrasted with "representative" forms of governance, it does not imply an outright rejection of this democratic form; rather, it implies the strengthening of the participatory dimensions of these processes and the creation of new institutions where the state and citizens can engage and redefine their relationship (Gaventa 2002). Reddel and Woolcock's (2004) summary of some of the principles and methodologies that are important to the practice of participatory governance illustrate its compatibility with community forestry. They include innovation; negotiation and transformative partnerships; the privileging of local and technical knowledge in the policy process; and the reinvention of government based on system-wide information exchange, knowledge transfer, decentralization of decision-making, and interinstitutional dialogue.

The community forestry literature, both in developing and industrialized contexts, is grappling with the question of how best to facilitate participatory governance. The intention of participatory governance is to solicit broad-based participation – or, as articulated by Moote et al. (2001, 99), to create "a process that proactively engages all community groups, providing a forum that allows everyone equal voice and participation, regardless of their relative size or influence within the larger community." However, in reality, there are limits to participation, in terms of both workability and public interest. Contemporary discussions therefore revolve around how to build a governance process that is sufficiently inclusive while also being accountable both downwards to the community and upwards to higher-level governance structures. A number of key concepts have emerged as essential attributes of citizen-based governance processes, including representation, inclusiveness, transparency, and empowerment (Wellstead et al. 2003; Larson and Ribot 2004; Leach 2004; Beckley, Parkins, and Sheppard 2005).

## Participatory Governance: Case Study Results

For the Creston Valley Forest Corporation (CVFC), the primary goal of the community forest is the protection of the community's watershed through the application of ecosystem-based management. The community forest was created in the mid-1990s under distinctly conflict-filled circumstances. For several decades, there had been division in the community over the issue of timber harvesting in the watershed. The Arrow Creek watershed, which lies directly adjacent to the community, provides the town's domestic and industrial water supply. Maintaining the quality of water is particularly important for the community because of the presence of several industries such as fruit orchards and a large brewery, which rely on high-quality water. The conservation-oriented contingent argued that harvesting in the watershed should be banned outright and called for the permanent protection of this forest area through some type of ecological reserve or protected area status. These residents had succeeded in securing an informal hiatus on logging from the 1970s onward, despite the fact that logging rights in Arrow Creek were allocated to a large forest company. The other contingent, more closely allied with forest industry interests, argued that logging and the protection of water quality were not incompatible and that forestry activity was necessary in order to bolster a declining forest economy in the region. When the watershed failed to be protected through subsequent land-use planning processes such as the provincial Commission on Resources and Environment (CORE) and the Protected Area Strategy in the 1990s, residents lobbied the government for a community-based solution. Some residents, with the help of a local MLA, persuaded the government to allocate the forest area to a community-based organization. The community forest was therefore conceived as a type of compromise solution – an organization that would be dedicated to the implementation of "careful" harvesting practices and would be responsive to local interests by including within its governance the very organizations that were opposed to logging.

In terms of participatory governance, the primary mechanism was conceived as the creation of a diverse board of directors, representing numerous forestry-related interests but with the all-important inclusion of those groups opposed to logging in the watershed. At the outset, five organizations came together to form the not-for-profit CVFC: the Town of Creston and the Regional District of Central Kootenay, both local government organizations; the Creston Development Authority, a public-private partnership organization promoting economic development; the Lower Kootenay Indian Band; and Wildsight, an environmental organization previously known as the East Kootenay Environmental Society. The local forest industry, composed primarily of medium-sized, family-owned businesses, made a parallel application for the community forest tenure but was turned down. In order to enhance the level of community participation, the five founding organizations chose to reserve an additional five seats on the board of directors for members of the community at large. According to CVFC policy, individuals wishing to join are required to make an application to the board of directors and are elected for a period of one year, with the possibility of renewal.

On the one hand, members of the board of directors describe this as a strong and diverse board structure and feel that it is sufficiently inclusive of the community's interests. On the other hand, the community forest has done relatively little to develop the public participation aspect of the organization. CVFC undertakes few formal public participation activities such as public presentations, advisory committees, or open houses, and there is no direct mechanism for the community at large to influence decisions through the election of board members. The primary means of gathering public views, beyond the input of the board, are through public review of management plans (a legislative requirement) and referral letters (also a legislative requirement) and through offering field tours to interested members of the public. One employee of the organization estimates that they provide, on average, a dozen field tours per year, although some of these are for people from outside the region. A sentiment expressed in interviews was that public participation is more an opportunity to enhance social acceptability of the organization through demonstrating its forest-management practices than an exercise in gathering public input. This may be due to the conflict-filled circumstances in which the organization was formed and the primary mandate of preserving ecological integrity. An employee describes the organization's philosophy about consultation: "A much better form of communication is first to live, to walk your talk, and then to take people out and walk them through the result. And you don't have to say anything. They can see for themselves."[1]

The impetus for the CVFC was to create a local governance structure that could bring together those interests in the community willing to forge a common vision of ecosystem-based forest management. The community forest has, to a large extent, remained wedded to this vision, focusing on the application of alternative forestry practices. Most harvesting is done using small equipment, and in the past, horse logging was attempted. The use of these alternative approaches appears to have had the desired effect of calming the debate around forestry to some extent. While in the early years the organization was under intense scrutiny from local citizens, public interest has died down considerably. After a veritable hiatus on forestry in the watershed in the 1970s and 1980s, the community forest has now conducted forestry operations without controversy for more than ten years.

However, according to interviews with several board members, it has not always been easy to maintain consensus around the vision of ecosystem-based forestry, and certain members, such as the general manager and the environmental group representative, have had to continually promote this vision among board members. Certain organizations, such as local government interests, question the lack of economic return to the community due to what they perceive as an inordinate focus on ecologically appropriate forestry practices.

The example of the CVFC illustrates what can be a tenuous balance within the "interest group" board structure between the need to forge and adhere to a common vision for the community forest and the need for decision makers to protect the particular interests of their organizations. This was also reflected in the dynamics that emerged between the community forest and one of its founding members, the Lower Kootenay Band. The First Nation originally joined the organization because its members shared a concern for protecting water quality in the Arrow Creek watershed; however, the relationship has been fraught with difficulties at a political level.

The Corporation de gestion de la Forêt de l'Aigle has a similar governance structure to that of the CVFC. The Forêt de l'Aigle is a shareholder-based corporation composed of organizations with a direct interest in forest management. Unlike CVFC, however, it was not formed under conflictual circumstances. The formation of the Forêt de l'Aigle was tied to the emergence of provincial policy in the mid-1990s that, albeit briefly, provided financial support to a series of "inhabited forest" pilot projects. The inhabited forest movement, which was endorsed briefly by the Government of Quebec, aims to bring lands adjacent to settlements under some form of collective management for the socio-economic benefit of those communities (Bouthillier and Dionne 1995). The forest, which had been under protected reserve status since the 1970s because of the presence of large stands of white pine, was offered up by the Ministère des Ressources naturelles et de la Faune as a site for experimentation around local participation and forestry innovation. A number of different local organizations recognized this as an opportunity and expressed an interest in managing the area. For the Institut québecois d'aménagement de la forêt feuillue (IQAFF), the forest area represented a place to experiment with new and innovative forestry practices and to generate organizational financing through forest harvesting. For the Kitigan Zibi Anishinabeg First Nation, the forest represented an opportunity for employment creation and a way to ensure the sustainable management of a forest adjacent to their own reserve lands. For the Societé sylvicole de la Haute-Gatineau, the forest represented an economic opportunity and a means to increase their forest-management expertise. However, rather than granting it to one group, the provincial ministry encouraged the organizations to adopt a collaborative management model. The governance structure of the Forêt de l'Aigle was purposely left open in order to include new members. Since its inception, five other organizations have joined, including two hunting and fishing organizations (ZEC Bras-Coupé-Désert, ZEC Pontiac), the municipality of Cayamant, a snowmobile organization (l'Association des motoneigistes les Ours Blancs), and a college (Collège des Outaouais). Therefore, while the organization does not have a specific mandate to reduce conflict, it was nonetheless conceived as an explicit attempt to foster collaboration on a regional level.

From the perspective of fostering regional-level cooperation and decision-making, the current structure appears to be working well. The organization has been described as a success story in the literature because of its ability to integrate the different values and priorities of its diverse membership, while staying tuned in to those of the local population (Gélinas 2001; Chiasson, Boucher, and Martin 2005).[2] According to interview respondents, the original board developed a strong set of organizational objectives from the outset, including a multiuse vision for forest management, strong citizen involvement, and the maximization of employment opportunities. The founding members were cognizant of the importance of generating public support for their initiative, particularly because this area had been open-access Crown land for several decades and therefore represented an important recreational zone for local residents. Although the Forêt de l'Aigle had no rights over non-timber activities, residents were nonetheless concerned that the new organization would interfere with their access to hunting, camping, and fishing. The Forêt de l'Aigle therefore made numerous efforts to disseminate information about the initiative and to solicit feedback, including an initial survey to gauge public interests and values towards forest management. This appears to have had the desired affect, as interviewees described the social acceptability of the community forest as having grown over time.

Interviews nonetheless revealed certain dissatisfactions with the functioning of the Forêt de l'Aigle's decision-making structure. The organization quickly built up its internal capacity for management through the hiring of a number of specialists such as a general manager, a recreation specialist, and a biologist. While this had the effect of building considerable internal capacity from a management perspective, interview respondents described a certain dynamic that has arisen whereby the board of directors has taken a back seat to employees in the decision-making process. Several employees described the board as insufficiently interested and engaged with the community forest and as overly reliant on employees for leadership. There has also been high turnover among board members.

As in the case of the CVFC, some board members, though not all, tend to perceive their participation in the community forest as an opportunity to ensure that the interests of their own organization are not compromised. For example, one of the founding members chose to withdraw from the board when the board decided to change the organization's bidding process for contracts. In describing this situation, one employee of the business put it this way: "That's why we were members. We figure that if we are a member of something, we have the right to expect certain things ... There is no longer any advantage to being a member of an organization like the Forêt de l'Aigle." Similarly, another board member's principal interest in participating was to ensure that the area remained open for local hunters and fishers. It seems that for some shareholder members, the relationship of accountability is more closely aligned with the constituents of their particular organization – whether this be the residents of a particular municipality or a certain group of recreational users – than with the well-being of the community as a whole.

The Forêt de l'Aigle has, however, been very proactive in soliciting public input. The community forest is strongly conceived by founding organizations as a "social project" that aims to incorporate the values and vision of the regional community in the decision-making process and the overall management approach. The organization quickly set up an office in the town of Maniwaki and, soon after, a visitor pavilion within the forest itself, providing the Forêt de l'Aigle with strong visibility and an infrastructure by which to interact with the public. The organization also has a strong public consultation regime, including annual open meetings at multiple locations to present its annual plan, harmonization agreements with other users of the territory, and field tours to discuss the potential impact of operations. A government employee contrasts the consultation approach of the Forêt de l'Aigle with that of other industrial licensees:

> They [forest companies] are more in the mode ... of putting an ad in the newspaper and say[ing], "Come and see us," whereas the Forêt de l'Aigle takes a different approach. They really publicize. They go into the municipalities to present their annual plan and their vision for development. So their approach brings them a lot closer to the population. Whereas the forestry companies, they just follow the law. That's it.

The other two case studies, the Matane Regional Council Municipality and the South Nation Conservation Authority, are both loosely based on the "local government" approach. Under this model, local elected officials, or some derivative thereof, are the primary decision-makers for the community forest. The Matane RCM is a regional municipal organization that is responsible for scattered parcels of Crown land within municipal borders, commonly referred to as intramunicipal lands. Responsibility for these lands was transferred from the Ministère des Ressources naturelles et de la Faune to the RCM in 1999 as part of a larger decentralization program on the part of the government, which extended forest-management responsibilities for intramunicipal lands to the level of RCM. Previous to this, responsibility for these forests was contracted out to joint management organizations. The objective of this program is to allow the regional municipalities to manage these forests more actively and thereby to contribute to the creation of new economic opportunities and the overall revitalization of local communities. Revenues generated through stumpage payments or the sale of certain lots contribute to a fund (Fond de mise en valeur) that can be used to hire a forest manager and to finance forestry or agricultural projects. The ministry offers start-up funds to RCMs in accordance with the size of the forest lands being transferred.

The Matane RCM, and this region more generally, has a history of mobilization around forests. The most recent widespread grassroots movement took place in the early 1990s, when a series of forestry committees formed throughout the municipalities of the RCM. These groups formed in response to what was seen as an unacceptably high level of corporate control over Crown land (80 percent of the RCM), coupled with the loss of processing opportunities and the perception of a growing degradation of the forest resource (Dionne and Saucier 1995). The vision of these forestry committees, which were supported by various community-based organizations including the CLSCs (centre local de services communautaires, or community-based health organizations) and the RCM, was to privatize municipal Crown lands and establish forest tenant farms (290 in total). According to a past director of the RCM, in the early 1990s, the RCM was much more politically active than at the present time and made efforts to extend regional influence over forest management, including attempts to pass bylaws in order to limit the extent of clearcutting and to protect certain endangered habitats.

> In the beginning it was really the appropriation of all the public lands within the regional county municipality. Not just the intramuncipal lands. We meant Crown lands in the large sense of the word. We were in a time when manual forestry was being replaced by harvesters, and people felt they were going to lose their jobs if they didn't do something – that they'd be replaced by machines. So people took charge. They organized themselves. Like I said, they formed a committee, and they did a lot of consultation – on the whole territory of the Matane RCM.

In 2000, the Matane RCM was granted responsibility for the management of thirteen thousand hectares of intramunicipal lands through a territorial management agreement tenure. These lands are scattered throughout the regional municipalities and are not considered highly profitable from a forestry perspective because of a history of exploitation and because they are small and surrounded by private lands (Roy

2006). Initially, the RCM set up a forest tenant-farming project on a portion of these lands, although on a much more modest scale than had been envisioned previously. Seven one-thousand-hectare farms were established and were operational between 2000 and 2005. However, these forest farms were not particularly profitable, in part because of the ineligibility of the forest farmers for employment insurance, which many other forest workers in the region relied on. The project was dissolved in 2004, and since then, the forests have been allocated to a municipal corporation.

The primary tenure holder, and thereby decision-making structure, for these forests is the RCM board, made up of mayors from each of the eleven municipalities within the RCM borders, whose responsibilities include a broad portfolio of political responsibilities in addition to forest management. The RCM has one forester on staff and has hired a biologist on contract through a government-sponsored program. As is permitted through its particular tenure arrangement, called a territorial management agreement, the RCM has opted to download many operational-level decisions to two municipal corporations, the Corporation de développement de Sainte-Paule and the Corporation des municipalités de la MRC de Matane (which took over after the forest farms folded), through the signing of forest management contracts with these organizations. The decision-making structure for community forest lands in the RCM of Matane is therefore essentially two tiered. The organization with the higher level of authority is the Matane RCM, with responsibility for "big picture" tasks such as planning for land use, drafting long-term management plans, ensuring financial accountability and complying with rules and regulations, and so on. The second tier is the board of directors of the two municipal corporations. The scope of their decision-making is restricted to operational issues such as harvesting, silviculture, and the marketing of timber.

This power-sharing relationship has set up a particular dynamic whereby the Matane RCM, as the administrator and overseer of the municipal corporation, is perceived by some members of the municipal corporations as more akin to another level of bureaucracy, much like the role the Ministère des Ressources naturelles et de la Faune played previously. This situation is complicated by the fact that one of the municipal corporations, the Corporation de développement de Sainte-Paule, had its own ambitions to obtain decision-making authority over the community forest lands and was already managing the forests for the provincial ministry before the establishment of the territorial management agreement. The Municipality of Sainte-Paule has a long history of political mobilization around forestry, including a community forestry experiment dating back to the 1970s and created as part of a popular movement to oppose government threats of closures of certain rural parishes. Therefore, the transfer of authority from the province to the regional municipality did not meet the expectations of this municipal organization for local control. One interview respondent from Sainte-Paule said, "In the beginning, it was pretty difficult because it wasn't what we were looking for. We wanted to manage our own territory, not to manage it for the RCM."

The perception of the Matane RCM playing a primarily bureaucratic role is compounded by the fact that the RCM, under its territorial management agreement tenure, must apply a regulatory regime that is widely considered to be ill adapted for the management of intramunicipal lands (Moranville 2005). Interviewees, as well as supporting documentation, described the regulatory regime, which is based on the model of the large-scale industrial tenure called the timber supply and forest management agreement, as unnecessarily complex and unduly expensive because it is based on the industry practice, which has greater opportunities for adding value through processing. Adherence to the regulatory regime also includes costly payments to provincial funds for fire and pest protection, road maintenance, and so on. Interview respondents described the intramunicipal lots as having more in common with private forest lands in Quebec than with large-scale Crown tenures. The RCM is therefore charged with administering these forests according to a complex set of rules and regulations designed by the provincial government and are not free to make adjustments at the request of the corporations. Moranville (2005, 6), writing about the intramunicipal lands program for all RCMs in the Lower St-Laurent region, describes this dynamic:

> The mandate delegated confers them a role akin to a support pillar for government, with little latitude in the application of laws and regulations on intramunicipal lands. The RCMs were expecting more power and autonomy to adapt the management process to the regional characteristics of the intramunicipal lands. From the perspective of the Ministry of Natural Resources, the RCMs have all the latitude they need to apply the delegation of management. Rather, the origin of the problem stems from the fact that they cannot extract themselves from the legal framework surrounding the management of lands in the public domain, since intramunicipal lands are considered to be entities akin to large public forests.

But interview respondents described the transfer of power from the central government to the RCM as also having had positive impacts, particularly in terms of opening up lines of communication and bringing greater accountability in forest-management decision-making.

In the case of the Matane RCM, relatively few resources have been invested in soliciting public input. The RCM does not make management plans available for public review, nor do they organize any other formal public participation activities. An advisory group made up of local forestry stakeholders provides regular input to the RCM, but this is a legal requirement of the territorial management agreement tenure. Participants described the advisory group as fairly inactive. Evident in interviews was the sentiment that there was a certain disconnect between the aspirations and motivations of the local population vis-à-vis forest management and the current manifestation of community forestry – namely, the intramunicipal lands. Perhaps because of the limited and parcelled nature of the forests themselves, or because of a restrictive regulatory regime, this community forest does not meet several of the interviewees' expectations of devolution.

The fourth case study, the South Nation Conservation Authority, is situated in what was historically an agricultural area close to the urban centres of Montreal and Ottawa. Like many other conservation authorities across Ontario, South Nation acquired abandoned and degraded farmland at highly subsidized rates in the first half of the twentieth century through the Agreement Forest Program offered by the Ontario Ministry of Natural Resources. The ministry managed these lands until the early 1990s, rehabilitating them back into forests through the establishment of plantations and through natural regeneration. However, as far as forest governance was concerned, South Nation was not in charge of the decision-making process.

While for some conservation authorities and counties this may have been a satisfactory arrangement, South Nation was motivated to take over management, both in order to increase their influence over forest-management practices and because they recognized an economic opportunity. As an organization, South Nation was well positioned to take over management of these forests: the organization had decades of experience working on environmental issues such as water protection, flood control, the management of conservation areas, and fish and wildlife conservation. It also had a small forestry department, which at the time worked primarily on private-land forest management. The organization also had a certain financial stability, stemming in part from access to financing from municipalities, a positive track record of obtaining grants, and its strategic positioning as a type of environmental consultant in the region. At the time that forest land management was transferred to South Nation, the organization had an existing staff of more than thirty and therefore came to the task of community forest management with considerably more expertise and capacity than most community forests.

South Nation also had a well-established governance structure. The board of directors is composed of thirteen representatives appointed from each of the municipalities within the watershed. In most cases, these are either elected or previously elected municipal officials. The approach to representation is therefore quasi-democratic, since board members are not, as a matter of process, elected by the residents of a municipality. According to Bullock (2007), the fact that directors are appointed rather than elected has long raised concerns for account-ability. The organization also has a permanent forestry committee, made up primarily of local stakeholders with a direct interest in forest management such as representatives from the Ontario Ministry of Natural Resources, forestry consultants, farmers, and rural landowners. The membership of the committee is open. The role of the committee is to help direct forestry activities and bolster the technical expertise of the board, a role that is particularly important considering that most municipal officials have little experience in forest management. According to interview respondents, the combined influence of the board and the forestry committee provides a balance in terms of ad-dressing concerns related to public opinion (e.g., local values), which are brought to the table by local politicians, and forestry concerns, which are raised by the forestry committee.

Interviewees described the board of directors as quite active. In addition to their regular duties, members are required to participate in a number of subcommittees such as forestry, fisheries, water, com-munications, and so on. There also appears to be broad consensus on the management approach adopted by South Nation, which is oriented towards the implementation of multiple forest use and towards serv-ing the recreational needs of the regional population. A wide variety of groups use South Nation's forest, including hunters, snowmobilers, dogsledders, and naturalists. South Nation also maintains an empha-sis on the ecological restoration of these forests, through both re-search practices and the implementation of low-impact harvesting practices. The forests are certified by the Forest Stewardship Council through a group certification coordinated by the Eastern Ontario Model Forest.

In terms of its level of decision-making authority, South Nation finds itself in a privileged position compared with the Crown land cases. Much like a private forest owner, South Nation controls the full suite of property rights on its land base. Although the organization is bound by the Ontario Conservation Authorities Act in terms of its legal status, mandate, and organizational structure, it nonetheless has the ability to control and regulate key aspects of land management such as determining harvest levels for all forest and non-timber forest products, managing wildlife and associated activities, placing infrastructure, controlling access, and selling or leasing of land. This makes it an interesting contrast to the Crown land cases, because it embodies a tenure arrangement that is much closer to what many perceive to be true devolution, a condition that many community forests aspire towards.

In terms of public outreach, this case study is interesting because, unlike Crown lands, there are no legislated consultation requirements; therefore, should decision makers choose, they could manage the forests with little or no input from the public. However, interview respondents from within the organization described themselves as strongly oriented towards public participation. As one employee emphasized, "You have to involve the local community, because it's their land. It's their forest."[3] South Nation has created advisory committees on its own forests and for the two county forests under its management. It also holds periodic open houses to consult on major management plans and holds meetings on an annual basis with other users, such as snowmobilers, adjacent landowners, and municipalities, in order to inform them about upcoming activities and work towards the harmonization of different activities. South Nation participates in numerous environmental partnerships within the region and provides many opportunities for local residents to volunteer with the organization.

## Discussion

Compared with the status quo in local involvement in forest management, these initiatives measure up fairly well. Most, if not all, Canadian provinces promote public participation activities, whether in the form of reviews of management plans or advisory committees, as the main mechanisms for local participation in forest-management decision-making. Most of these activities are consultative in nature and treat public requests in isolation rather than inviting citizens to participate in the decision-making process from the outset. While there are a few exceptions, such as co-management boards, these are few and far between. Community forests are therefore unique in that they sanction the creation of a decision-making body made up entirely of local citizens that has been accorded some level of management control over an actual public land base, albeit a small one compared with most industrial licensees.

However, as has been recounted time and time again in the literature on devolution and common property, these types of bodies are ineffectual if they are not granted sufficient rights, management scope, and authority in decision-making (Schlager and Ostrom 1992; Agrawal and Ostrom 2001). Booth (1998) questions whether initiatives can be characterized as "community forests" if they must comply with government-set regulations such as annual harvest levels. All three of the Crown-land case studies face similar struggles in terms of operating within a legal structure, in terms of both tenure arrangements and regulatory regimes that are designed by provincial agencies and are based almost exclusively on a "timber tenure" blueprint (Ross and Smith 2002). Even in the case where greater administrative responsibility has been devolved to regional governments, as in the case of the Matane RCM, scope for intervention in the area of rule making was described as limited.

How this impedes a community forest's ability to achieve its vision varied from one case study to the next, depending on an organization's objectives. For the CVFC and the Matane RCM, obstacles relate mostly to inflexible or maladapted rules surrounding forest management, leading to unnecessary costs and red tape. For the Forêt de l'Aigle, the issue was first and foremost one of gaining rights to non-timber activities and thereby being able to implement a truly integrated approach to forest management. Although interviewees recognized the importance of having a regulatory regime in place and protecting the forest-related interests of all citizens, they were looking for greater autonomy, flexibility, and regional differentiation in the way the regulatory regime is imposed. This will be an important step towards building successful community forests in the Canadian context. South Nation, the only case study operating on its own land base, faced few of these constraints. Although there is no guarantee that an unregulated situation such as South Nation will result in sustainable forestry practices, in this case, the organization has been proactive in drawing on available guidelines, building up forestry expertise, and demonstrating its effectiveness through independent certification.

Representation is another critical issue in the governance of community forestry (Larson and Ribot 2004; Leach 2004). The ultimate goal of representation is described by Leach (2004, 2) as "ensuring that the interests of all affected individuals are effectively advocated." These four case studies have adopted two approaches: the interest group approach and the local government approach. All four cases tend to draw on "established" groups or individuals in the community that bring a certain measure of expertise to the table – whether this be administrative experience or direct forest-management experience. In the case of CVFC and the Forêt de l'Aigle, the adoption of the interest group approach has allowed these organizations to draw on pre-existing forestry expertise and human resources – for example, through the participation of research, education, and First Nation organizations. However, it brings to mind an observation made by Moote et al. (2001) that so-called community-based management in North America tends to be dominated not so much by communities as by agencies, commodity groups, conservation groups, and the like. In the case of Matane RCM and South Nation, power has been entrusted to elected or appointed municipal officials who, although they have little direct forest-management experience, are strong in financial and administrative skills. Therefore, these four community forests do not address the concern expressed in the community-based management literature of bringing excluded or marginalized groups, such as women or low-income groups, into decision-making or of working explicitly towards the alleviation of poverty (Ostrom 1990; Smoke 2001; Larson and Ribot 2004). Rather, they tend to draw on and support existing capacity in the community.

These case studies also reveal a certain inherent tension between the mandates of particular interest groups and the overall mandate of the community forest. In both the Forêt de l'Aigle and CVFC, instances occurred when an organization chose to withdraw or cease its participation within the board of directors because of a perception that its own interests were at odds with those of the community forest. The theory of representative democracy posits that under ideal circumstances, the interest groups sitting at the table will represent the overall distribution of interests in society (Overdevest 2000). In reality, however, the number of organizations participating in these community forests is relatively limited – encompassing, at best, five or six organizations. This raises the question of whether the governance of a community forest adequately serves the community by involving organizations that are purposefully oriented towards protecting a specific spectrum of community interests.

Public participation represents an important mechanism for broadening community input in the governance process. In some cases, such as South Nation, the organization is making a deliberate attempt to interact with a diverse array of people that extends beyond traditional conceptions of "forestry stakeholders." Local stewardship projects are particularly instrumental in connecting the organization with different groups, such as Aboriginal communities and retirees. In other cases, however, such as the Matane RCM and CVFC, extending the scope of community participation beyond the input of local decision-makers was not considered to be a central priority. Much like other Crown-land licensees, these community forests recognized the need to make room for specific views and requests through the legal consultation process, and in the case of CVFC, there was an openness to sharing the organization's experience through field tours. However, actual interactions between the community forest organizations and the public at large were relatively limited. The different manifestations of participation are indicative of the fact that not all community forests perceive their "public" role in the same way. Some community forests may remain relatively unknown in the community, while others have a much higher profile. This raises a number of related questions: Is empowering a handful of community-derived individuals or interest groups as decision makers enough to merit the label "community forestry"? Is democratic representation essential? Do we need to further refine our definitions of community forestry to assure some type of minimal level of community participation?

### Notes

1  Interview with forest manager, June 6, 2006.
2  After the time of research, the organization experienced serious challenges, including financial bankruptcy, and has since been restarted under the name Coopérative de solidarité de la Forêt de l'Aigle.
3  Interview with forest manager, June 6, 2006.

### References

Agrawal, A., and E. Ostrom. 2001. "Collective action, property rights, and decentralization in resource use in India and Nepal." *Politics and Society* 29 (4): 485–514. http://dx.doi.org/10.1177/0032329201029004002.
Albert, S. 2007. "Transition to a bio-economy: A community development strategy discussion." *Journal of Rural and Community Development* 2 (2): 64–83.
Anderson, N.B., and W. Horter. 2002. *Connecting lands and people: Community forests in British Columbia.* Victoria, BC: Dogwood Initiative.

Ayeva, T. 2003. *Gouvernance locale et renforcement des capacités: Quelques pistes de réflexion pour un développement territorial durable des collectivités rurales.* Montreal: Initiative sur la nouvelle économie rurale, Centre de recherche sur la gouvernance rurale.

Beckley, T. 1998. "Moving toward consensus-based forest management: A comparison of industrial, co-managed, community, and small private forests in Canada." *Forestry Chronicle* 74 (5): 736–44. http://dx.doi.org/10.5558/tfc 74736-5.

Beckley, T.M., J.R. Parkins, and S.R.J. Sheppard. 2005. *Public participation in sustainable forest management: A reference guide.* Edmonton, AB: Sustainable Forest Management Network.

Booth, A.L. 1998. "Putting 'forestry' and 'community' into First Nations' resource management." *Forestry Chronicle* 74 (3): 347–52. http://dx.doi.org/10.5558/ tfc74347-3.

Bouthillier, L., and H. Dionne. 1995. *La forêt à habiter, la notion de forêt habitée et ses critères de mise en oeuvre: Rapport final aux Ressources naturelles Canada.* Rimouski: Université du Québec à Rimouski et Université Laval.

Bradshaw, B. 2003. "Questioning the credibility and capacity of community-based resource management." *The Canadian Geographer* 47 (2): 137–50. http://dx.doi.org/10.1111/1541-0064.t01-1-00001.

Bullock, R. 2007. "Two sides of the forest." *Journal of Soil and Water Conservation* 62 (1): 12A–15A.

Carrier, M., and B. Jean. 2000. "La reconstruction de la légitimité des colléctivités rurales: Entre gouvernment et gouvernance." In *Gouvernance et territoires ruraux: Éléments d'un débat sur la résponsabilité du développement*, edited by M. Carrier and S. Côté, 41–64. Ste-Foy: Presses de l'Université du Québec.

Chiasson, G., J.L. Boucher, and T. Martin. 2005. "La forêt plurielle: Nouveau mode de gestion et d'utilisation de la forêt, le cas de la Forêt de l'Aigle." *Vertigo* 6 (2): 115–25.

Coon, D. 2004. *Give us back our forests.* Report written for the Conservation Council of New Brunswick.

Dionne, H., and C. Saucier. 1995. "Intervention sociale et développement local: La Coalition urgence rurale du Bas-Saint-Laurent." *Nouvelles Pratiques Sociales* 8 (1): 45–61. http://dx.doi.org/10.7202/301304ar.

Duinker, P.N., P. Matakala, F. Chege, and L. Bouthillier. 1994. "Community forestry in Canada: An overview." *Forestry Chronicle* 70 (6): 711–20. http://dx.doi.org/ 10.5558/tfc70711-6.

Fung, A., and E.O. Wright. 2001. "Deepening democracy: Innovations in empowered participatory governance." *Politics and Society* 29 (1): 5–41. http:// dx.doi.org/10.1177/0032329201029001002.

Gaventa, J. 2002. "Towards participatory governance." *Currents* 29: 29–35.

Gélinas, N. 2001. *La gestion partenariale: Un nouveau mode de gestion pour les forêts québécoises.* Québec, QC: Faculté de foresterie et de géomatique, Université Laval.

Kearney, J., F. Berkes, A. Charles, E. Pinkerton, and M. Wiber. 2007. "The role of participatory governance and community-based management in integrated coastal and ocean management in Canada." *Coastal Management* 35 (1): 79–104. http://dx.doi.org/10.1080/10.1080/08920750600970511.

Larson, A., and J. Ribot. 2004. "Democratic decentralisation through a natural resource lens: An introduction." *European Journal of Development Research* 16 (1): 1–25. http://dx.doi.org/10.1080/09578810410001688707.

Leach, W.D. 2004. *Is devolution democratic? Assessing collaborative environmental management*. Sacramento: Center for Collaborative Policy, California State University.

McCarthy, J. 2006. "Neoliberalism and the politics of alternatives: Community forestry in British Columbia and the United States." *Annals of the Association of American Geographers* 96 (1): 84–104. http://dx.doi.org/10.1111/j.1467 -8306.2006.00500.x.

McIlveen, K., and B. Bradshaw. 2005–6. "A preliminary review of British Columbia's Community Forest Pilot Project." *Western Geography* 15–16: 68–84.

Moote, M.A., B.A. Brown, E. Kingsley, S.X. Lee, S. Marshall, D.E. Voth, and G.B. Walker. 2001. "Process: Redefining relationships." In *Understanding community-based forest ecosystem management*, edited by G.J. Gray, M.J. Enzer, and J. Kusel, 97–116. New York: Food Products Press.

Moranville, D. 2005. *Rapport d'évaluation: Entente specifique sur la gestion et la mise en valeur du territoire public intramunicipal (TPI) du Bas-Saint-Laurent*. Presented to the Table de concertation TPI du Bas-Saint-Laurent.

Newig, J., and O. Fritsch. 2009. "Environmental governance: Participatory, multi-level – and effective?" *Environmental Policy and Governance* 19 (3): 197–214. http://dx.doi.org/10.1002/eet.509.

Ostrom, E. 1990. *Governing the commons: The evolution of institutions for collective action*. Cambridge, UK: Cambridge University Press. http://dx.doi. org/10.1017/CBO9780511807763.

Overdevest, C. 2000. "Participatory democracy, representative democracy, and the nature of diffuse and concentrated interests: A case study of public involvement on a National Forest District." *Society and Natural Resources* 13 (7): 685–96. http://dx.doi.org/10.1080/08941920050121945.

Québec. MRNF (Ministère des Ressources naturelles et de la Faune). 1999. *Allocution de monsieur Jacques Brassard. Ministre des Ressources naturelles, lors de la signature de l'Entente spécifique sur la gestion et la mise en valeur du territoire public intramunicipal du Bas-Saint-Laurent. Quebec.*

Reddel, T., and G. Woolcock. 2004. "From consultation to participatory governance? A critical review of citizen engagement strategies in Queensland." *Australian Journal of Public Administration* 63 (3): 75–87. http://dx.doi. org/10.1111/j.1467-8500.2004.00392.x.

Reed, M., and K. McIlveen. 2006. "Toward a pluralistic civic science? Assessing community forestry." *Society and Natural Resources* 19 (7): 591–607. http:// dx.doi.org/10.1080/08941920600742344.

Robinson, D., M. Robson, and R. Rollins. 2001. "Towards increased citizen influence in Canadian forest management." *Environments* 29 (2): 21–41.

Ross, M.M., and P. Smith. 2002. *Accommodation of Aboriginal rights: The need for an Aboriginal forest tenure.* Edmonton, AB: Sustainable Forest Management Network.

Roy, M.É. 2006. "Des fermes forestières en métayage sur le territoire public québécois: Vers un outil d'évaluation pour les communautés." Master's thesis, Faculté de foresterie, de géographie et de géomatique, Université Laval, Ste-Foye, QC.

Schlager, E., and E. Ostrom. 1992. "Property rights regimes and natural resources: A conceptual analysis." *Land Economics* 68 (3): 249–62. http://dx.doi.org/10.2307/3146375.

Schusler, T.M., D.J. Decker, and M.J. Pfeffer. 2003. "Social learning for collaborative natural resource management." *Society and Natural Resources* 16 (4): 309–26. http://dx.doi.org/10.1080/08941920390178874.

Sierra Club of Canada. 2004. *Tenure reform and community forests.* Fact sheet. Ottawa: Sierra Club of Canada.

Smoke, P. 2001. "Fiscal decentralization in developing countries: A review of current concepts and practice." Democracy, Governance and Human Rights Programme Paper Number 2, United Nations Research Institute on Social Development.

Solidarité rurale. 2008. *L'avenir des communautés forestières par le prise en charge de leur ressource comme levier d'une développement durable: Mémoire déposé à la Commission de l'économie et du travail de l'Assemblée nationale du Québec.* Nicolet, QC: Solidarité Rurale.

Teitelbaum, S. 2014. "Criteria and indicators for the assessment of community forestry outcomes: A comparative analysis from Canada." *Journal of Environmental Management* 132: 257-67.

Teitelbaum, S., T. Beckley, and S. Nadeau. 2006. "A national portrait of community forestry on public land in Canada." *Forestry Chronicle* 82 (3): 416-28.

Turnhout, E., S. van Bommel, and N. Aarts. 2010. "How participation creates citizens: Participatory governance as performative practice." *Ecology and Society* 15 (4): art. 26. http://www.ecologyandsociety.org/vol15/iss4/art26/.

Wellstead, A.M., R.C. Stedman, and J.R. Parkins. 2003. "Understanding the concept of representation within the context of local forest management decision making." *Forest Policy and Economics* 5 (1): 1–11. http://dx.doi.org/10.1016/S1389-9341(02)00031-X.

# PART 3
## Community Forestry:
## Looking towards the Future

Chapter 12   **Stronger Rights, Novel Outcomes**
Why Community Forests Need More
Control over Forest Management

*Erik Leslie*

Community forests operating on provincial Crown land in Canada have
limited rights to the forests they manage. Generally, they can remove
timber and can develop and implement operational plans to do so. However, management-planning rights are subject to government regulation and approvals, and community forests cannot convert the land to
other uses or transfer rights to the land. When considering the scope
and strength of resource-management rights, it is helpful to think
in terms of disaggregated "bundles of rights" (Schlager and Ostrom
1992). In this context, community forests generally have rights of
access and withdrawal, but they do not have rights of exclusion and
alienation.[1] Management rights – including the right to set priorities
and rules for resource use and the right to transform the resource
(Agrawal and Ostrom 2001) – are shared with provincial governments.
Access and withdrawal rights are simple operational-level rights, while
management, exclusion, and alienation rights are strategic decision-
making rights that allow the rights holder to set rules and standards
for exercising power. In community forests, the issue of who has the
authority to make strategic forest-management decisions is often contested (Pinkerton et al. 2008).

In the global context, community forestry rights are situated within
larger discussions of decentralization and/or devolution of central
government power (Ribot 2002; Larson and Ribot 2004; Larson, Barry,
and Dahal 2010) and within conversations about governance and metagovernance (Jessop 2003; Parkins 2008). Decentralization reforms are
generally based on the proposition that better outcomes will occur if
representative and accountable local bodies have the rights and power

to make decisions regarding public resources than if central authorities make those decisions (Ribot 2004). However, this proposition
cannot be tested if only very limited rights are conferred. Community
forests around the world struggle with many constraints placed on the
scope and strength of their forest-management rights and are often
saddled with inappropriate and burdensome central government regulations (Ribot 2004; Pulhin, Larson, and Pacheco 2010; Cronkleton,
Pulhin, and Saigal 2012). Constraints on community forestry management rights typically include the inability to set higher-level direction
and priorities for the land base, the lack of right to manage for both
timber and non-timber values, limited duration of tenure term, and
onerous reporting and compliance requirements. Apparent devolution
of government control to local groups takes place in the context of lingering bureaucratic hierarchies and state oversight (Parkins 2008).

In this chapter, I argue that stronger community forest-management
rights can lead to innovative forest management and community outcomes. I illustrate my argument with an analysis of how the Harrop-
Procter Community Forest has worked for fifteen years to expand
the scope and strength of its management rights and how stronger
management rights have led to unique management approaches and
facilitated continued engagement with the local community. Having
worked since 2006 as Harrop-Procter's forest manager, I have participated in over one hundred community forest meetings, have engaged
in dozens of community forest planning and decision-making processes, and have been involved in negotiations with the provincial
government. As a front-line community forestry practitioner, I have
seen first-hand how stronger rights and enhanced decision-making
power can lead to novel management approaches and to more robust
and durable public engagement.

The descriptive material presented in this chapter derives from five
types of experiences: extensive discussions with current and former
Harrop-Procter Community Cooperative (HPCC) and Harrop-Procter
Watershed Protection Society (HPWPS) board members; participation,
from 2005 to 2015, in many HPCC board meetings where management
issues were debated and discussed extensively; participation, from
2005 to 2015, in at least eight well-attended open public meetings related to Harrop-Procter's management priorities, management plan,
and harvest levels; a review of findings of Harrop-Procter community
surveys completed in 1998, 2006, and 2012; and discussions with BC's
Ministry of Forests management and staff between 2005 and 2015.

The theoretical framework that guided my empirical analysis was based in the scholarship on common property (Schlager and Ostrom 1992; Agrawal and Ostrom 2001) and the literature regarding devolution of management rights (Ribot 2002; Larson and Ribot 2004; Cronkleton, Pulhin, and Saigal 2012).

## Rights Delegation in BC's Community Forests

British Columbia's community forests have, in the Canadian context, relatively strong and secure forest tenure rights that provide exclusive, long-term timber rights to a clearly defined area. However, these rights are still relatively narrow and heavily attenuated by provincial government regulation and oversight. Despite some recent administrative streamlining, community forests remain largely subject to the same complex suite of forestry legislation and regulations as are large industrial forest companies. Detailed regulations govern forest-planning requirements, survey methods, regeneration obligations, waste assessments, timber-scaling requirements, and five-year cut control requirements, to name just a few examples. Perhaps most importantly, the scope of the tenure is limited by an almost exclusive focus on timber.[2] Ambus and Hoberg (2011) argue that BC's community forests have a very low level of authority to make strategic management decisions, such as the power to engage in land-use planning, establish standards of practice, or set timber-harvest levels.

In British Columbia, the Community Forest Agreement (CFA) is the primary tenure document that sets out the terms under which community forests operate. When the community forest program was first piloted in the late 1990s, many expected that a customized tenure document would be developed. The Community Forest Advisory Committee (CFAC), which had been appointed by the minister of Forests to help guide the development of community forest legislation, had recommended that community forests be provided with comprehensive resource rights, including rights to manage timber, botanicals, firewood, recreation, range, and gravel extraction, as well as eventually being given the rights to manage fish, wildlife, and water (CFAC 1998). However, the legislation that followed was limited to timber rights, and a customized community forests regulation and a broader tenure document were never developed. As Ambus and Hoberg (2011) point out, the CFA tenure document is largely indistinguishable from tenure documents for large private timber companies operating in BC, such as the Tree Farm Licence. Thus, from the perspective of tenure structure, community forests are a slightly modified form of small-scale industrial tenure.

While the CFA is the primary tenure document, community forest rights are also expressed in a comprehensive suite of regulations, approved planning documents, and administrative processes. One could perhaps argue that there has been a gradual process of de facto evolution of tenure rights and devolution of power through recent incremental administrative "streamlining" efforts, such as the removal of provincial requirements for standard cutting permit applications. The move to simplify stumpage rates to account for the increased costs of community forestry has freed community forests from the onerous and restrictive requirements of the BC stumpage appraisal system. However, the reduction of administrative barriers, while it has enabled more independence in operational management, does not constitute devolution of strategic management control or a fundamental increase in management rights. It has been largely driven by a common interest in reducing administrative transaction costs. While community forests may be taking small de facto steps towards more operational management independence, there has been no central government commitment to devolution of higher-level management rights.

The BC Community Forest Association (BCCFA) has advocated for many years for stronger management rights for community forests. At the time of writing, the BCCFA represents fifty community forest organizations in British Columbia. The association's guiding principles include "community responsibility for land use and allocation decisions," and one of the organization's objectives is to enlarge the scope of community forests to include resources other than timber (BCCFA 2012). On behalf its members, the BCCFA works closely with the provincial government to influence community forest policy and advocates to loosen onerous regulatory restrictions, lower barriers to innovation, and remove one-size-fits-all tenure requirements. Following the CFAC's recommendations, the BCCFA has often argued that community forests should have a broader tenure document and a customized community forest regulation.

## The Allowable Annual Cut and BC's Community Forests

The management plan for a community forest outlines the goals and resource-management objectives of the community forest and proposes an allowable annual cut (AAC). An AAC is a target average annual harvest level that community forests are generally expected to meet.[3] Once approved by the Ministry of Forests, the management plan becomes part of the Community Forest Agreement, and thus part of the tenure arrangement with the province.[4]

The determination of an AAC, which is often presented as a technical calculation, is actually a key strategic forest-management decision. While an AAC determination requires a technical analysis of timber supply, it also requires weighing and balancing a range of ecological and social factors. Timber supply analyses are based on a series of key assumptions – including assumptions regarding natural disturbances, timber growth rates, economic operability, and safe locations for roads. They also imply an acceptance of various levels of risk – including the risks of landslides, loss of biodiversity, reduced timber growth rates, wildfire, and negative socio-economic impacts. They often require an analysis of multiple scenarios in an uncertain future. Ultimately, all AAC determinations are choices based on values, judgments, and assessments of benefits and risks. They require assessing and balancing land-management priorities. When determining an AAC for a large timber supply area, BC's chief forester must consider the provincial government's political priorities as expressed by the minister of forests. Likewise, in theory, community forest organizations have the responsibility and the right to consider local public values and local priorities, as well as environmental impacts, when setting their AAC.

While community forest harvest levels should only be established after a detailed land base analysis and a broad community discussion that assesses relative resource risks and benefits, the creation of new community forests in British Columbia is driven by an initial offering from the Province of an AAC, not a land base.[5] Because community forests are offered an AAC apportionment from a timber supply area instead of a land base to manage, the CFA tenure is framed from the very beginning in terms of a predetermined harvest level. This unduly narrows the focus and effectively limits the management scope for community forests.

## The Case of Harrop-Procter: Management Rights and the AAC

In this section, I discuss the history of the Harrop-Procter Community Forest and how the autonomous determination of its AAC has highlighted issues related to strategic forest-management rights. I briefly discuss the early days of the Harrop-Procter AAC debate and tell the story of how Harrop-Procter's AAC was changed when its management plan was rewritten in 2012. I then tie the Harrop-Procter AAC issue to forest-management rights generally.

Harrop and Procter are small rural communities in BC's southern interior that have a long history of social activism regarding local forestry issues (Elias 2000). The Harrop-Procter Community Forest was born in the 1990s out of BC's so-called War in the Woods, which pitted environmental and social activists against the provincial government and the forest industry in a fierce debate regarding the environmental and social sustainability of forest management and the lack of community input into resource decision-making. Many Harrop-Procter residents had long opposed conventional industrial logging in the watersheds from which they drew their drinking water and had spent many years trying to obtain control over their local forests. In the mid-1990s, many local residents advocated strongly for a provincial park in their watersheds. When the park option was denied, the Harrop-Procter Watershed Protection Society applied for one of the newly advertised community forest pilot agreements. Their objective was to practice low-impact ecosystem-based forestry, in which timber extraction is secondary to ecosystem and water protection. The initial Harrop-Procter Ecosystem-Based Plan (1999) called for a much lower AAC than that proposed by the Ministry of Forests.[6] Harrop-Procter was eventually awarded a community forest pilot agreement in 2000 because the minister of forests, under considerable public pressure, had committed to testing a wide range of forest-management approaches under the new community forest tenure.[7] Harrop-Procter's pilot agreement incorporated the lower AAC derived from the organization's ecosystem-based plan.

The Harrop-Procter Community Forest is run by the Harrop-Procter Community Co-operative (HPCC). As a cooperative of local residents without any municipal or forest industry affiliation, HPCC's volunteer board of directors is elected directly from its membership. The board is composed primarily of non-technical people without any forestry background. HPCC prides itself on its direct, unmediated relationship with Harrop-Procter residents.

Why does the AAC matter so much to Harrop-Procter residents? In Harrop-Procter, the AAC represents a de facto expression of community values and priorities. Other aspects of forest management – such as planning, road building, and harvest and regeneration methods – are also regarded as important and have also been contested. However, the AAC is the issue to which people have paid the most attention. As such, the AAC synthesizes and also symbolizes a forest-management approach and has long been a political issue in Harrop-Procter.

HPCC's lower AAC in its ecosystem-based plan represented a clear assertion of strategic control over local forest management. This assertion brought the cooperative into a protracted power struggle with the Ministry of Forests. At its core, this power struggle was about the strength of HPCC's forest-management rights – in other words, about who has the right to set the strategic forest-management direction for the community forest. Through the process of negotiating the tenure agreement and management plan, the HPCC had some success in effectively enlarging the scope of its forest-management rights. As the HPCC's forest-management approach has proven viable over time, the formerly contested broader scope of HPCC's management rights has become more entrenched.

Over the past thirteen years, the HPCC has at times been able to leverage the Ministry of Forests' (often begrudging) acceptance of its ecosystem-based plan and lower AAC to advocate for flexibility in the ministry's review and approval of its operational plans. For example, the HPCC implements larger riparian reserves than other BC forestry companies despite provincial regulatory requirements to "not unduly reduce" timber supply. It also permanently retains more trees in its cutblocks than legally required, again despite timber impacts. The HPCC has also been granted exceptions to standard restocking requirements in some areas, allowing the cooperative to begin to develop fuel breaks that will not be densely restocked with conifers. After a decade-long process of relaxing control as it developed trust in the HPCC's board and management, the Ministry of Forests has accepted that the organization is "different" and will take a unique, community-based approach to forestry. The HPCC's assertion of strategic forest-management rights has thus created some additional space in which to operate. The AAC has proven to be a key strategic management lever.

In 2012, the HPCC reopened the strategic forest-management discussion, since a new management plan and AAC were required. While the cooperative's non-timber values and objectives and "alternative" forest-management practices were reaffirmed in its new management plan without any debate, a lengthy discussion ensued regarding the AAC. It was widely recognized that the 1999 AAC calculation was out of date and that new information needed to be considered. The organization had also matured and was ready to consider emerging issues such as climate change and wildfire risks.

In early 2012, the HPCC completed a new timber supply analysis. This analysis was done internally with no involvement by the Ministry of Forests.[8] It indicated that a significantly higher AAC might be appropriate for several reasons, including an expanded land base, a new forest inventory, and improved terrain stability information.[9] A new argument was also introduced: a higher AAC could allow a more adaptive approach to climate change, improve ecosystem resilience at the landscape level, and reduce the risk of catastrophic wildfire.[10] The uniformly mature coniferous forests in Harrop-Procter, while currently valuable for both water protection and timber, could also be seen as vulnerable, ultimately unsustainable, and a potential liability from a wildfire perspective.

This information and new perspective were brought to the board of directors by the HPCC's forest manager and then was shared with the public. To discuss the AAC, a series of board and public meetings were held, and a community survey was conducted. Technical information was translated into language that the HPCC board and the public could understand – and that would facilitate informed decision-making. The decision was not considered just a technical decision based on timber supply: it needed to reflect community priorities and perceived risks.

The HPCC board was fully engaged throughout the ten-month deliberation process and requested additional analyses of AAC scenarios. Some public concerns and priorities regarding the HPCC's land base and protected areas were brought forward during this process. Many public concerns were primarily non-technical in nature, including the importance of wilderness and spiritual values. A new approach to the AAC emerged through these discussions, based in part on climate change considerations. The board settled on an approach that used a conservative estimation of the timber-harvesting land base – only 30 percent of the land base would be available for logging – but that promoted a slightly accelerated harvest rate on this land base. This approach acknowledged that accelerating harvest rates in areas near the communities of Harrop and Procter can provide increased protection from wildfire and can promote a more resilient forest in the face of climate change.[11]

The HPCC board applied its own collective judgment and took full responsibility for the proposed higher AAC, acknowledging and weighing local socio-economic and environmental risks. The board's decision was backed by the results of the community survey, which indicated that over 75 percent of respondents supported the HPCC's approach to forest management and to setting an AAC. Though the management plan and AAC were ultimately subject to government approval, the Ministry of Forests did not question the cooperative's approach and the social priorities it expressed, and the ministry approved the management plan and AAC without any changes. The board, which still includes many of the same people who were involved in the early days of the community forest and the fight for a low AAC, was willing to reconsider past decisions, and in the end, it surprised itself – and the Ministry of Forests – by choosing a higher AAC of ten thousand cubic metres.

By 2012, most of the Harrop-Procter residents were willing to consider a significantly increased AAC because of a relatively high level of trust in the HPCC's board and management. Public trust was higher because the board is strongly independent of government and industry and remains a grassroots, non-technical group of people who speak from the heart about how their perspectives have evolved over the years. The organization's structure as a community cooperative has facilitated this trust.

How do conceptual arguments regarding forest-management rights help explain the recent developments in Harrop-Procter? And, conversely, what does the Harrop-Procter story tell us about concepts of forest-management rights? I believe that the HPCC's mature and largely uncontested control over strategic forest-management decisions was a necessary precondition for the development of the cooperative's new approach to its AAC. The HPCC board and most Harrop-Procter residents would not have supported a significantly increased AAC if they had been compelled to do so by the Ministry of Forests. They were only willing to change course because they felt they had enough autonomy and control over the AAC process and outcome. This autonomy and control provided the "space" to engage in an open, nuanced community discussion and to take difficult steps to proactively address complex climate change–related issues and risks.

What is important for this discussion is not whether the HPCC's AAC went up or down. What matters is that the community had the right to determine harvest levels and that control over harvest levels can be a lever to address other forest-management issues such as regulating water flows, reducing forest fuels, or adaptively managing forests for climate change. Moreover, having strategic management decision-making rights is what engages the community at a deeper level in the community forest. Setting the AAC is just one strategic forest-management right, but in the current Canadian context, it is a critical one. Other strategic rights such as setting land-use priorities, engaging in land conversion, and managing and charging for non-timber resources including carbon may be just as – or more – important as timber harvest levels in the future.

For Harrop-Procter, more control over strategic resource decision-making has allowed the HPCC's forest management to be somewhat more innovative and adaptive. For example, the organization has developed new silviculture regimes to prioritize ecosystem resilience and community wildfire protection over maximizing timber harvest volumes. Increased control over strategic decision-making has also been a critical factor in maintaining community engagement and support. When community members feel like they actually have meaningful rights and ultimate control, a subtle but profound societal shift occurs and new approaches can emerge. As small, area-based, long-term, multiple-value, and multiple-use public tenures, community forests such as Harrop-Procter can lead the way in experimenting with new forest-management approaches, if they have the space – and the rights – to do so.

Any broad inferences from the HPCC's experience of contesting forest-management rights are subject to several caveats. First, the HPCC remains in a structurally weak position vis-à-vis the Ministry of Forests in that local strategic forest-management decisions are still ultimately subject to central government oversight. The requirements for the content of management and stewardship plans are prescribed in regulation, and plans must be approved by the provincial government. Community forests must still negotiate exceptions to government requirements regarding, for example, five-year harvest controls and silvicultural stocking standards. The regulatory system does not generally permit community forests to take a landscape-level approach (instead of a uniform hectare-by-hectare approach) to forest regeneration. Allowing the proliferation of non-commercial hardwood or shrub species that may improve ecosystem diversity and lower wildfire hazards is discouraged because of anticipated timber-supply impacts. The system is biased against partial-cutting approaches that do not fit into legal reporting systems. All deviations from standard government rules and procedures require applications for exemptions, which are time consuming and at risk of being refused.

Second, the HPCC's recent success in enlarging the scope and strength of its forest-management rights must be seen in the context of a steady erosion of the BC government's command-and-control approach to forest management over the last ten years. The Liberal government's "results-based" regulatory approach and the increased use of "professional reliance" in forestry matters has led to a lesser role for government in the affairs of all BC forest tenure holders. The BC government is no longer considered the "steward" of most forest lands (Reader 2006, 179) – this role is now assumed by forest industry licensees, including community forests. While focused primarily on operational forestry matters, the "results-based" and "professional reliance" environment has inadvertently provided space for community forests to undertake initiatives with potential strategic-level significance.

## Strengthened Rights for Community Forests: Risks and Liabilities

### Reduced Accountability and Control by Special Interests

One argument against stronger community control of forest management is the risk of limited accountability and democratic control (Bradshaw 2003). While democratically elected central governments must maintain the support of a broad range of constituents and are at least nominally accountable for their decisions, small local groups can potentially be controlled by powerful local elites or special interest groups (Reed 1995).

As discussed above, community forests in British Columbia remain upwardly accountable in many respects to the BC government. If upward accountability to, and control by, the central government is to be reduced, then downward accountability to the community becomes more critically important.[12] While upward accountability is clearly defined and relatively easy to assess, downward accountability is much harder to assess systematically. Currently, community forest governance structures in British Columbia are highly diverse, and governance is largely unregulated. While community forest governance structures include societies, cooperatives, First Nation governments, and local government–First Nations partnerships, the most common governance structure for new community forests is the municipally controlled corporation, with the municipality as the sole shareholder. Municipally controlled corporations, like other corporations, have minimal legal reporting and public consultation requirements and limited downward accountability (Tyler, Ambus, and Davis-Case 2007).

Requirements regarding the downward accountability of community forests are relatively weak. In British Columbia, the initial awarding of a CFA requires an application that demonstrates community support. Once a CFA is awarded, a basic level of annual community reporting is required. However, the Ministry of Forests is ill-equipped to assess local support, governance, and downward accountability. In the pilot phase of the CFA program, when community forest tenures were competitively awarded and a five-year probationary period was in place, community forests were required to demonstrate an ongoing high level of local engagement and support. The CFAC, which had broad membership outside of the Ministry of Forests, was responsible for assessing community support and providing advice to the minister of forests. However, with the dissolution of the CFAC and the end of the five-year probationary period for CFAs, the downward accountability of CFAs is no longer systematically assessed after the CFA is awarded. This is problematic.

Community forests should follow good governance principles (e.g., Graham et al. 2003), and community forest governance structures must ensure strong downward accountability (Ribot 2004). Instead of focusing on upward accountability, the BC government could require consistent implementation of good governance principles such as legitimacy, participation, responsiveness, equity, transparency, and accountability for community forest boards and management: for example, an openly elected board, open meetings, structured ongoing citizen involvement (e.g., advisory committees), formal community surveys, and so on.[13] Assessments of community forest governance structures and processes could be made by an independent committee with experience in community-based governance and accountability.

## Poor Environmental Outcomes

It has been argued that unregulated local control of forest management can lead to unsustainable practices and poor environmental or ecological outcomes. This risk is real: history has shown that local communities have not always been good stewards of local resources (see examples cited in Bradshaw [2003]). However, this risk should not be overstated in the BC context. The small size, long-term area-based nature, legislated professional stewardship requirements, and inherent diversity of community forest values and priorities all reduce ecological risks. Good governance structures and accountability processes also help mitigate against unbalanced management approaches.

With regard to larger-scale environmental issues such as loss of biodiversity and climate change, provincial and regional ecological requirements should continue to apply to community forests. For example, community forests should continue to abide by regulations regarding the protection of wildlife species at risk and old-growth forests. However, community forests should have more control over management decisions that are primarily local or site specific, such as determining local priorities for the use of public land and requirements for forest regeneration.

Community forests have shown that they can take the lead on complex environmental issues independently. Notwithstanding central government initiatives on climate change, several community forests in British Columbia are at the forefront of developing climate change adaptation and mitigation strategies. For example, the Cheakamus (Whistler) Community Forest is developing carbon sequestration strategies, the Esk'etemc Community Forest is restoring natural grassland ecosystems, and both the Slocan and Creston Community Forests have developed alternative silviculture stocking standards that address ecosystem resilience and community protection from wildfires in dry forests.

## Poor Socio-economic Outcomes

It could be argued that unregulated community forest decision-making could lead to poor socio-economic outcomes such as job losses or negative impacts on tourism values due to poorly planned logging operations. In any such discussion, the governance structure and mandate of community forests is again of critical importance. Unlike for large multinational forestry companies, there are relatively few social "externalities" for community forests. From an economic perspective, the internalization of social costs and benefits is an argument in favour of community forest rights.

For small, long-term, area-based tenures, the provincial government has a very limited role to play in attempting to determine socio-economic outcomes. Many community forests throughout British Columbia have actively supported local and regional value-added wood manufacturing, as well forest-based recreation and tourism projects and broader economic diversification initiatives. As long as downward accountability structures are in place, community forests should have the right to assess and balance the socio-economic outcomes, risks, and benefits of local resource use.

## Conclusions

Community forest rights need to be considered within the context of a broader discussion regarding devolution of central government authority over natural resources. Ribot (2004), Cronkleton, Pulhin, and Saigal (2012), and others have argued that in the global community forestry context, very limited rights are being transferred to communities. Charnley and Poe (2007, 325) highlight the wide gap between community forest theory and practice and point out that "without real devolution of power, the goals of community forestry will be difficult to achieve because they are premised on this transfer." Following Schlager and Ostrom's (1992) concept of "bundles of rights," the most critical property rights for Canadian community forests operating on public land are resource management rights, which include the right to regulate use patterns and the right to transform the resource.

In the context of Canadian public forest management, community forests are uniquely capable of implementing new and adaptive management and governance approaches that integrate a wide range of community interests and benefits. Community forest management can help our ecosystems and our communities become more diverse and resilient. However, socio-economic and environmental transformation through community-based management cannot occur if only limited management rights are transferred. As long as community forests rights are narrow and heavily attenuated, new ideas, local commitment, and public engagement will be suppressed. Community forests need space to innovate; central governments have to be willing to relinquish key aspects of strategic forest-management control.

Over the past several years in British Columbia, there has been some de facto incremental devolution of administrative responsibility to community forests but no fundamental discussion of, or broad progress towards, devolution of strategic-level management rights. Community forests should have the rights to prioritize and manage the land base for non-timber values and benefits, including water quality and quantity and carbon sequestration. Strategic decisions such as setting harvest rates are inevitably tied to water, carbon, wildlife habitat, wildfire protection, and other values. In Harrop-Procter, the right to set the AAC has been a proxy for the right to manage for these other resource values.

Although community forests have a mandate to manage for a wide range of resource values, they have limited rights and no economic incentives to manage for these other values.[14] As a community forest manager, I struggle with this incongruity every day. Rather than a narrow tenure system based on paying stumpage for timber as it is harvested, a more appropriate tenure system may be a land "rent" model based on area, not timber. In discussions of community forest program development and expansion, it is inappropriate to frame community forests as simply another tenure group seeking access to timber. Rather, we need to be thinking about new ways to facilitate integrated management of forest resources.

The issues of scope and strength of tenure rights are tied to issues of community governance and engagement. Despite considerable outside interest and support, operating community forests in British Columbia often have a hard time maintaining local community engagement in their forestry plans and operations unless a controversial issue arises. In general, as long as forest management is timber-centric and the central government retains strategic control, few local people are likely to remain engaged in community forest management. Most people relate more strongly to non-timber values, so a broader approach will engage more people, and this will improve community forest governance. The experience in Harrop-Procter has demonstrated that community engagement is stronger when the discussion goes beyond timber and when local residents know that they have control over strategic decisions.

### Notes

1  An access right is the right to enter a property. A withdrawal right is the right to take "products" (e.g., timber, water, fish) from a resource. An exclusion right is the right to determine who will have an access right and how that right may be transferred. Finally, an alienation right is the right to sell or lease rights to the resource (Schlager and Ostrom 1992).

2  While BC's community forests nominally have the rights to manage non-timber forest products (NTFPs), the lack of enabling government regulation for NTFPs renders the "rights" to manage NTFPs largely irrelevant (McIlveen and Bradshaw 2005–6; Ambus and Hoberg 2011).

3  Theoretically, an AAC sets a maximum permitted timber harvest level and there is no minimum harvest requirement; however, in the current BC context, AACs are often interpreted as a target harvest level. A sustained timber yield approach based on a target AAC is problematic from the perspective of ecosystem sustainability, adaptability, and resilience (Luckert and Williamson 2005).

4  The name of the BC government ministry responsible for forests has changed several times in the past decade. The current name is the Ministry of Forests, Lands, and Natural Resource Operations. In this chapter, I will simply use "Ministry of Forests" to refer to this ministry.

5  The community forest land base is assumed to come from somewhere near the community, but the land base is determined later.

6  The Ministry of Forests had calculated an AAC of 6,000 to 10,000 cubic metres, while Harrop-Procter's plan called for an initial AAC of 2,603 cubic metres.

7  Harrop-Procter's success in gaining approval of its ecosystem-based plan and lower AAC relied, to a considerable degree, on social and political conditions that were unique to British Columbia in the 1990s. Because forestry and environmental issues were hot political topics at the time, Harrop-Procter was able to make effective use of the media and exert political pressure at a provincial level.

8  HPCC simply notified the ministry of its methodology and its preliminary results.

9  The new forest inventory indicated that the community forest has higher coniferous timber volumes and increased growth rates compared to previous estimates.

10  The Harrop-Procter land base comprises nearly continuous hundred-year-old fire-origin forests. The uniformity of the forests is a historical anomaly resulting from very large settlement-era fires in the early 1900s, followed by decades of active fire suppression.

11  Logged areas can either be partially cut to increase structural diversity in the ecosystem, or they can be logged and regenerated using a mix of natural regeneration and planting of species and genetic provenances better adapted to the hotter, drier summers expected over the next century. Some logged areas can also be converted to shrub or deciduous species to develop landscape-level fuel breaks. The HPCC has begun implementing all of these strategies.

12  Regarding concepts of "upward" and "downward" accountability, see Ribot (2002).

13  In Harrop-Procter, in addition to open membership, open meetings, and direct election of HPCC directors, there is also a separate community environmental oversight organization (the Harrop-Procter Watershed Protection Society) that provides a check and balance for the HPCC through overlapping board membership.

14  For example, community forests do not "own" and cannot "sell" carbon or water rights despite the fact that they directly or indirectly manage carbon and water resources.

### References

Agrawal, A., and E. Ostrom. 2001. "Collective action, property rights, and decentralization in resource use in India and Nepal." *Politics and Society* 29 (4): 485–514. http://dx.doi.org/10.1177/0032329201029004002.

Ambus, L., and G. Hoberg. 2011. "The evolution of devolution: A critical analysis of the community forest agreement in British Columbia." *Society and Natural Resources* 24 (9): 933–50.

BCCFA (British Columbia Community Forest Association). 2012. *Strategic Plan 2012–2017.* http://bccfa.ca/index.php/about-us/strategic-plan.

Bradshaw, B. 2003. "Questioning the credibility and capacity of community-based resource management." *The Canadian Geographer* 47 (2): 137–50. http://dx.doi.org/10.1111/1541-0064.t01-1-00001.

CFAC (Community Forestry Advisory Committee). 1998. *Final recommendations on attributes of a community forest tenure.* Report to Minister of Forests, Victoria, BC.

Charnley, S., and M.R. Poe. 2007. "Community forestry in theory and practice: Where are we now?" *Annual Review of Anthropology* 36 (1): 301–36. http://dx.doi.org/10.1146/annurev.anthro.35.081705.123143.

Cronkleton, P., J.M. Pulhin, and S. Saigal. 2012. "Co-management in community forestry: How the partial devolution of management rights creates challenges for forest communities." *Conservation and Society* 10 (2): 91–102. http://dx.doi.org/10.4103/0972-4923.97481.

Elias, H. 2000. "Harrop and Procter: How a persistent community fashioned its own forestry future." *Ecoforestry* 15: 218–26.

Graham, J., B. Amos, and T. Plumptre. 2003. "Principles for good governance in the twenty-first century." Policy brief no. 15. Ottawa, ON: Institute on Governance.

Jessop, B. 2003. "Governance and metagovernance: On reflexivity, requisite variety, and requisite irony." In *Governance as social and political communication*, edited by H.P. Bang, 142–72. Manchester, UK: Manchester University Press.

Larson, A.M., D. Barry, and G.R. Dahal. 2010. "New rights for forest-based communities? Understanding processes of forest tenure reform." *International Forestry Review* 12 (1): 78–96. http://dx.doi.org/10.1505/ifor.12.1.78.

Larson, A.M., and J. Ribot. 2004. "Democratic decentralization through a natural resource lens: An introduction." *European Journal of Development Research* 16 (1): 1–25. http://dx.doi.org/10.1080/09578810410001688707.

Luckert, M.K., and T. Williamson. 2005. "Should sustained yield be part of sustainable forest management?" *Canadian Journal of Forest Research* 35 (2): 356–64. http://dx.doi.org/10.1139/x04-172.

McIlveen, B., and B. Bradshaw. 2005–6. "A preliminary review of British Columbia's community forest pilot project." *Western Geography* 15–16: 66–84.

Parkins, J. 2008. "The metagovernance of climate change: Institutional adaptation to the mountain pine beetle epidemic in British Columbia." *Journal of Rural and Community Development* 3 (2): 7–26.

Pinkerton, E., R. Heaslip, J.J. Silver, and K. Furman. 2008. "Finding 'space' for comanagement of forests within the neoliberal paradigm: Rights, strategies, and tools for asserting a local agenda." *Human Ecology* 36 (3): 343–55. http://dx.doi.org/10.1007/s10745-008-9167-4.

Pulhin, J.M., A.M. Larson, and P. Pacheco. 2010. "Regulations as barriers to community benefits in tenure reform." In *Forests for people: Community rights and forest tenure reform*, edited by A.M. Larson, D. Barry, G.R. Dahal, and C.J.P. Colfer, 139–59. Washington, DC: Earthscan.

Reader, R. 2006. *The expectations that affect the management of public forest and range lands in British Columbia: Looking outside the legislation.* Victoria: Government of British Columbia.

Reed, M. 1995. "Co-operative management of environmental resources: An application from northern Ontario, Canada." *Economic Geography* 71 (2): 132–49. http://dx.doi.org/10.2307/144355.

Ribot, J. 2002. *Democratic decentralization of natural resources: Institutionalizing popular participation.* Washington, DC: World Resources Institute.

–. 2004. *Waiting for democracy: The politics of choice in natural resource decentralization.* Washington, DC: World Resources Institute.

Schlager, E., and E. Ostrom. 1992. "Property rights and natural resources: A conceptual analysis." *Land Economics* 68 (3): 249–62. http://dx.doi.org/10.2307/3146375.

Tyler, S., L. Ambus, and D. Davis-Case. 2007. "Governance and management of small forest tenures in British Columbia." *BC Journal of Ecosystems and Management* 8 (2): 67–78.

# Chapter 13  Whither Community Forests in Canada? Scenarios of Forest Governance, Adaptive Policy Development, and the Example of Nova Scotia

*Peter N. Duinker and L. Kris MacLellan*

Community forests represent a paradox in Canada. On one hand, our experiences with them, if we interpret the concept broadly, are relatively rich and reach back over a century. On the other hand, they still play only a minor role in the forest policies of the nation's provinces. Perhaps that is because forest policy is overwhelmingly about the forest-products industry – a commodity-focused, export-oriented sector of the economy – and the Crown-owned forest lands administered by the provinces. Perhaps it is because the domain of forest policy has, for many decades, been dominated by industrial interests and provincial bureaucrats who might see community forests as hobby-scale enterprises that do not really fit the industrial model on which Canada's economy was based from the late 1800s right through the twentieth century. Things are changing, though, as the forest sector experiences profound transition in the early twenty-first century (Standing Committee on Natural Resources 2008).

A comprehensive historical account of how various types of community forests have developed in Canada does not exist. Broad summaries of the recent state of play vis-à-vis community forests across the country have been offered (e.g., Duinker et al. 1994; Teitelbaum, Beckley, and Nadeau 2006). Papers on experiences with specific community forests in Canada (e.g., Allan and Frank 1994) and on community-forest concepts and experiences in specific regions (e.g., Bullock and Hanna 2007; Bullock, Hanna, and Slocombe 2009; Ambus and Hoberg 2011; Matakala and Duinker 1993; Duinker, Matakala, and Zhang 1991; Harvey and Hillier 1994) are growing in number as scholars pay increasing attention to the phenomenon. Indeed, entire books generated from Canada (e.g., Bullock and Hanna 2012, and this volume) are beginning to appear and capture audience attention. What we lack, though, are prospective analyses about how community forests might continue to develop through the twenty-first century in this vast forested country.

Our mission, therefore, with this chapter is twofold: (a) to offer some alternative conceptions on how the community-forest theme might play out in Canada over the next several decades, and (b) to outline some of our hopes for that development and some principles that should deliver the learning we need about the most promising models for future community forests. On the first objective, we take inspiration from a recent project of the Sustainable Forest Management Network (SFM Network), a national research network that was active between 1995 and 2010 and was based at the University of Alberta. On the second, we dig into our previous writings and betray our scholarly biases for strong learning processes based on adaptive management and policy development.

## Scenarios of Forest Governance to 2050: Where Do Community Forests Fit?

In 2007, the SFM Network (funded jointly by the Government of Canada, several provincial governments, a host of forest industry companies, and a range of non-government organizations) launched the Forest Futures Project, a two-year research endeavour dedicated to using scenario analysis to try to increase our understanding of how forest and forest-sector policy makers could improve their long-term expectations and redirect policy more strongly towards sustainability. The project was highly participatory. First, a core group of eight university-based scholars (the SFM Network's research-area leaders) created an initial set of long-range scenarios for Canada's forests and forest sector (see Box 13.1). The SFM Network's Research Planning Committee and Partnership Committee (together numbering several dozen academics and stakeholder representatives) then reviewed the scenarios for plausibility and utility for policy analysis. Finally, hundreds of interested and knowledgeable Canadians participated in workshop-based conversations across the country on the implications of those scenarios for immediate policy development.

||||||||||||||||||||||||||||||||||||||||||||||||||||||||||||||||||||||||||||||||||||||||||||||||||||||||||||||||

**BOX 13.1**

## SCENARIO A: GOODS FROM THE WOODS

*Summary*

The forest sector has been able to make good on some of the issues of the day in the 1990s and 2000s, but not on others. Forest ecosystem diversity is not shameful, even if the species-at-risk situation is. Climate change – touch wood – has so far been tolerable, and even helpful in some respects. The strong global timber and forest-products markets have repositioned Canada as a profitable player on the world stage. The energy situation has brought new emphasis to bioenergy, much to the delight of the business community and governments. Local and community-based input and control on forests has waned in light of globalization and a return to stronger provincial governance and the influence of large multinational corporations. Conflicts in and over forest use are abundant and rarely expeditiously resolved. Aboriginal communities, while not much better off politically, have certainly improved their lot economically.

*Governance*

There has been continued strengthening of global integrated forest-products companies. The most illustrative example is Exron Global Enterprises, founded in 2027 and involved in fossil fuels, forest products, real estate, shipping, and investment fund management. Exron holds timber licences on fully one-third of Canada's publicly owned timber landbase. Some interesting changes have taken place in forest ownership. In 2019, the federal government made the three territories into provinces, giving a big drop in federal forest lands in the middle of the period. On top of that, the last two decades have seen a pair of divestments from provincial governments – to the private sector as well as to Aboriginal communities.

## SCENARIO B: PEACE IN THE WOODS

*Summary*

The forest sector has been able to make good on many of the issues of the day in the 1990s and 2000s. Forest ecosystem diversity has improved, and the species-at-risk situation has turned the corner. Climate change – fingers crossed – has so far been tolerable, and even helpful in some respects. The weak global timber and forest-products markets have repositioned Canada as a low-volume, value-added player, as much on the domestic stage as on the world's. The energy situation has brought little emphasis to bioenergy, despite the initial enthusiasms. Local and community-based input and control on forests have blossomed. Whatever conflicts have been occurring in and over the woods have been fairly peacefully resolved. Aboriginal communities are, by their own admission, much better off both politically and economically.

### Governance

The dominance of provincial governments and industry over Canada's forests back in 2000 has given way to profound shifts. One is the deeding of provincial Crown land to local, forest-based communities. In 2000, communities across Canada owned a total of 52,000 hectares of timber-producing forest land. Today, that number is 25 million hectares. The second shift is the granting of comprehensive forest-management licences to communities in places where the old industries have either left town or become low-volume, high-tech manufacturers. The timber licences of the past have been replaced with broad resource-management licences, giving holders rights to timber, water, wildlife, surface minerals and recreation. The new resource-management licences, as before, require extensive public consultations during the preparation of management plans. The third shift is a huge expansion of Aboriginal lands in Canada.

## SCENARIO C: TURBULENCE IN THE WOODS

### Summary

The forest sector has been able to make good on rather few of the issues of the day in the 1990s and 2000s. Forest ecosystem diversity is shameful. Climate change is overwhelming all types of ecosystems. Despite strong global timber and forest-products markets, Canada is a marginal player on the world stage. The energy situation has brought new emphasis to bioenergy. Local and community-based input and control on forests has waned in light of a strange return to a kind of forest feudalism. Abundant conflicts occurring in and over the woods are rarely peacefully resolved – strife is the order of the day. Aboriginal communities are worse off politically and economically. All in all, one could long for a return to the conditions of 2000 – most indicators were more favourable at that time. The future can only get better – we hope!

*Governance*

Forest governance today seems neo-feudal, almost nineteenth-century-like. The provinces experimented with transfer of Crown-forest tenures to local interests such as municipalities, thinking that community forests would emerge. However, most community-forest administrations devolved into fiefdoms for harvest contractors and other timber entrepreneurs. Forests across Canada are in such a sorry state that people seem to care little for them, and they are 'managed' largely by timber barons. The industry is not entirely unhappy, because wood has kept flowing in abundance. Forest ownership has changed little. Aboriginals across the boreal north now own 5 percent of Canada's forests.

## SCENARIO D: RESTORATION IN THE WOODS

*Summary*

The forest sector has been able to make good on rather few of the issues of the day in the 1990s and 2000s. Forest ecosystem diversity is shameful. Climate change is overwhelming all types of ecosystems. Global timber and forest-products markets are flat, and Canada is a marginal player on the world stage. Continually increasing energy demand has led to growth in supply from all sources except biomass. Local and community-based input and control on forests has flourished, with strong global institutions. What conflicts occur in and over the woods are peacefully resolved. Aboriginal communities are better off politically but not economically. All in all, it's a time of considerable social and economic progress despite forest ecosystems having fallen into a degraded state. Creative and energetic political leadership has helped shape a country where people are working together nationwide to try to restore better conditions in the woods. A few high-profile community collapses in the 2020s and 2030s also helped people understand the profound importance of collaborating, rather than competing, to secure their continued prosperity and the sustainability of their communities. Canada continues as a forest nation, and Canadians finally are behaving as a forest people.

*Governance*

the forest-products industry in Canada has lost considerable interest in managing the wood-supplying public forests. Three important developments are noteworthy here. One is the devolution of authority over public forests to co-management partnerships involving non-Aboriginal and Aboriginal communities working together. The second development relates to an abundance of settled land claims for Aboriginal peoples. The third development is a function of the rise of several new and influential worldwide forest-governance institutions. Global institutions have managed to capture strong participation from local forest interests across Canada, and the passage of helpful information and advice from local settings to the global stage, and vice versa, is exceptional. As a result, forest-land ownership patterns have changed substantially.

||||||||||||||||||||||||||||||||||||||||||||||||||||||||||||||||||||||||||||||||||||||||||||||||||||||||||||||||||||

Creation of the scenarios, numbering four in total, followed, in general, the protocols of Ogilvy and Schwarz (2004) and was supported by a series of working papers analyzing the major drivers of change on Canada's forests and forest sector. Topics covered by those papers included global climate change, global forest products demand and Canadian wood supply, geopolitics, global energy, technology, governance, Aboriginal empowerment, air pollution, conflict over resources, society's forest values, demographics, and industry profitability.

The four scenarios explore possible trajectories for Canada's forests and the forest sector to the year 2050. They were developed with parallel thematic structure but dramatically different constellations of outcomes. The structure includes not only the themes covered in the driver papers but also a suite of so-called response themes, grouped into three categories: (a) ecological themes – biodiversity, ecosystem condition and productivity, water and soil, and carbon; (b) social and community themes – amenity values, participatory processes, and forest-related employment; and (c) economic themes – wood harvests, wood-processing industries, harvests of non-wood products, and markets for forest services.

Each scenario was written as a substantial essay outlining the future state of Canada's forests and forest sector. For our purposes here, we have summarized each scenario and recounted the relevant passages from the governance section of each one (Box 13.1). These scenario texts were written as though it were the year 2050, and the perspective of the account is to look back to the turn of the century and relate the major developments associated with the Canadian forests and forest sector. For this chapter, the texts have been abridged from the original scenario texts to direct our attention more strongly to community forests in this chapter.

A word on scenario construction and use is in order. Scenarios are defined as descriptions of alternative future outcomes, states, or pathways (see Cornish 2005). They are built to illuminate possible pathways and outcomes for the purpose of exploring future possibilities for important topics as well as promising policy options for preventing undesirable outcomes and fostering desirable ones. They are not laden with the baggage of probability, because confident assignment of likelihoods of outcome is both impossible and unnecessary (see Frittaion, Duinker, and Grant 2010, 2011). In short, the scenarios are selected stories that, if prepared well, serve as strong conversation starters about possible and desirable states of the world as bounded by project scope. Written as historical narratives, they cannot but contain the values of their creators. It is hoped, however, that each scenario highlights a different constellation of values such that value diversity characterizes the entire scenario set. The scenario materials are reproduced in Box 13.1, with text taken verbatim from the full scenarios as written for the project in 2007–8 (Forest Futures Project 2008a; Forest Futures Project 2008b; Forest Futures Project 2008c; Forest Futures Project 2008d).

## Emerging Messages for Community Forests

Needless to say, the scenarios above are all fictions. Participants in the cross-Canada workshops all had their own views about the relative preferences for each. The point of these scenarios is to explore not what is probable but what is possible for the future of Canada's forests and the forest sector of the economy. In the context of this chapter, the point is to examine what might be possible for community forests. The scenarios, in aggregate, suggest that under some circumstances, community forests could thrive in Canada. Scenario B, Peace in the Woods, posits a huge increase in community-run forests, by both non-Aboriginal and Aboriginal communities. Scenario D, Restoration in the Woods, posits a similar future, but with an added twist of unprecedented cooperation among non-Aboriginal and Aboriginal communities.

In contrast, the two other scenarios (Scenarios A and C) imagine futures with few or no community forests. In the Goods from the Woods scenario, people just want to go back to the easier days of government and industrial control under a thriving and stable economy based on natural resources. In the Turbulence in the Woods scenario, community forests were attempted but were unsuccessful in terms of their governance structures, and the forest sector became a kind of free-for-all Wild West.

The bulk of our workshop participants across Canada leaned towards the desirable features of the Peace in the Woods scenario. This preference was doubtless shaped by the better outcomes along both environmental and social dimensions and by an acceptance of a smaller forest industrial sector. However, many participants were not so optimistic about environmental outcomes, expecting the worst in terms of climate change and social outcomes and fearing less socially progressive behaviours both within Canada and abroad. Participants who were predisposed to favour the concept of community forests were cheerleaders for the Peace in the Woods scenario (and the Restoration scenario, if that were the direction for environmental change). Those who feared that Canada was headed towards the Goods or Turbulence scenarios (A and C, respectively) were not overtly against the concept of community forests; rather, they were skeptical about the prospects of successful implementation.

It is pointless to single out one scenario and say, "Well, that will never happen!" In socio-economic systems, so much of what does happen results from people making it happen. This brings to mind the old adage "Where there is a will, there is a way." So, the right approach to scenario analysis is to ask "What would we prefer to happen?" If a nation with abundant, diverse, successful community forests from coast to coast is desired – and we believe that huge numbers of forest-sector participants do want this – then how do Canadians collectively make that happen?

There is no nationwide empirical research revealing the strength of preferences of Canadians, either lay or professional, for community forests. The SFM Network scenario analysis leads us to the following characterization of Canadian thought about community forests: few are against them, many are skeptical, and many more are enthusiastic. This situation suggests the following strategies for building up the overall community forest enterprise across Canada:

- Make efforts to convince the skeptics that community forests can be successful. This requires aggressive documentation and communication about successful – and, of course, less-than-successful – experiences with implementation.
- Commit to continuing with implementation and experimentation.

- Experiment with forms of forest tenure that go well beyond conventional timber-harvest rights and responsibilities.
- Diligently oppose socio-economic forces that thwart possibilities for successful community-forest ventures.

Scenario practice tells us that the future is ours (May 1996), meaning that we have considerable ability to choose a specific future and control our actions to work for its eventual manifestation. The SFM Network scenario project revealed a strong leaning across Canada for a future characterized by more community-controlled forests. Given this, we conclude that community forests can and should be expanded substantially across Canada. We turn now to some concepts for strong implementation, with a focus on Nova Scotia, where initial signs are promising.

### Innovations and Creative Arrangements: A Prospectus for Nova Scotia

Our firm starting point is this: many people who are keen participants in forest and forest-sector affairs in Canada would like to see various community forest models implemented and tested. We have several experiences that provide at least some evidence for this claim. First, many participants in the Forest Futures Project expressed hope for the outcomes of scenarios that include a rich array of community forests across the country. Second, we see a general expansion of interest in community forests across Canada as sensed through increased experimental ventures, reinterpretation of alternative tenure models as community forests, and increased literature on the topic. Third, we note, with great delight and surprise, that the Government of Nova Scotia recently declared its interest in community forests, and it is to that prospect that we now turn. Our surprise comes from the lack of forewarning that community forests were to be featured in the government's recent natural resources strategy, titled *The Path We Share* (Nova Scotia, Natural Resources 2011). Our delight is that we live in Nova Scotia and that our senior author, Peter Duinker, has advocated for and researched community forests in Ontario, his previous province of residence (e.g., Duinker, Matakala, and Zhang 1991).

More than three-quarters of Nova Scotia's land base is forested, with about half of forest land ownership in small private holdings and roughly a quarter in large industrial holdings and another quarter in government Crown parcels. The forest products industry, once thriving, has recently endured debilitating shutdowns and through-put reductions, both in pulp and paper and in lumber. Forest policy in the province is very much a matter of supporting and regulating activities on private land. That policy has been turbulent of late, as the province went about developing a new forest strategy (Duinker 2012).

We are unable to explain why the concept of community forest has gained essentially no traction in Nova Scotia until now. Perhaps there is something in the notion that to date, community forests seem to have flourished best under conditions where there is abundant Crown land. After all, most of Canada's community forests are in British Columbia, Ontario, and Quebec, which, as it turns out, are Canada's most populous provinces. But that argument loses ground when one considers that most of Ontario's community forests – especially the county forests and conservation authority forests – are in the heavily populated south of the province. Perhaps when the forest products industry is thriving, people are less inclined to feel the need to imagine alternative forms of forest tenure. Perhaps when small private fee-simple holdings represent the predominant form of land ownership, forest policy generally is dominated by their issues. Regardless, little was said or written about community forests in Nova Scotia until 2012 (MacLellan and Duinker 2012a, 2012b; MacLellan 2013), and now the tide seems to be turning.

### Forum Project

The delight and surprise alluded to above came in response to the following government commitment in *The Path We Share* (Nova Scotia, Natural Resources 2011, 38): "Explore ways to establish and operate working community forests on Crown land." Under the aegis of the Nova Forest Alliance and with enthusiastic financial support from the Department of Natural Resources, we ran a stakeholder forum in June 2012 with the goal of bringing a wide range of interested people together to explore the concept of community forests and, specifically, their possible implementation in Nova Scotia. Thus, our forum objectives were to identify and discuss a range of ideas on what kinds of community forests people in the province would like to see developed and applied and to document the results of the discussions (MacLellan and Duinker 2012a, 2012b).

Based on the results of the forum, MacLellan and Duinker (2012b) concluded the following:

1. The forum participants were strongly excited and optimistic about the prospects of experimenting with a range of community forest ventures across the province.
2. The forum participants expect strong leadership, initiative, and facilitation from the Government of Nova Scotia regarding experimental ventures.
3. Early and strong leadership also needs to be forthcoming from the various parts of the forest products industry (e.g., pulp and paper companies, sawmillers), from non-government organizations (e.g., Nova Forest Alliance, Ecology Action Centre, Mersey Tobeatic Research Institute), and from municipal governments (e.g., municipal councillors).
4. The forum represented an important but insufficient first step in building both momentum and understanding around the community forest concept. It is paramount that we foster both of these factors through ongoing dialogue involving diverse, interested participants across the province.

### Recommendations to Government
In order to directly encourage the Government of Nova Scotia to take appropriate actions without delay, Duinker and MacLellan (2012) forwarded the following recommendations to the minister of natural resources:

- That the Department of Natural Resources begin immediately to develop a program of assistance, guidance, and encouragement for the establishment of community forests across the province
- That, in its policy and program development for community forests in the province, the Department of Natural Resources embrace a diverse array of possible models and proceed with full-scale experimentation of the chosen models
- That the Department of Natural Resources mount a vigorous program of stakeholder and citizen engagement in association with policy and program development on community forests
- That the Department of Natural Resources take immediate steps to establish its leadership of, and support for, implementing community forests in the province
- That the Department of Natural Resources organize an annual review mechanism for community forest policy and programming

whereby all interested parties will have a chance to examine critically the accomplishments and failures of program implementation and plot promising avenues forward.

## Recent Developments

The year 2013 was a year of both further conversation and community forest program implementation. On the conversation side, two initiatives are noteworthy. First, another forum took place in June 2013. Its objectives were to review initial attempts to prepare a guidebook on community forests in Nova Scotia (MacLellan 2013) and to discuss two key themes in community forest design and implementation: governance and entrepreneurship. Second, the Department of Natural Resources contracted the Nova Forest Alliance to establish and facilitate a Community Forest Advisory Group. The group, comprising individuals from both government and non-government organizations, met monthly throughout the year to develop guidance for ongoing program development.

Associated with the work of the advisory group was the production of the *Guide to Community Forests in Nova Scotia* (MacLellan 2013). The guide is meant to help those involved in the forest sector in Nova Scotia to orient their thinking towards the main concepts and issues associated with setting up a community forest in the province.

Finally, of greatest significance was an announcement from the Department of Natural Resources that negotiations would begin early in 2014 to establish the first community forest on Crown land – the Medway Community Forest (Nova Scotia, Natural Resources 2013). This venture covers fifteen thousand hectares in the Medway River area of southwestern Nova Scotia. The land will be managed for an initial three-year period by the Medway Community Forest Cooperative. The Department of Natural Resources also indicated that it was "exploring opportunities for a Mi'kmaq forestry initiative with the Assembly of Nova Scotia Mi'kmaq Chiefs" (Nova Scotia, Natural Resources 2013).

## Prospects for Community Forests

The key to success with community forests is to engage in adaptive policy development (Ontario Forest Policy Panel 1993), itself a successor to the broad concept of adaptive management (Lee 1993). Without using that moniker, De Young and Kaplan (1988) offer a powerful three-pronged approach to sustainable development that they call "adaptive muddling." The first prong is exploration: that is, experimenting, at scales beyond the lab bench, with as wide a range of alternative resource-management models as people can come up with. The second is stability in the policy framework under which experimentation occurs. While some experiments will succeed, others will not. Policy stability is needed to provide the safety net for local entrepreneurs to take risks. It also provides the protracted time frame needed to allow local experimentation to bear fruit, or not, with reliable knowledge as to the factors behind success or failure. The third prong is distributed leadership, which recognizes the immense energies and talents of local people and the facilitative character of local empowerment.

Based on the concepts of adaptive policy development and adaptive muddling, a robust policy and program for the development and implementation of community forests in Nova Scotia (and probably anywhere) needs the following characteristics:

- A strong and stable enabling policy framework providing senior government oversight for the broader public interest, a safety net for failures, and resources such as land, expertise, and money
- A reliance on distributed leadership that taps into the creative energies and intellects of local people
- A receptivity to wide diversity in the range of sizes, governance structures, and objectives for community forest management
- A boldness to abandon prevailing conventions and step smartly, with eyes wide open, into novel territory.

Let us step back for a moment and be reminded of the basic tenets of adaptive management. The description by Duinker and Trevisan (2003) is helpful here. At its technical roots, adaptive management consists of the following:

- Making formal statements of future expectations (i.e., hypotheses) of system performance under specified management strategies
- Implementing the specified strategies, complete with a monitoring scheme that address both action implementation and system responses in terms of indicators as projected in the hypotheses
- Rigorously comparing, at an appropriate future time, the hypotheses and the monitoring data to discern divergences between

expectations and realities, with an eye to understanding the divergences and repositioning the strategies for improved system performance in the next management cycle.

For the purposes of this chapter, we propose that each new community forest should adopt a formal program of adaptive management, a concept that, as described above, has clear applicability at the level of managing forests in a physical sense. For the community forest program as a whole, we advocate adoption of adaptive policy development as embodied in De Young and Kaplan's (1988) conception. Key to both levels of application of adaptivity is the avoidance of premature abandonment of initiatives that seem to bear little fruit at the beginning: clear signals of success, whether in managing forests or developing provincial policy for forest management, take considerable time to manifest.

The principle of diverse experimentation is, we believe, crucial for advancement of learning. What kinds of community forest concepts and models might be developed, refined, and implemented to give Nova Scotians a strong sense of progress and learning? Here are some suggestions for consideration; no doubt many others could be conceived and designed.

- Demonstration forests: These would probably be relatively small (e.g., measured in terms of hundreds of hectares) but should be numerous, given the educational power of working examples of advanced approaches to forest management and conservation across the province. One such community forest is now in operation – the Otter Ponds Demonstration Forest (Prest 2010).
- Watershed conservation forests: Many communities in Nova Scotia, including the Halifax Regional Municipality, draw water from surface sources in forest watersheds. These are ideal settings for experimentation with the community forest concept. Experience from British Columbia, where many of the new community forests have potable water supply as a top priority in forest management, could provide useful lessons for application in Nova Scotia (see Bullock and Hanna 2012). Watershed community forests might be measured in terms of thousands of hectares (e.g., Halifax Regional Municipality 2012; HPFP 2015).
- Timber-production forests: Notwithstanding the vagaries of the commercial timber market in Nova Scotia, and in much of the rest

of Canada, timber production is still a likely major income source to pay the bills of forest management. The full range of conventional forest products (e.g., high-quality saw logs, stud wood, pulp wood) and less-conventional materials (e.g., commercial-scale energy wood) can be considered under the aegis of community forest. These enterprises might well be suited for forests measuring in the tens of thousands of hectares.

- Protected areas and recreation forests: We are not averse to the idea that communities could be appropriate managers of national parks, provincial parks, designated wilderness areas, and other conservation-oriented forests. Such community forests could be of any extent.

## Conclusions

As ardent advocates for community forests, we relish the thought of engaging in what Boyer (1990) calls the scholarship of application. In that form of scholarship, we become involved in a formal way in the learning processes needed to propel the community forest concept towards a promising future. It is exciting to think of community forests becoming exemplars of Lee's (1993) "compass and gyroscope," in which the science of adaptive management and the politics of principled negotiation combine to achieve real progress in sustainability (Duinker 2012). If ever an opportunity for such progress were laid at our feet, it is community forests now. In returning to the Forest Future Project's scenarios, we profess our affinity for a scenario where community forests flourish across Canada. It is indeed exciting to imagine a rich suite of partnerships involving forest practitioners, business people, Aboriginal communities, rural (and even urban) residents, and scholars, among others, coming together over the coming decades in numerous locations across Canada to give the community forest concept a chance to prove itself.

## References

Allan, K., and D. Frank. 1994. "Community forestry in British Columbia: Models that work." *Forestry Chronicle* 70 (6): 721–24. http://dx.doi.org/10.5558/tfc70721-6.

Ambus, L., and G. Hoberg. 2011. "The evolution of devolution: A critical analysis of the community forest agreement in British Columbia." *Society and Natural Resources* 24 (9): 933–50. http://dx.doi.org/10.1080/08941920.2010.520078.

Boyer, E.L. 1990. *Scholarship reconsidered: Priorities of the professoriate.* San Fran-
cisco: Carnegie Foundation for the Advancement of Teaching, and Jossey-Bass.

Bullock, R., and K. Hanna. 2007. "Community forestry: Mitigating or creating
conflict in British Columbia." *Society and Natural Resources* 21 (1): 77–85.
http://dx.doi.org/10.1080/08941920701561007.

Bullock, R., K. Hanna, and S. Slocombe. 2009. "Learning from community forestry
experience: Challenges and lessons from British Columbia." *Forestry Chronicle*
85 (2): 293–304. http://dx.doi.org/10.5558/tfc85293-2.

Bullock, R.C.L., and K.S. Hanna. 2012. *Community forestry: Local values, conflict
and forest governance.* Cambridge, UK: Cambridge University Press. http://
dx.doi.org/10.1017/CBO9780511978678.

Cornish, E.E. 2005. *Futuring: The exploration of the future.* Bethesda, MD: World
Future Society.

De Young, R., and S. Kaplan. 1988. "On averting the tragedy of the commons."
*Environmental Management* 12 (3): 273–83. http://dx.doi.org/10.1007/BF
01867519.

Duinker, P.N. 2012. "In search of 'compass and gyroscope': Where were adaptive
management and principled negotiation in Nova Scotia's forest-strategy process?" *Dalhousie Law Journal* 35 (1): 31–46.

Duinker, P.N., and L.K. MacLellan. 2012. *In support of community forests: Recommendations to the Government of Nova Scotia.* Halifax, NS: School for Resource
and Environmental Studies, Dalhousie University.

Duinker, P.N., P. Matakala, F. Chege, and L. Bouthillier. 1994. "Community forests
in Canada: An overview." *Forestry Chronicle* 70 (6): 711–20. http://dx.doi.org/
10.5558/tfc70711-6.

Duinker, P.N., P.W. Matakala, and D. Zhang. 1991. "Community forestry and its
implications for northern Ontario." *Forestry Chronicle* 67 (2): 131–35. http://
dx.doi.org/10.5558/tfc67131-2.

Duinker, P.N., and L.M. Trevisan. 2003. "Adaptive management: progress and
prospects for Canadian forests." In *Towards sustainable management of the
boreal forest: Emulating nature, minimizing impacts and supporting communities,* edited by V. Adamowicz, P. Burton, C. Messier, and D. Smith, 857-92.
Ottawa, ON: NRC Press.

Forest Futures Project. 2008a. "Goods from the woods: A history of Canada's
forests and forest sector, 2000-2050." Scenario A of the Forest Futures
Project. Sustainable Forest Management Network, University of Alberta,
Edmonton, AB. 17 pp. http://www.sfmn.ales.ualberta.ca/en/Research/Forest
Futures/ForestFuturesDocuments.aspx.

–. 2008b. "Peace in the woods: A history of Canada's forests and forest sector,
2000-2050." Scenario B of the Forest Futures Project. Sustainable Forest
Management Network, University of Alberta, Edmonton, AB. 17 pp. http://
www.sfmn.ales.ualberta.ca/en/Research/ForestFutures/ForestFutures
Documents.aspx.

–. 2008c. "Turbulence in the woods: A history of Canada's forests and forest sector, 2000-2050." Scenario C of the Forest Futures Project. Sustainable Forest

Management Network, University of Alberta, Edmonton, AB. 17 pp. http://www.sfmn.ales.ualberta.ca/en/Research/ForestFutures/ForestFutures Documents.aspx.

–. 2008d. "Restoration in the woods: A history of Canada's forests and forest sector, 2000-2050." Scenario D of the Forest Futures Project. Sustainable Forest Management Network, University of Alberta, Edmonton, AB. 17 pp. http://www.sfmn.ales.ualberta.ca/en/Research/ForestFutures/ForestFutures Documents.aspx.

Frittaion, C.M., P.N. Duinker, and J.L. Grant. 2010. "Narratives of the future: Suspending disbelief in forest-sector scenarios." *Futures* 42 (10): 1156–65. http://dx.doi.org/10.1016/j.futures.2010.05.003.

–. 2011. "Suspending disbelief: Influencing engagement in scenarios of forest futures." *Technological Forecasting and Social Change* 78 (3): 421–30. http://dx.doi.org/10.1016/j.techfore.2010.08.008.

Halifax Regional Municipality. 2012. *Halifax water: Pockwock Lake and Tomahawk Lake watersheds.* Halifax, NS: Halifax Regional Municipality. http://www.halifax.ca/hrwc/PockwockLakeandTomahawkLakeWatersheds.html.

Harvey, S., and B. Hillier. 1994. "Community forestry in Ontario." *Forestry Chronicle* 70 (6): 725–30. http://dx.doi.org/10.5558/tfc70725-6.

HPFP (Harrop-Procter Forest Products). 2015. "Deep Community Roots." *HPFP.* http://www.hpcommunityforest.org .

Lee, K.N. 1993. *Compass and gyroscope: Integrating science and politics for the environment.* Washington, DC: Island Press.

MacLellan, L.K. 2013. *Guide to community forests in Nova Scotia.* Stewiacke, NS: Nova Forest Alliance.

MacLellan, L.K., and P.N. Duinker. 2012a. *Community forests: A discussion paper for Nova Scotians.* Stewiacke/Halifax, NS: Nova Forest Alliance/School for Resource and Environmental Studies, Dalhousie University.

–. 2012b. *Advancing the conversation on community forests in Nova Scotia: Proceedings from the June 2012 Forum on Community Forests.* Stewiacke/Halifax, NS: Nova Forest Alliance/School for Resource and Environmental Studies, Dalhousie University

Matakala, P., and P.N. Duinker. 1993. "Community forestry as a forest-land management option in Ontario." In *Forest dependent communities: Challenges and opportunities,* edited by D. Bruce and M. Whitla, 26–58. Sackville, NB: Rural and Small Town Research and Studies Program, Mount Allison University.

May, G.H. 1996. *The future is ours: Foreseeing, managing and creating the future.* Praeger Studies on the 21st Century. Westport, CT: Praeger.

Nova Scotia. Natural Resources. 2011. *The path we share: A natural resources strategy for Nova Scotia.* Halifax: Department of Natural Resources.

–. 2013. *Province moving forward on community forests.* News release, 10 October. http://novascotia.ca/news/release/?id=20131018004.

Ogilvy, J., and P. Schwarz. 2004. *Plotting your scenarios.* Emeryville, CA: Global Business Network.

Ontario Forest Policy Panel. 1993. *Diversity: Forests, peoples, communities.* Toronto: Ministry of Natural Resources, Government of Ontario.

Prest, W. 2010. *Making progress: The Otter Ponds Demonstration Forest is now a reality.* Guest editorial. *Atlantic Forestry*, September.

Standing Committee on Natural Resources. 2008. *Canada's forest industry: Recognizing the challenges and opportunities.* Ottawa: House of Commons, Government of Canada.

Teitelbaum, S., T. Beckley, and S. Nadeau. 2006. "A national portrait of community forestry on public land in Canada." *Forestry Chronicle* 82 (3): 416–28. http://dx.doi.org/10.5558/tfc82416-3.

# Chapter 14 Towards an Integrated System of Communities and Forests in Canada

*Ryan Bullock and Maureen G. Reed*

Previous chapters in this volume have illustrated the capabilities and diversity that exist within localized community forestry efforts in Canada. Earlier examples point to challenges that can arise at local to national levels: for instance, challenges associated with captured policy networks, access to markets, limitations of knowledge systems, and the "reach" of grassroots initiatives. In our view, current disconnects limit the flow of ideas and resources, and these disconnects have, in turn, constrained the overall success of community forestry. This view is consistent with Berkes (2009), who, with respect to community-based natural resource management in general, points out that local initiatives are often constrained for the same reason as top-down ones – a failure to recognize the need for achieving integration across subsystems. Hence, we argue that there is a need and an opportunity for greater integration among the many localized community forestry movements developing across Canada, including stronger linkages among local and national organizations, practitioners, and advocates. In this chapter, we reflect on factors influencing community roles in forest development to suggest what changes could help to generate a well-supported and integrated system of locally managed forests in Canada.

We begin by discussing how the historic rise of natural resource industries created isolated communities and what we refer to here as a "myth of self-reliance" that permeates the collective mindsets of many decision makers and residents in resource towns to this day. The well-being of resource communities has been heavily dependent on single industries and sectors and on the policy choices of senior governments. Our discussion draws on complex systems concepts and Canadian regional planning to characterize resource communities under both industrial and community forestry approaches, in order to understand the dynamics that shape governance and development opportunities. We present the conventional industrial model and its links to the core-periphery model of regional development to demonstrate how the creation of *community-company systems* reinforced isolationist views among communities and constrained collaboration and stability. We then contrast these systems with *community-forest systems*, which draw on ecosystem-based management and territorial planning approaches (M'Gonigle 1997; Hodge and Robinson 2002). We argue that these community-forest systems are more appropriate to support the formation of multiple interdependent community forests. Finally, we identify opportunities to improve regional network development, coordination, and capacity requisite to advancing functioning systems that might more effectively support community forestry in Canada.

## Conceptualizing Communities, Regions, and (Inter)Dependency in Natural Resource Systems

### Retrospective: Community-Company Systems

There is a rich and romantic cultural history associated with natural resource development and town planning in Canada – one reinforced by images and narratives that portray self-reliant northerners and isolated communities eking out an existence in rugged Canadian landscapes, surrounded by pristine environs and natural resources containing immeasurable wealth. Geographers have highlighted how such scenes and ideas have been important to shaping both a national consciousness (e.g., Wallace 1987; Bone 1992) and provincial identities linked to a staples economy (e.g., Hayter 2000; Dunk 2003; Davis 2011). Scenic and "wild" landscapes depicting forests, water, and rocky terrain have been celebrated for more than a century, as in the works of Canada's prominent Group of Seven painters from the early 1900s. Ongoing characterizations, whether by artists or governments, have portrayed interior and northern regions as rich Crown resource storehouses largely devoid of Aboriginal and non-Aboriginal peoples (Lawson, Levy, and Sandberg 2001; Ballamingie 2009). The process of forming isolated settlements across an "open frontier" was important to national and provincial development. However, some suggest that propagating

a myth of self-reliance was also important to garner support for development agendas that would actually produce isolated, dependent
settlements to serve distant interests (Clement 1997).[1] Most notable
of these are natural resource towns and Indian reserves.[2]

It is ironic that so many self-reliant people became so dependent
on single employers within the remote towns that were, by design,
administered from afar and tied to foreign commodity markets. By
the late 1800s, provincial governments were encouraging resource development and private investment in resource-rich areas, which gave
rise to company towns in the forestry, mining, and transportation sectors (see Lucas 1971). Many company towns were closed systems for
workers and residents in that "the Company" provided all the essen-
tials required for life and work (Saarinen 1986). In addition to natural
resource processing and energy-generating facilities, forest and mining companies built roads, infrastructure, and housing and provided
services and recreation opportunities to attract and retain labour
in remote and rural areas (Saarinen 1986; Robson 1988; Goltz 1992).[3]
Physical isolation – coupled with limited transportation, communication infrastructure, and technology – greatly restricted the flow of
people and ideas among resource communities, as well as between
those communities and urban centres.

Under this model, labourers, town administrators, and companies
helped to form what could be considered remote industrial forestry
systems rather than place-based communities. A community-company
compact was formed that set out roles and relations for residents as
labourers, forests as fibre sources, and town sites as service hubs for
the main industry. The particular resource at the centre of development produced a "temporarily stable and self-sustaining pattern of
relations" (Barnes, Hayter, and Hay 2001, 2130). Natural resource extraction and companies became the centre of local culture, work, and
economies (Bullock 2013). This arrangement also produced in the settler population a local acceptance of, and loyalty to, extractive resource
development and utilitarian landscapes, even if some groups of residents (e.g., women) were largely excluded from or marginalized within
the local workforce (Reed 2003). As Saarinen (1986, 245) points out,
"Resistance to the status quo was muted by the reliance of residents
upon the company for housing and employment." Above all, this system
created a local dependence on the company, and resource extraction
more broadly produced a strong relationship among social relations,
community planning, and natural resource development. However, these

benefits were not provided to Aboriginal peoples, who were usually the original inhabitants of these regions. Still reliant on these resource territories, Aboriginals were often present on the fringes of resource communities.

As community-company systems have depended on constant and large exchanges (both outflows and inflows) of energy, material, and information with distant power centres to maintain their structure and function, they are largely open systems (Kay et al. 1999, 723). Examples of vital inflows include physical capital (e.g., machinery), human capital (e.g., people and their skills and knowledge), private investment, government transfers and development funding (e.g., FedNor), management decisions, and information from corporate offices and central government branches in major centres, as well as technical expertise (e.g., consultants) and other professional services (e.g., financial) fed in from afar to support forest planning and management decision-making and mill operations.

Under the conventional model, some benefits from fishing, hunting, non-timber forest products, and tourism flowed to the community through a small local market or direct interactions with the forest. However, community-company systems have relied on high volumes of energy, materials, and information *flowing out*, mainly in the form of raw resources or semi-finished products and resource rents, in order to receive basic returns needed to maintain structure. The majority of forest materials and profits have long flowed out, bypassing local entities (e.g., municipalities, local enterprise, households) before benefits from forest extraction and processing could return to the community, primarily in the form of employment, corporate taxes, and reinvestment for the provision and maintenance of infrastructure and services provided by companies and/or senior governments.

While some feedbacks occurred within these systems (e.g., small businesses in forestry support services), experience and theory have demonstrated that service jobs are generally insufficient by themselves to generate and maintain core system structure. The export focus and built-in dependency on external controls means that significant external inputs are needed to maintain buildings and processing facilities, infrastructure and services, human populations, and managed forests and lands (i.e., through inputs for silviculture and intensive forest management, decisions, and innovation). In Canada, research has demonstrated that there is generally a negative association between forest dependence and community well-being (Clapp 1998), particularly in logging, where "the extraction and supplying of inputs for others to process does not result in [substantial and long-term] economic development" (Stedman, Parkins, and Beckley 2005). Cases such as Port Alberni, British Columbia (forestry; see Barnes, Hayter, and Hay 2001), Dubreuilville, Ontario (forestry; see Bullock 2010), and Schefferville, Quebec (mining; see Bradbury and St-Martin 1983) provide clear examples of how community-company systems can decline with the loss of the major industry.

Community-company systems remain stable so long as significant disturbances are not introduced from the outside. These disturbances might include decisions to reduce timber extraction, policy changes or natural disturbances that limit timber supply, market changes that dramatically reduce demand, or corporate restructuring causing job loss. Each of these examples would reduce inputs necessary to maintain system structure and function. In Canada, it is widely held that resource communities have had limited control over their role in the economy or in forest-management and planning decisions. Resource management and planning have been handled by external actors (i.e., forest companies and government agencies) that historically have controlled public forest resources. External forces such as new public or private investment strategies or technological innovations may invoke changes at the local level, causing the system to "flip" (Walker and Salt 2006). Such flips in forest economies and social conditions have long been driven by positive, reinforcing feedback loops that create the boom-and-bust cycle.[4]

Historically, communities have been tied to a political economy of resource extraction focused on exports to core urban areas (i.e., concentrated on flows to a dominant centre; see, e.g., M'Gonigle [1997]) (Figure 14.1). This arrangement demonstrated limited functional integration and political organization (Wallace 1987) among resource-based communities *within* a region. Furthermore, local governance cultures within extractive-resource settings did not foster entrepreneurialism or cooperation at a regional level (Hayter 2000). While communities were characterized by an initial period of rapid growth, their "maturation" was characterized by a more stunted form and character. Reflecting on town-planning experiences in northern Ontario over the past century, Saarinen (1986) observed that creating one-off resource-based settlements eventually produced towns with common structural characteristics and limitations, including small populations, slow growth rates, isolation, limited hinterlands, a narrow economic base, and poorly developed physical and socio-cultural infrastructures. With respect to company towns in Canadian resource

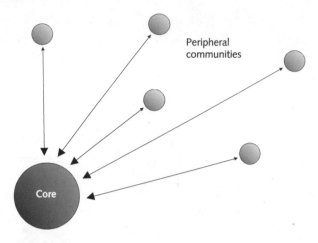

**Figure 14.1**  Basic core-periphery structure of company-community systems under the centrist model.
*Note:* The comparative sizes of the arrowheads indicate that more of the benefits are flowing from the peripheral communities to the core. Observe the absence of connections among peripheral communities.

regions, Saarinen (1986, 228) concluded that "the functioning of the urban network as a whole has been seriously influenced by powerful exogenous forces, instability and uncertainty, and weak spatial interaction among the constituent communities."

The associated social contract was formed by an "iron triad" of industry, government, and labour unions. It was predicated on the assumption of the male breadwinner who secured high wages for what was frequently considered to be dirty, dangerous, physical labour (Hayter 2000; Dunk 2003; Reed 2003). Governments provided favourable policy conditions for resource exploitation, and industry gained profits and provided high wages secured through union contracts. The associated division of labour was highly segregated into primary, secondary, and tertiary sectors and was highly gendered, providing greater opportunities for secure, well-paid employment and income to Euro-Canadian men than to women and/or Aboriginal inhabitants (Reed 2003; Mills 2006). Associated with these wages were supports for local communities such as recreational facilities and infrastructure, already described. Frequently missing, however, was a strong sense of community and identity built on diversified employment, social supports, and community infrastructure (e.g., medical and social services, well-serviced public schools, communities designed for social interaction) that might assist communities in becoming "fully functional" through all stages of the employee's life course.

According to Barnes, Hayter, and Hay (2001), the resultant structure of dispersed satellite settlements also created intellectually constrained regions that further limited prospects for industrial and research diversification. It is widely accepted that provincial ministries and private firms have controlled Crown-resource development through centralized and distant institutions and decision-making supported by professional disciplines such as forestry, planning, surveying, and engineering (e.g., Saarinen 1986; Baskerville 1995; Nelles 2005; Robson 2010). With respect to forestry, university-produced technical expertise (and associated research funding and centres) has most commonly emanated from core regions (e.g., Toronto, Vancouver, Edmonton). Some advocates point out that local and traditional forms of knowledge, practice, and leadership in resource towns have long been overlooked by universities (Bullock, Armitage, and Mitchell 2012). Public forestry advisory committees are still relatively recent, and they remain dominated by the values and interests of industrial forestry (Parkins et al. 2006; Parkins and Davidson 2008; Reed and Davidson 2011).

In summary, dominant flows have allowed for system exchanges to take place between the core and the periphery, generating a dependency of the periphery on the core. Such relationships have not provided capacity for the building of effective regional networks across communities in the periphery in order to strengthen regional economic, political, cultural, and social well-being, resilience, or self-reliance. Instead, a pattern of external and centralized control has shaped relations among forests, communities, private investors, and senior governments. While companies, senior governments, and some groups of workers have benefited financially from such arrangements, there remains very little local control over industries that have such a dominant influence on resource town viability (Bullock, Armitage, and Mitchell 2012). Yet as conditions and stereotypes of resource and rural communities continue to evolve, there may be opportunities for endemic forms of development and transformation (Randall and Ironside 1996; Parkins and Reed 2013). One such opportunity lies with community forest systems, as described below.

**Prospectus: Community Forest Systems**
In contrast to the historic or *retrospective* deconstruction of socioeconomic relations in resource towns presented above, the following

section is deliberately *prospective*. Since community forestry is often heralded as an alternative to the conventional model discussed in the previous pages, in this section we conceptualize community forestry as it might unfold in Canada, describing key visible parts of the existing situation and using relevant theory to hypothesize what *could* develop if community forestry were permitted to thrive and proliferate.

Community forestry is an increasingly popular and sometimes idealized concept, described as having the potential to alleviate many of the challenges of conventional industrial forestry and resource-dependent communities described above. The approach is aligned with the goals and values of community sustainability, community economic development, and civic environmentalism (Baker and Kusel 2003; Teitelbaum and Bullock 2012). It seeks to account for a broader collection of values and benefits, to address social and environmental justice issues, to build resilience and self-reliance, and to utilize ecologically minded practices that together enable forest constituents and societies to flourish rather than be at the whim of external interests and forces (Bullock and Hanna 2012).

Experience in Canada indicates that many community forestry practitioners and supporters embrace an ecosystem-based approach to planning and management. The recent movement to establish community forests in British Columbia provides examples of local groups that have developed (e.g., on Denman Island and Malcolm Island) and, in some cases, actually implemented ecosystem-based plans (e.g., in Cheakamus, Creston, and Harrop-Procter). In another example, Ontario's thirty-six conservation authorities were founded on principles that align with an ecosystem-based approach (e.g., watershed management unit, local initiative, and partnership; see Shrubsole 1996). As scholars have indicated (e.g., Dunster 1989; Hammond 1991; Slocombe 1993a, 1993b; M'Gonigle 1997; Beckley 1998), the ecosystem-based approach differs from the conventional industrial approach in some important ways (see Table 14.1). The ecosystem-based approach to planning focuses on the regional or landscape level of scale and seeks to maintain ecosystem processes and structures that support communities of people and ecosystems. As such, it is based on broad, long-term objectives targeted to sustain whole ecosystems rather than emphasizing the narrow objective of economic growth characterized by short-term profit for a few private interests. Ecosystem-based approaches tend to utilize the boundaries of ecological systems (i.e., watersheds) rather than arbitrary planning divisions dictated by government administrations and power relations among multinational firms. When coupled

**TABLE 14.1   Comparison of conventional and ecosystem-based approaches**

|  | Conventional industrial approach | Ecosystem-based approach |
|---|---|---|
| Focus objectives | Firms, market conditions<br>Narrow: economic growth | Communities, ecosystems<br>Broad: economic, social, cultural, and environmental sustainability |
| Values | Maximum efficiency and short-term profits from production to a group of shareholders | Effective long-term provision of forest ecosystem services and processes for well-being of most residents and whole ecosystem |
| Boundaries | Arbitrary political-administrative lines to maintain spheres of influence | Based on ecological and traditional cultural territories to account for context |
| Decision-making | Top-down, centralized, command and control | Bottom-up, decentralized, collaborative |
| Nature of flows | Linear, direct, accumulative | Circular, indirect, distributive |
| Relevant knowledge and information | Science-based, disciplinary, professional and technical expertise | Place-based, interdisciplinary, experiential, lay knowledge, civic science |
| Motivation for, and application of, knowledge and information | Collect information about market, timber inventory, and production for internal use to maximize competitiveness and profits | Generate place-based research and knowledge for education, awareness, and capacity building to enable development and empowerment |
| Social contract | Male breadwinner, high wages, company-provided infrastructure and services | Diversified labour force and income sources, community-provided infrastructure and services |

*Sources:* Adapted from Dunster (1989); Hammond (1991); Slocombe (1993a, 1993b); M'Gonigle (1997); Beckley (1998).

with grassroots initiatives, they seek to integrate local and traditional knowledge with landscape-level scientific knowledge and technical expertise. Typically, this requires extensive community-engagement activities, an approach that contrasts with the command-and-control style of centralized decision-making that has often failed to acknowledge the merit of local knowledge.[5] In theory, following an ecosystem-based approach helps to produce a plan for sustainable management that respects the unique conditions and attributes of the regional context based on scale-appropriate knowledge, information, and objectives. Once completed and implemented, such plans further support education and build community capacity to participate in forestry decision-making and development, thereby helping communities and ecosystems to be more resilient (see Bullock, Hanna, and Slocombe 2009).

When viewed as a system, community forestry seeks to alter flows of energy, information, and material and to fundamentally change feedbacks among communities, forests, and external forces for sustainability. Achieving such a system requires revising patterns of control and creating tighter feedbacks for local information inputs, developing pathways to circulate locally generated benefits within and among forest communities in a region, and maximizing value created and captured within communities and nearby forests to enable local reinvestment for ongoing development and well-being. Community forestry seeks to achieve long-term self-sufficiency by increasing local access to wood, local value-added processing, development and retention of local talent, and investment and entrepreneurialism. Community forestry also implicitly embeds a social contract that has different elements from those associated with the industrial forestry model. In a sense, the iron triad that describes the actor group in industrial forestry and the range of social benefits is broadened. Sources of income are more diverse, as are actual wage rates. Economic reliance shifts from a single (or narrow) employer base to a broader range of employers undertaking a more diverse set of activities within a forested ecosystem. Activities may include timber harvesting, but they may also include activities related to other forest products such as traditional botanicals and/or ecotourism opportunities, as well as professional outfitting and guiding for hunting, fishing, and trapping; thus, more people are likely to be employed through value-added employment opportunities, and local innovation is encouraged. (See Auden [1944] for an example of diversification through a fully integrated community forest plan.) We still do not have complete information regarding the full variety and value of commercial and non-commercial non-timber forest products (NTFPs) that are already harvested in North American forests. Yet the diversity of products, markets, and groups linked to NTFP management (both formally and informally) and the economic and cultural importance of NTFPs is evident (Jones and Lynch 2008). Together, these sources of income are less likely to provide the high wages associated with the industrial forestry model, but they are designed to provide greater opportunity across a range of community members. Increasing diversity also promises greater resilience during economic downturns. For example, recent research in British Columbia found that when large mills closed during the recent downturn in the Canadian forest sector, small-scale value-added wood processors – those typically associated with small-scale community forests – remained operational because of their ability to adapt and retool and their demonstrated commitment to place (Pinkerton and Benner 2013).

Those previously excluded from the industrial economy (e.g., women and Aboriginal peoples) may also find opportunities in this more diverse economy. Ideally, through more inclusive and flatter decision-making structures and processes, local opportunities for leadership will arise for groups previously marginalized from the governance system, and opportunities will increase for local investment and empowerment previously untapped or underdeveloped. Emery and Flora (2006) suggest that such initiatives generate a shift towards more strategic, integrated, and systemic change that relies on and contributes to new leadership styles, new actors, and new ways of doing things.

In theory at least, community forestry is concerned with maintaining and improving the structure and processes of the community forest system in a manner that is consistent with local values and desires and can adapt over time to stresses and changes, whether they originate internally or are imposed from outside. In contrast to the company- and export-oriented centralized industrial forestry systems outlined above, community forest systems "assume that ecological and community processes are circular" (M'Gonigle and Parfitt 1994). The model counts on decentralized decision-making to use place-based natural and social capital in a manner that can (re)generate community structures and that functions in a self-sustaining manner. The system is not closed, however, and resource flows are intentionally circulated back into local processes to increase local capacities and decrease outside dependency.

In the Canadian context, residents and local representatives are encouraged to become involved in forest planning and management decision-making through membership on boards and advisory committees. Community forests may also hire private consultants to assist residents in identifying and pursuing development options (Mallik and

Rahman 1994; Teitelbaum and Bullock 2012). The process of creating, collecting, and incorporating local knowledge for community forests is a central strategic aspect for developing locally owned resource inventories, forest-management plans, and community-based land-use mapping (Bullock, Hanna, and Slocombe 2009). The initial period of organization and planning can be greatly facilitated by technical and networking support from provincial governments, regional associations, and community-based research organizations (e.g., BC Community Forestry Association, Eastern Ontario Model Forest, Nova Forest Alliance in Nova Scotia). In theory, under community forestry, both the province and large private companies would have less decision-making control, nor would they be primarily responsible for providing information and resource support for management planning. Practical examples of community forests in Canada show that the need for scale-specific information (e.g., environmental, social, and economic) and technical and management expertise can be effectively produced through local processes and actors with supplemental state-level support programs. For example, communities in British Columbia have generated their own forest inventories but have also received technical advice from provincial resource agencies (Bullock, Hanna, and Slocombe 2009).

In addition to community forestry's links to ecosystem-based management, previous scholars (e.g., M'Gonigle 1997) observe that it has similarities with the territorial planning model that focuses on regional interdependency (as depicted in Figure 14.2 below) rather than with a functional or centrist model focused on linear flows between a dominant core and its periphery, which has been shown to entrench local dependency. In theory, a regional network of community forests would have the majority of forest resources and benefits redistributed through the community and within the territory first, and only then out to the system environment (i.e., larger urban and external markets). Under the territorial model, each community is intended to gain more control over its local economy and its role in the regional economy. There is also maximum intraterritorial integration of local actors, information, and resource flows. This arrangement would also serve the values of community forests, which are also expected to contribute significantly to education and awareness, training and retention of local talent, and local pride and identity and, in doing so, to support local economic development, ecological sustainability, and social equity (Reed and McIlveen 2006; Teitelbaum and Bullock 2012).

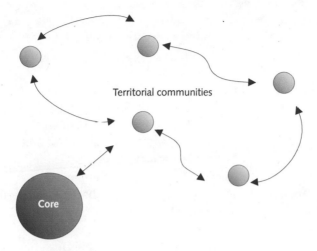

**Figure 14.2**   Basic schema of a prospective regional integrated network of community forests under the territorial model.

*Note:* Benefits of this model are more evenly distributed between the core and territorial communities. In addition, there are connections among the territorial communities.

To summarize, in theory, under the community forestry model, each community would interact more directly with surrounding forests and other communities than it would in the conventional industrial model (in terms of decision-making and management practice), and each community would thereby generate and receive increased benefits. This model holds that communities are bound regionally by a shared commitment to place and diversification, shared ecological and economic functions, and a culture of local empowerment, enterprise, and responsible resource use.

We recognize that the model offers an almost utopian view of community and social relations. In practice, there will be bumps in the road. Integration of communities and forests will require improved forms of networking if they are to avoid the fate of top-down approaches articulated by Berkes (2009) at the beginning of this chapter. In the next section, we turn to the issue of integration.

## Integrating Communities and Forests

Berkes (2009) notes that failure to integrate across subsystems has constrained top-down management approaches and is likely to constrain community-based resource-management approaches as well. Given the affinities among community forestry, ecosystem-based management, and the territorial planning approach discussed above, there are opportunities to better apply systems concepts to guide community forestry in order to close identifiable gaps for improved integration.

A key challenge with community forestry in Canada, as we see it, is the absence of a fully functioning network to coordinate and advocate initiatives across local, regional, and national scales. Conceived as a system, community forests are without a well-developed supporting network through which resources, information, and power can flow. The development of such a network is partly hampered by either the spatial concentration, or "pockets," of community forests in certain parts of some provinces (Ontario), or the dispersion of a few small community forests across large areas, in other provinces (British Columbia). Isolation, whether produced by physical or social conditions, interrupts communication, learning, and resource sharing and poses real challenges to collaboration. As the number of community forests rises, the question remains: Could a larger coordinating network effectively support interaction and collaboration that might surpass local capacity?

Support for the utility of such networks exists in regional associations (i.e., BC Community Forest Association) and non-government community-based forestry groups (e.g., Northern Ontario Sustainable Communities Partnership, Nova Forest Alliance, and Nova Scotia Community Forest Advisory Group). Such networks have emerged in British Columbia, Ontario, and Nova Scotia. There is little doubt that groups like these are currently making headway in linking community forestry initiatives across Canada (see Palmer et al. 2013). However, horizontal and vertical links to address cross-scale and cross-level issues have developed very slowly and organically rather than with intent and careful design. While Model Forests and the Canadian Model Forest Network can also be seen to support the values of community forests at the regional and national levels, respectively, these organizations face an uncertain future because of federal funding cuts in 2012. Other groups are more international in focus (e.g., Community Forestry International) or have community forestry as one program area of many (e.g., Conservation Council of New Brunswick). Linking supporting organizations, funding opportunities, relevant government agencies, research organizations, training opportunities, public forums and processes, and forest lands is essential to creating an integrated system of communities and forests in Canada.

Building and linking these systems is an important and necessary part of fostering strategic efforts to change the thinking about the value of networks and integration. As discussed above, the historical model of resource development explicitly worked to limit regional organization and resource towns frequently considered (and still considered by themselves) as competitors within a regional setting rather than as partners (Nelson, Duxbury, and Murray 2013; Davis and Reed 2013; Bullock 2013). To enable regional integration, various assets and resources – whether material, physical, socio-political, socio-cultural, or psychological – must be *collectively* harnessed and mobilized (Emery and Flora 2006).

Leadership and experience (or memory) are key in this regard. For example, Davis and Reed (2013) illustrate that some of the successes attributed to regional coalitions in addressing the pine beetle infestation in British Columbia's interior forests in the mid-2000s were attained because of previous experience in regional planning associated with land-use planning by the BC Commission on Resources and Environment (CORE) in the 1990s. Hence, CORE's "latent" benefits may actually reside in the capacity that it built for regional integration going forward. Emery and Flora (2006) observe that engaging youth in leadership positions helped a waning regional resource economy "spiral up" to enjoy gains in human, economic, and social capital at a regional scale. They emphasize that "without changes in the traditional leadership structures and actors, the community could not have mobilized citizens to support changes" (25). Likewise, when the 2007 downturn in the forest sector hit the Northeast Superior Region of northern Ontario (closing five mills across six towns), local leaders drew on their existing de facto regional network to pool resources and political influence in order to access several government programs and develop new business partnerships (Bullock, Armitage, and Mitchell 2012). During turbulent times, such transitory regional associations provide leaders and residents with forums where problems, solutions, and collective identity can be reframed as the basis for new power structures and visions for regional development and cooperation. Thus, both leadership and experience can help forge capacity and shift the collective and isolating culture of dependence towards one that seeks opportunity through regional integration.

Community forestry practice and research has remained a backwater to mainstream forestry research (Bullock and Lawler 2015), perhaps in part because community forestry is inherently bottom-up and environmental governance research has been predominantly case-based (Reed and Bruyneel 2010).[6] As discussed above, the legacy of resource dependency and separation, as well as the more recent fixation on securing tenure rights locally, has perhaps directed attention away from national collaboration and the development of well-resourced regional networks that would link actors and establish forums to exchange ideas and resources. As explained in this chapter, this is partly because of how individual sites, government programs, and community forestry practices have evolved, and also because of differences in perceptions among those pressed with solving integration challenges. However, more and more rural and resource-based communities are recognizing that they are actually part of a regional and national fabric of interdependent economies, social networks, and ecosystems involving complex exchanges and interrelated forces that influence their ability to make a living (Bullock, Armitage, and Mitchell 2012; Nelson, Duxbury, and Murray 2013; Parkins and Reed 2013). There is a need for research and practice to examine how governance mechanisms that support community forestry scale down (to communities), scale up (to cross provincial, national, and international borders), scale out (across regions and actor groups), and scale backward and forward in time (to better understand context and history; Reed and Bruyneel 2010). Bridging structures, processes, and theories are needed to better integrate and engage the community aspects of forestry practice and research with their wider geographic, social, political, environmental, and economic counterparts.

## Notes

1  Lockhart (1991, 25) warned that "product-oriented concepts of self-reliance" that have informed hinterland development policy in northern Canada must be recognized as propaganda in that they actually create dependency and depletion. A focus on human development and community-based process orientation instead would help generate self-reliance and renewal in northern communities.

2  For a full treatment of the Indian reserve system and the abhorrent conditions of forced settlement under colonialism in Canada, see Harris (2002) and Bartlett (1990). Briefly, the system of Indian reserves in Canada ensured that Aboriginal people were excluded from rights and responsibilities associated with resource extraction, use, and export. Aside from reserves, Aboriginal people who migrated to resource towns for seasonal work were typically excluded from townsites through discriminatory local and provincial regulations. Where they were employed, it was only for the most economically and physically vulnerable occupations. For examples, see Keeling (2010), about settlement, and Mills (2006), about employment.

3 Based on experiences in regions such as northern Manitoba, Ontario, and British Columbia, scholars (Saarinen 1986; Robson 1988; Gill 1990) point out that conditions in resource towns steadily improved through three phases of development: an initial period when companies provided essential services only needed to produce the resource in unplanned camps, a second period of company-government cooperation that saw marked improvements through additive planning, and a third period when competition in the labour market and regional planning encouraged planned communities to better support healthy lifestyles and development.

4 For example, Freudenburg (1992) suggests that volatile resource prices have given rise to cyclical, yet well-paid employment within the resource sector. Because the "highs" of price and employment are so compelling, they reinforce a positive addiction to resource extraction, discouraging workers and policy makers from identifying the signs of long-term resource decline (Davidson, Williamson, and Parkins 2003). Once signs of depletion appear, workers and local politicians pressure senior government officials to subsidize further extraction to the point of depletion, often jeopardizing opportunities for successor industries and steering the system into an undesirable state from which recovery is unlikely (Freudenburg 1992; see also Clapp 1998).

5 A notable example of this sort of community-based planning approach is in the planning work supported by Silva Forest Foundation over the past two decades (http://www.silvafor.org/).

6 Much research on community forestry has been conducted with or by community groups and for specific project-based purposes and objectives. As a result, much of it is descriptive, case specific, and applied rather than concerned with generalization and theory development.

## References

Auden, A. 1944. "Nipigon Forest Village: A prospectus." *Forestry Chronicle* 20 (4): 209–61. http://dx.doi.org/10.5558/tfc20209-4.

Baker, M., and J. Kusel. 2003. *Community forestry in the United States.* Washington, DC: Island Press.

Ballamingie, P. 2009. "First Nations, ENGOs, and Ontario's Lands for Life consultation process." In *Environmental conflict and democracy in Canada*, edited by L. Adkin, 84–102. Vancouver: UBC Press.

Barnes, T., R. Hayter, and E. Hay. 2001. "Stormy weather: Cyclones, Harold Innis, and Port Alberni, BC." *Environment and Planning A* 33 (12): 2127–47. http://dx.doi.org/10.1068/a34187.

Bartlett, R. 1990. *Indian reserves and Aboriginal lands in Canada – A homeland: A study in law and history.* Saskatoon: Native Law Centre, University of Saskatchewan,.

Baskerville, G. 1995. "The forestry problem: Adaptive lurches of renewal." In *Barriers and bridges to the renewal of ecosystems and institutions*, edited by L. Gunderson, C.S. Holling, and S. Light, 37–102. New York: Columbia University Press.

Beckley, T.M. 1998. "Moving toward consensus-based forest management: A comparison of industrial, co-managed, community, and small private forests in Canada." *Forestry Chronicle* 74 (5): 736–44. http://dx.doi.org/10.5558/tfc74736-5.

Berkes, F. 2009 "Evolution of co-management: Role of knowledge generation, bridging organizations, and social learning." *Journal of Environmental Management* 90 (5): 1692–702. http://dx.doi.org/10.1016/j.jenvman.2008.12.001.

Bone, R. 1992. *The geography of the Canadian North: Issues and challenges.* Toronto: Oxford University Press.

Bradbury, J., and I. St. Martin. 1983. "Winding down in a Québec mining town: A case study of Schefferville." *The Canadian Geographer* 27 (2): 128–44. http://dx.doi.org/10.1111/j.1541-0064.1983.tb01468.x.

Bullock, R. 2010. "A critical frame analysis of Northern Ontario's 'forestry crisis.'" PhD diss., Department of Geography and Environmental Management, University of Waterloo, Waterloo, ON.

–. 2013. "'Mill town' identity crisis: Reframing the culture of forest resource dependence in single industry towns." In Parkins and Reed, *Social transformation in rural Canada*, 269–90.

Bullock, R., D. Armitage, and B. Mitchell. 2012. "Shadow networks, social learning, and collaborating through crisis: Building resilient forest-based communities in northern Ontario, Canada." In *Collaborative resilience: Moving through crisis to opportunity*, edited by B. Goldstein, 309–37. Cambridge, MA: MIT Press.

Bullock, R., and K. Hanna. 2012. *Community forestry: Local values, conflict, and forest governance.* Cambridge, UK: Cambridge University Press. http://dx.doi.org/10.1017/CBO9780511978678.

Bullock, R., K. Hanna, and D.S. Slocombe. 2009. "Learning from community forestry experience: Challenges and lessons from British Columbia." *Forestry Chronicle* 85 (2): 293–304. http://dx.doi.org/10.5558/tfc85293-2.

Bullock, R., and J. Lawler. 2015. Community forestry research in Canada: A bibliometrics perspective. *Forest Policy and Economics.* 59: 47-55.

Clapp, R.A. 1998. "The resource cycle in forestry and fishing." *The Canadian Geographer* 42 (2): 129–44. http://dx.doi.org/10.1111/j.1541-0064.1998.tb01560.x.

Clement, W. 1997. *Understanding Canada: Building on the new Canadian political economy.* Montreal and Kingston: McGill-Queen's University Press.

Davidson, D., T. Williamson, and J. Parkins. 2003. "Understanding climate change risk and vulnerability in northern forest-based communities." *Canadian Journal of Forest Research* 33 (11): 2252–61. http://dx.doi.org/10.1139/x03-138.

Davis, E.J. 2011. "Resilient forests, resilient communities: Facing change, challenge, and disturbance in British Columbia and Oregon." PhD diss., Department of Geography, University of British Columbia, Vancouver.

Davis, E.J., and M.G. Reed. 2013. "Governing for transformation and resilience: The role of identity in renegotiating roles for forest-based communities of

British Columbia's interior." In Parkins and Reed, *Social transformation in rural Canada*, 249–68.

Dunk, T. 2003. *It's a working man's town: Male working class culture in northwestern Ontario*. 2nd ed. Montreal and Kingston: McGill-Queen's University Press.

Dunster, J. 1989. "Concepts underlying a community forest." *Forest Planning Canada* 5 (6): 5–13.

Emery, M., and C. Flora. 2006. "Spiraling-up: Mapping community transformation with community capitals framework." *Community Development* 37 (1): 19–35. http://dx.doi.org/10.1080/15575330609490152.

Freudenburg, W. 1992. "Addictive economies: Extractive industries and vulnerable localities in a changing world economy." *Rural Sociology* 57 (3): 305–32. http://dx.doi.org/10.1111/j.1549-0831.1992.tb00467.x.

Gill, A. 1990. "Enhancing social interaction in new resource towns: Planning perspectives." *Tijdschrift voor Economische en Sociale Geographic* 81 (5): 348–63. http://dx.doi.org/10.1111/j.1467-9663.1990.tb00719.x.

Goltz, E. 1992. "The image and the reality of life in a northern Ontario company-owned town." In *At the end of the shift: Mines and single industry towns in Northern Ontario*, edited by M. Bray and A. Thomson, 62–91. Toronto: Dundurn Press.

Hammond, H. 1991. *Seeing the forest among the trees: The case for wholistic forest use*. Winlaw, BC: Polestar Press.

Harris, C. 2002. *Making Native space: Colonialism, resistance, and reserves in British Columbia*. Vancouver: UBC Press.

Hayter, R. 2000. *Flexible crossroads: The restructuring of British Columbia's forest economy*. Vancouver: UBC Press.

Hodge, G., and I. Robinson. 2002. *Planning Canadian regions*. Vancouver: UBC Press.

Jones, E., and K. Lynch. 2008. "Integrating commercial nontimber forest product harvesters into forest management: Opportunities and challenges." In *Forest community connections*, edited by E. Donoghue and V. Sturtevant, 143-61. Washington, DC: Resources for the Future.

Kay, J., H. Regier, M. Boyle, and G. Francis. 1999. "An ecosystem approach for sustainability: Addressing the challenge of complexity." *Futures* 31 (7): 721–42. http://dx.doi.org/10.1016/S0016-3287(99)00029-4.

Keeling, A. 2010. "Born in an atomic test tube": Landscapes of cyclonic development at Uranium City, Saskatchewan. *The Canadian Geographer* 54(2): 228-52.

Lawson, J., M. Levy, and A. Sandberg. 2001. Perceptual revenues and the delights of the primitive: Change, continuity, and forest policy regimes in Ontario. In *Canadian forest policy: Adapting to change*, edited by M. Howlett, 279–315. Toronto: University of Toronto Press.

Lockhart, A. 1991. Northern development policy: Hinterland communities and metropolitan academics. In *Social relations in resource hinterlands*, edited by T. Dunk, 22–28. Vol. 1, Northern and Regional Studies Series. Thunder Bay, ON: Centre for Northern Studies, Lakehead University.

Lucas, R. 1971. *Minetown, milltown, railtown: Life in Canadian communities of single industry.* Toronto: University of Toronto Press.

M'Gonigle, M. 1997. "Reinventing British Columbia: Towards a new political economy in the forest." In *Troubles in the rainforest: British Columbia's forest economy in transition*, edited by T. Barnes and R. Hayter, 37–52. Victoria, BC: Western Geographical Press.

M'Gonigle, M., and B. Parfitt. 1994. *Forestopia: A practical guide to the new forest economy.* Madeira Park, BC: Harbour.

Mallik, A., and H. Rahman. 1994. "Community forestry in developed and developing countries: A comparative study." *Forestry Chronicle* 70 (6): 731–35. http://dx.doi.org/10.5558/tfc70731-6.

Mills, S. 2006. "Segregation of women and Aboriginal people within Canada's forest sector by industry and occupation." *Canadian Journal of Native Studies* 26: 147–71.

Nelles, H. 2005. *The politics of development: Forests, mines, and hydro-electric power in Ontario, 1849–1941.* 2nd ed. Montreal and Kingston: McGill-Queen's University Press.

Nelson, R., N. Duxbury, and C. Murray. 2013. "Cultural and creative economy strategies for community transformation: Four approaches." In Parker and Reed, *Social transformation in rural Canada*, 368–86.

Palmer, L., P. Smith, and R. Bullock. 2013. "Community Forests Canada: A new national network." *Forestry Chronicle* 89 (2): 133–34. http://dx.doi.org/10.5558/tfc2013-028.

Parkins, J., and D. Davidson. 2008. "Constructing the public sphere in compromised settings: Environmental governance in the Alberta forest sector." *Canadian Review of Sociology* 45 (2): 177–96. http://dx.doi.org/10.1111/j.1755-618X.2008.00009.x.

Parkins, J., L. Hunt, S. Nadeau, J. Sinclair, M. Reed, and S. Wallace. 2006. *Public participation in forest management: Results from a national survey of advisory committees.* Northern Forestry Centre Information Report NOR-X-409. Edmonton, AB: Northern Forestry Centre, Canadian Forest Service, Natural Resources Canada.

Parkins, J., and M.G. Reed. 2013. *Social transformation in rural Canada: New insights into community, cultures, and collective action.* Vancouver: UBC Press.

Pinkerton, E.W., and J. Benner. 2013. "Small sawmills persevere while the majors close: Evaluating resilience and desirable timber allocation in British Columbia, Canada." *Ecology and Society* 18 (2): art. 34. http://dx.doi.org/10.5751/ES-05515-180234.

Randall, J., and R. Ironside. 1996. "Communities on the edge: An economic geography and resource-dependent communities in Canada." *The Canadian Geographer* 40 (1): 17–35. http://dx.doi.org/10.1111/j.1541-0064.1996.tb00430.x.

Reed, M.G. 2003. *Taking stands: Gender and the sustainability of rural communities.* Vancouver: UBC Press.

Reed, M.G., and S. Bruyneel. 2010. "Rescaling environmental governance, rethinking the state: A three-dimensional view." *Progress in Human Geography* 34 (5): 646–53. http://dx.doi.org/10.1177/0309132509354836.

Reed, M.G., and D. Davidson. 2011. "Terms of engagement: The involvement of Canadian rural communities in sustainable forest management." In *Reshaping gender and class in rural spaces*, edited by B. Pini and B. Leach, 199–220. Aldershot, UK: Ashgate.

Reed, M.G., and K. McIlveen. 2006. "Toward a pluralistic civic science? Assessing community forestry." *Society and Natural Resources* 19(7): 591-607.

Robson, M. 2010. "Sustainable forestry in the 21st century: Multiple voices, multiple knowledges." *Forestry Chronicle* 86 (6): 667–68.

Robson, R. 1988. "Manitoba's resource towns: The twentieth century frontier." *Manitoba History* 16 (Autumn). http://www.mhs.mb.ca/docs/mb_history/16/resourcetowns.shtml.

Saarinen, O. 1986. "Single-sector communities in northern Ontario: The creation and planning of dependent towns." In *Power and place: Canadian urban development in the North American context*, edited by G. Stelter and A. Artibise, 219–64. Vancouver: UBC Press.

Shrubsole, D. 1996. "Ontario conservation authorities: Principles, practice, and challenges 50 years later." *Applied Geography* 16 (4): 319–35. http://dx.doi.org/10.1016/0143-6228(96)00017-3.

Slocombe, D.S. 1993a. "Environmental planning, ecosystem science, and ecosystem approaches for integrating environment and development." *Environmental Management* 17 (3): 289–303. http://dx.doi.org/10.1007/BF02394672.

–. 1993b. "Implementing ecosystem-based management: Development of theory, practice, and research for planning and managing a region." *Bioscience* 43 (9): 612–22. http://dx.doi.org/10.2307/1312148.

Stedman, R., J.R. Parkins, and T. Beckley. 2005. "Forest dependence and community well-being in rural Canada: Variation by forest sector and region." *Canadian Journal of Forest Research* 35 (1): 215–20. http://dx.doi.org/10.1139/x04-140.

Teitelbaum, S., and R. Bullock. 2012. "Are community forestry principles at work in Ontario's county, municipal, and conservation authority forests?" *Forestry Chronicle* 88 (6): 697–707. http://dx.doi.org/10.5558/tfc2012-136.

Walker, B., and D. Salt. 2006. *Resilience thinking: Sustaining ecosystems and people in a changing world*. Washington, DC: Island Press.

Wallace, I. 1987. "The Canadian Shield: The development of a resource frontier." In *Heartland and hinterland: A geography of Canada*, 2nd ed., edited by L. McCann, 443–81. Scarborough, ON: Prentice-Hall Canada.

# Contributors

**Lisa Ambus** has been deeply engaged in community-based natural resource management over many years and in various capacities, including as a coordinator of a global network of community forest organizations, director of the Wetzin'kwa Community Forest Corporation, member of the BC Community Forest Association, and through her graduate studies at the University of British Columbia, which focused on the community forest program in British Columbia.

**Thomas Beckley** has been working in the field of natural resource sociology in Canada for twenty years. He teaches in the Faculty of Forestry and Environmental Management at the University of New Brunswick in Fredericton. His research topics have included social problems and issues in forest dependent communities, public participation in resource management and policy, community sustainability, and adaptability and environmental values. Recently Tom has branched out into explorations of energy issues and climate change.

**Ryan Bullock** is an assistant professor in the Department of Environmental Studies and Sciences, and member of the Centre for Forest Interdisciplinary Research, at the University of Winnipeg. His research focusses on conflict and cross-cultural collaboration in emerging multi-level environmental resource governance systems, as well as community-based research approaches. Dr. Bullock is lead author of the recent book *Community Forestry: Local Values, Conflict and Forest Governance* (Cambridge Press, 2012), which examines the promises and pitfalls of local forest-management arrangements in Canada.

**Sara Carson** is originally from New Brunswick and now calls Newfoundland home. Sara been fortunate enough to work in many forest-dependent communities across Canada from the Coast Mountains of British Columbia to the coastlines of Newfoundland. She completed an undergraduate and Master's degree in Forestry at the University of New Brunswick, which focussed on the potential for community forestry in Newfoundland. Sara is a Registered Professional Forester in the province of Newfoundland and Labrador and is the silviculture forester for the Avalon Peninsula.

**Guy Chiasson** teaches political science and regional development at the Université du Québec en Outaouais in Gatineau, Québec. His research is mainly concerned with issues of local governance in peripheral regions. Since 2013, he has been the scientific director of the Centre de recherche sur le développement des territoires, a network of scholars on regional development.

**Alan Diduck** is a professor in the Department of Environmental Studies and Sciences at the University of Winnipeg and a member of the Centre for Forest Interdisciplinary Research. Prior to joining the University of Winnipeg, he was a lawyer and the executive director of the Community Legal Education Association, a social-profit organization providing public legal education and information services. His research program focuses on citizen involvement in environmental governance, the learning implications of involvement, and the consequences for sustainability.

**Peter Duinker** has been a professor since 1988, and his major areas of scholarly interest are forests and environmental assessment. He has researched and advocated for community forests since the early 1990s. Peter was active in various processes leading to the establishment of the first community forest in Nova Scotia – the Medway Community Forest Cooperative.

**Erin Kelly** is an assistant professor in Forest Policy, Economics, and Administration at Humboldt State University in Arcata, California. Her research has focused on forest governance and policy, with an emphasis on rural community capacity and community forestry. Erin earned her PhD from the College of Forestry at Oregon State University, then moved to Corner Brook, Newfoundland, for two years for a postdoctoral fellowship at the Grenfell Campus of Memorial University.

**Édith Leclerc** is an associated researcher at the Chaire sur la franco-phonie et les politiques publiques at the Ottawa University. She finished her doctoral thesis on the regional organization of forest governance in Quebec at the Université du Québec en Outaouais. Her main reasearch concern is related to local and regional governance of Quebec's forest territories. Geographic scales, environmental policies, and public action are at the heart of her concerns.

**Erik Leslie** is a forestry consultant and the forest manager for the Harrop-Procter Community Co-operative. He is also the president of the BC Community Forest Association. Erik has worked for community organizations, regional and provincial governments, industry, and First Nations on projects from Haida Gwaii to Labrador. He has extensive experience in forestry planning and operations, community consultation, small business management, and forest certification.

**L. Kris MacLellan** has lived along Canada's Pacific Coast, above the Arctic Circle, against the Bay of Fundy, and on Cape Breton Island. He holds a Master of Resource and Environmental Management from Dalhousie University, where he focused on community-based resource management and contributed to the Halifax Urban Forest Master Plan. Today he contributes to the development of tidal power in the Bay of Fundy with Halifax-based Minas Energy. Kris also serves as a volunteer director at Nova Scotia's Ecology Action Centre.

**Kirsten McIlveen** worked as a forest technician in the Lakes Forest District and teaches geography and women's and gender studies at Capilano University, as well as a first-year social sciences course at a federal women's prison.

**Solange Nadeau** is a senior forest sociologist with the Canadian Forest Service of Natural Resources Canada. Her research concerns issues such as social values and attitudes related to forest management and policy, forest-based communities, and public participation.

**Teika Newton** is the community research coordinator for the Common Ground Research Forum. Based in her hometown of Kenora, Ontario, Teika is passionate about community, environmental, and social justice. She chairs the City of Kenora's Environmental Advisory Committee, is a local representative to the Rainy-Lake of the Woods Watershed

Board's Community Advisory Group, and she is the executive director of Transition Initiative Kenora, an environmental non-profit that works with local people, government, and other community groups to build the skills necessary to transition our economy to clean, renewable energy and to deal with the many impacts of climate change.

**Bram Noble** is a professor in the Department of Geography and Planning at the University of Saskatchewan. His research is focused on environmental assessment, resource development and regional land-use planning.

**Lynn Palmer** is a PhD candidate in Forest Sciences in the Faculty of Natural Resources Management at Lakehead University and holds a BSc and MSc in forestry. The focus of her PhD research is the ability of Ontario's new forest tenure policy framework to enable community forestry. She is a founding member of the Northern Ontario Sustainable Communities Partnership, an NGO that advocates for community forests in northern Ontario, and Community Forests Canada, a new national network that aims to bridge practice, research, and advocacy for community forests in Canada.

**John R. Parkins** is a professor in the Department of Resource Economics and Environmental Sociology at the University of Alberta. His recent work on forestry includes an analysis of media framing related to mountain pine beetle management in Alberta, forest governance, neoliberal strategy within model forest programs in Canada, and a comparative analysis of socio-economic conditions in forested regions of the Philippines.

**Evelyn Pinkerton**, professor in the School of Resource and Environmental Management, Simon Fraser University, is a maritime anthropologist specializing in political ecology and co-management of adjacent natural resources by rural communities traditionally dependent on them. She has a long-term interest in community forests and continues to follow up on the research conducted in this study.

**Maureen G. Reed** is a professor in the School of Environment and Sustainability at the University of Saskatchewan. She is particularly concerned with the social dimensions of environmental and land use policies as they affect rural places. Her research is focused on how participatory decision-making approaches, working conditions, gender relations, and socio-cultural change affect the capacity of rural communities to work towards sustainability and resilience.

**Lauren Rethoret** is a researcher at the Columbia Basin Rural Development Institute at Selkirk College in the Kootenay region of British Columbia. The research described in her chapter was completed as part of the requirements for a Masters degree, which Lauren obtained from the School of Resource and Environmental Management at Simon Fraser University. Her research interests include community based resource management, environmental planning, rural policy, and sustainable infrastructure management.

**Michelle Rhodes** is an associate professor of Geography and the Environment at the University of the Fraser Valley. Her research and teaching focuses on natural resources geographies and on the relationships between economic activities and landscape morphology.

**James Robson** is an assistant professor at the School of Environment and Sustainability, University of Saskatchewan. His research interests include the study of collective action arrangements for governing shared environmental resources (commons). He has been involved with the Common Ground Research Forum in Kenora since 2011, conducting research into people's understandings of "common ground" and, more recently, local uses of Tunnel Island trails as an emergent recreational commons.

**Murray Rutherford** is an associate professor of Environmental Policy and Planning in the School of Resource and Environmental Management at Simon Fraser University. He is a policy scientist whose research focuses on the human dimensions of environmental policy and planning, and particularly on the values, perspectives, and institutions that shape and govern people's relationships with the environment. He has studied and written about ecosystem-based management, large carnivore conservation policy, water planning, environmental impact assessment, and human values and attitudes towards nature and the conservation of biological diversity.

**Chander Shahi** is an associate professor in the Faculty of Natural Resources Management, Lakehead University. He specializes in Natural Resource Economics, and his research emphasis has been on developing models for socio-economic impact asssessement due to the transitions taking place in the resource-based industry. His emphasis in quantitative research has been on building models for socio-economic and environmental impact assessment of the emerging bio-economy initiatives, supply chain and value-chain optimization, reviving the resource-based industry in northwestern Ontario, market sector analysis, and value-added products. Whereas in qualitative research, his major focus is the economic development of local and First Nation communities.

**A. John Sinclair** is a professor at the Natural Resources Institute, University of Manitoba. His main research interest focuses on community involvement and learning in the process of resource and environmental decision-making. His applied research takes him to various locations in Canada, as well as East Africa and Asia.

**M.A. (Peggy) Smith** is an associate professor in Lakehead University's Faculty of Natural Resources Management and a Registered Professional Forester. Her research interests focus on the social impacts of natural resources management, including Indigenous peoples' involvement and community forestry. She considers herself privileged to be a part of the growing number of people of Indigenous ancestry (Cree) who are working in the field of natural resources management, conservation, and development in Canada.

**Sara Teitelbaum** is an assistant professor in the Sociology Department at the University of Montreal. Her research focuses on the social dimensions of forest management including community forestry, Indigenous rights, forest certification, and public participation. Sara teaches courses in environmental sociology as well as qualitative research methods.

**Mya Wheeler** is a PhD student at the Natural Resources Institute of the University of Manitoba and also teaches as contract faculty for the University of Winnipeg in the Geography Department. Her research looks at place-based inquiry within resources management as a way to encourage participation and reflection on decision-making in development. This research will be applied to cases of oil and gas expansion in Canada, specifically hydraulic fracturing technology.

# Index

Printed and bound in Canada by Friesens

Set in Syntax and Cambria by Artegraphica Design Co. Ltd.

Copy editor: Joyce Hildebrand

Proofreader: Ryan O'Connor

Indexer: Noeline Bridge

Cartographer: Eric Leinberger